D1060125

THE NEW NATURALIST LIBRARY

A SURVEY OF BRITISH NATURAL HISTORY

SHIELDBUGS

*

The aim of this series is to interest the general reader in the wildlife of Britain by recapturing the enquiring spirit of the old naturalists. The editors believe that the natural pride of the British public in the native flora and fauna, to which must be added concern for their conservation, is best fostered by maintaining a high standard of accuracy combined with clarity of exposition in presenting the results of modern scientific research.

THE NEW NATURALIST LIBRARY

SHIELDBUGS

RICHARD JONES

This edition published in 2023 by William Collins,
an imprint of HarperCollins*Publishers*

HarperCollins*Publishers*
1 London Bridge Street
London SE1 9GF

WilliamCollinsBooks.com

HarperCollins*Publishers*
Macken House, 39/40 Mayor Street Upper
Dublin 1, D01 C9W8, Ireland

First published 2023

Illustrations by Verity Ure-Jones

A CIP catalogue record for this book is available
from the British Library.

Set in Nexus Serif Pro and Nexus Mix Pro

Edited and designed by
D & N Publishing
Baydon, Wiltshire

Printed in Bosnia-Herzegovina by GPS Group

Hardback
ISBN 978-0-00-833489-5

Paperback
ISBN 978-0-00-833491-8

Contents

Editors' Preface

Among other aspects of natural history, the New Naturalist Library has a distinguished history of publishing titles on individual insect groups. The very first volume in the series, published in 1945, was E. B. Ford's *Butterflies,* and over the years this was followed by books on honeybees, moths, bumblebees, dragonflies, ants, ladybirds, grasshoppers and crickets, beetles, and most recently solitary bees. Whereas most of these groups were already popular when the books were written, shieldbugs currently have a smaller fan club and are less well known to non-specialists.

Following his excellent book on *Beetles,* we are delighted that Richard Jones has been willing to introduce us to this new group. We expect that this book, with its lively and informative text and its colourful illustrations, will make shieldbugs attractive and accessible to many more naturalists. The keys and illustrations, supported by notes on habitats and food plants, will be of great assistance in species identification, and the inclusion of recent additions to the British shieldbug fauna, and others likely to follow, challenges us to watch out for new arrivals from the continent.

This book will bring great pleasure and satisfaction to those who are inspired to study shieldbugs seriously, but even readers who have no intention of grubbing around the base of food plants or sniffing the bugs on their broad beans will enjoy the vivid colours and curious antics presented in this book. Perhaps in the future we will look back at its publication as the trigger for a fresh surge of interest in British shieldbugs.

Author's Foreword and Acknowledgements

Shieldbugs are lovely. Why else would we give them such an elegant heroic heraldic name? Sure enough, they are glossy, chunky, often large and brightly coloured, and suitably shield-shaped. Shieldbugs have a regal and aristocratic air about them. My stylish pentatomid-design enamel lapel badge often draws comments because its obvious shield shape looks as if it could be the emblem of some noble family. It is, of course, but not necessarily in the way most people imagine.

Shieldbugs walk with a friendly clockwork gait and take to the wing with a solid model-aircraft rattle. Each year common Green Shieldbugs (*Palomena prasina*) emerge from hibernation in my garden and disport themselves on the garden fence in the early spring sunshine. At that time, March or April, they are a rich brownish purple, seemingly well camouflaged against the brown wooden panels, but soon they will physiologically ferment themselves back to their original emerald. My family are inured to my childish enthusiasm, and politely look on whenever I eagerly rush inside with one cupped in my hands to show them.

Having said that, in North America especially they are called, less appealingly perhaps, stink bugs. True, they smell a bit, but I've always thought their aroma rather pleasant – a kind of oily marzipan scent. After I release that inevitable first show-and-tell *Palomena*, I can't resist sniffing my fingers. It's funny; as a child I couldn't stand marzipan, and always peeled that layer off my slice of Christmas cake, and yet the shieldbug 'stink' is terribly evocative of my childhood. Maybe my psychotherapist could make something of that. On a more positive note, some very brightly coloured tropical species of shieldbug (especially those in the family Scutelleridae) are called jewel bugs, for their mesmerising colours and bright patterns.

One of the most appealing things about shieldbugs is that they are seen as non-threatening. Even the biggest do not bite, and their short broad feet do not scratch as they walk up the outstretched finger. Sharp spines on the pronotum are not regarded as a danger to life or limb. The cats aren't scared of them – well, the cats aren't interested in them, but that's cats. Importantly they are not seen as significant garden, farm or forest pests. They feed, reassuringly (and usually discreetly), on a variety of wild plants. Mind you, there was that time I got into trouble for blithely dismissing them as 'harmless' in a wildlife article in a gardening magazine. A reader wrote an irate letter to the editor complaining of my indifference, and pointing out (with photos, if memory serves) that they were decimating her broad beans. A suitable apology was issued. On the whole, though, they rarely get out of hand in the garden. And mostly they are, indeed, harmless. Mostly.

In the autumn of 1992 my partner and I took a last-minute adventurous holiday to Costa Rica. A few days in, and we had driven our small hire car to the tiny town of Quepos on the Pacific coast with a view to visiting the Parque Nacional Manuel Antonio the next day in search of tree sloths, agouti, white-faced monkeys and hummingbirds. We saw them all, but being an entomologist my lingering memory of the place (shortly after the sloth sighting) is of a loud rattling buzz as a large insect flew over my head. Without a thought, I jumped up and caught it in my bare hands. It was a huge orange shieldbug, probably one of the *Edessa* species so diverse there in the Neotropics. I held it gently, as I had done for many large insects, but it did not recognise my friendly behaviour and immediately exuded a copious amount of defensive chemical from its thoracic glands. Ordinarily I would have savoured the delicate rancid marzipan smell I knew from shieldbugs back home in England, but I was surprised to see my thumb completely stained a rich ochre brown. There was none of the accompanying pungent smell that I had come to associate with shieldbugs, but my skin was marked, immovably, for the next five days. It was quite unnerving to think of the chemical power from this insect, particularly as I'd always thought of plant bugs as being at the mild end of the insect danger spectrum. I'd like to be able to say that I was chastened by this episode, more circumspect in my future dealings with large tropical insects, but this was not the case; I continue with my cavalier pick-it-up-and-see-what-happens attitude.

A few years later, demonstrating insects to small children at a local environmental event at the annual Nunhead Cemetery Open Day, I took great delight in getting my visitors to smell the mating pairs of the small bronze shieldbug *Eysarcoris venustissimus* that many of them were bringing to the bug-identification stall. Holding them in the palm of my left hand I gave the still-conjoined shieldbug couples a rude poke with my right index finger and held them out to the noses of my audience. Sometimes I was given a screwed-up face of disgust; sometimes an amazed eyes-wide-open recognition from those who knew what a cocktail of almonds and diesel smelled like. But mostly what I got, by the end of the day, was a memory of that Costa Rican encounter when I noticed that following the repeated prod-and-sniff performances my left palm was marked with streaks of vague brown stain that would not be washed away. The home-grown shieldbugs, smaller, less loaded with bodily secretions, nevertheless had the same ability to tan my hide, albeit in a more subdued and gentle fashion. Despite these assaults on my epidermis, I remain unafraid of shieldbugs and continue to handle them as gently as I can. Plenty of the photos in this book are of shieldbugs crawling over my hand, or held gently between finger and thumb. Thankfully my fingers have never been tested so again.

The North American suspicion of shieldbugs may be down to the possibility that North American pentatomids stink more than European species, though it is more likely to be a general distrust of insects there. We can be rather blasé about potential insect nuisances in the gentle temperate and oceanic climate we enjoy here in north-west Europe, but North Americans have had more than their fair share of significant invasive alien insect pests – from Gypsy Moths (*Lymantria dispar*) and Tiger Mosquitoes (*Aedes albopictus*) to Emerald Ash-borer beetles (*Agrilus planipennis*). One of the latest is the Brown Marmorated Shieldbug (*Halyomorpha halys*), a pentatomid originally from East Asia that is now a widespread orchard and soft-fruit pest in the USA. It is not just farmers and market-gardeners who notice this large and striking insect; it has a propensity to enter homes to hibernate in clusters, and the appearance of large numbers indoors is alarming and olfactorily offensive. Reports of many thousands invading houses, each buzzing about 'like an angry overweight wasp', are commonplace. A prominent article on them in the *New Yorker* clearly demonstrates that tabloid spluttering has now been replaced by mainstream angst (Schulz 2018). Incidentally, this is the Marmorated Shieldbug, not the Marmite-rated Shieldbug as suggested by the autocorrect on my phone.

Halyomorpha halys arrived in Europe (in Liechtenstein) in 2004 and now occurs sporadically in Switzerland, Italy, Germany and France. Its arrival into the British Isles has been widely anticipated; indeed, it has been intercepted at British ports in imported goods and passenger luggage, and live specimens have recently turned up in pheromone monitoring traps. If *Halyomorpha* does become well established here maybe its agricultural afflictions and household invasions will cause people to shy away from the benign aura that presently surrounds shieldbugs.

Before shieldbugs become blighted with bad press, however, I'd like to re-emphasise their usual harmlessness, their striking and attractive forms, their interesting behaviours and physiology, and their amenable accessibility to the field entomologist. I therefore humbly offer up this book in the furtherance of shieldbug appreciation.

ACKNOWLEDGEMENTS

During 2020 and 2021, it became a slightly facetious joke that everyone should be using the downtime – lockdown, curfew, furlough, hibernation, whatever you care to call it – to write a book. From the privileged armchair of the middle-class, middle-aged, white man with a university education I consider myself very fortunate that this is exactly what I was able to do. Although initiated in 2018 through reading and research, the writing began in earnest in May 2020, and has followed the on/off time I have had available. There have been stoppages and slow-downs, but by early December of that year I was back on track again. As ever, my partner Catrina Ure helped me find a safe place – physically, mentally, spiritually, and emotionally – to get on with the task. She can only guess at the importance I place on my entomology and what sociologist Harriet Martineau so eloquently described as the 'need of utterance' to write down and get out what is on my mind. It has been both cathartic and calming to sit writing about shieldbugs, amidst the raging tempests of shambolic domestic and international politics and a tragic medical calamity unfolding on a world scale. I hope that my prose has not been too distorted to either rant or whimsy, and that I have been able to steer a steady and neutral course through what, for many, has been a time of confusion, chaos, and personal heartache.

The text is only half this book, and the illustrations also needed considerable work. Many shieldbugs are common and striking and relatively easily identified from photographs, so they feature heavily in the several stock photo agencies through which I have been able to trawl for the usual portraits and even some behavioural shots. However, I have not always agreed with the identifications offered. I've done my best, but if there are errors in the names then I accept full responsibility. Thankfully I have also been able to call on several specialist entomologist photographers to fill the many gaps and provide spectacular photographs of some of the more difficult and obscure species, and also their unusual behaviours – Tristan Bantock, Yvonne Couch, Maria Justamond, Penny Metal and Simon Robson, you are all stars. My daughter Verity Ure-Jones skilfully produced all the line illustrations for the identification key, coping admirably with my sometimes garbled instructions and ill-formed ideas.

Roger Hawkins kindly read through some of the text, particularly the long list of British shieldbug species in Chapter 8; he made many helpful suggestions and contributed some interesting personal observations. Hugh Brazier then patiently edited my original text using subtle variations on a theme of 'did you really mean this?' when he had spotted a complete boo-boo. And thanks to designers Namrita and David Price-Goodfellow for arranging the sometimes complex page layouts that resulted when all the disparate pictures, verbose fact boxes and my obsessive footnotes needed to be slotted in together sensibly.

Many other people have helped with contributions, suggestions, photographs and more: Sybil Baldwin, Max Barclay, Ian Beavis, Nathaniel Blair, Steve Covey, Alan Watson Featherstone, Graham Fisher, Jim Flanagan, Will George, Ian Harding, Joanne Hatton, Graham Hopkins, Finley Hutchinson, Richard Lewington, Jerzy Lis, Graeme Lyons, Emma Nicholls, Matthew Oates, Simon Oliver, Colin Purrington, Stuart Reed, Jony Russell, Matt Shardlow, Arabella Sock, Stras Strekopytov, Lillian Ure-Jones, Leon van der Noll, Sue Vincent, Simon Warry, David Williams, Gary Williamson, Ron Woollacott, Aidan Wren, Antony Wren, Elaine Wright. To all of you go my grateful thanks.

CHAPTER 1

What is a Shieldbug?

FIRST – WHAT IS A BUG?

'Bug' is a rather vague term. Today it can mean a disease-causing germ from virus to bacterium, any invertebrate from tardigrade to giant squid, a computer coding problem, a mechanical glitch or an electronic failure. In natural history, though, it has mainly become a flexible and broad term for creepy-crawlies in general. I became Bugman Jones when I started working with local schoolchildren who knew that I studied the small insects, spiders, woodlice, centipedes and snails that they just lumped all together as 'bugs'. This trend is sometimes lamented as unfortunate, and for some people 'minibeast' is an even more unspeakable heresy. Yet these are inclusive, friendly and non-threatening terms in a science often mired in jargon, and they all have their place when used properly. Meanwhile, for the more academic entomologist, a 'true bug' is a member of the insect order Hemiptera. This is a large and diverse collection of creatures including aphids, whitefly, scales, jumping plant lice, leafhoppers, froghoppers, cicadas, water boatmen, back-swimmers, skaters, capsids and shieldbugs.

This was not always the case. Although 'bug' had sometimes been used disparagingly to mean a small nondescript insect before about 1622 (according to the *Oxford English Dictionary*),[1] it had previously mostly meant something completely different – a bugbear, boggard, bogey, hobgoblin or some other general night fear – and variants include bugabo and buglarde. The etymological origins are unclear, but it may have some connection with the Welsh *bwg* or

1 The earliest example I've been given (by Matt Shardlow, Twitter exchange, 2022) is from Rogers (1642): 'Rather let us be like him … who never trod upon a bug or worme to kill it.'

bwgan – ghost (again, the *OED* is my source here). This ill-defined night dread is actually the meaning used in the oft-misquoted (by entomologists) Psalm 91, verse 5, from the Matthew Bible (1537), 'So that thou shalt not nede to be afrayed of eny bugges by nyght.' For the 1611 King James edition, though, this was retranslated: 'Thou shalt not be afraid for the terror by night.' In other words, creepy, but not crawly.

FIG 1. The glory of shieldbugs, shown in examples of some of the groups covered in this book. (a) Brassica Bug (*Eurydema oleracea*, Pentatomidae). (b) Box Bug (*Gonocerus acuteangulatus*, Coreidae). (c) Firebug (*Pyrrhocoris apterus*, Pyrrhocoridae). (d) Cinnamon Bug (*Corizus hyoscyami*, Rhopalidae). (e) Common Tortoise Bug (*Eurygaster testudinaria*, Scutelleridae). (f) Bordered Shieldbug (*Legnotus limbosus*, Cydnidae). (Photos by Penny Metal)

However, in 1730 Londoner John Southall firmly fixed the name to the blood-sucking Bedbug (*Cimex lectularius*) when he wrote his *Treatise of Buggs*, mainly to promote his nonpareil liquor (2 shillings a bottle) to get rid of the damn things (Fig. 2). He was also appointed 'buggman' to Sir Thomas Coke, Lord Lovell, later first Earl of Leicester, to control the vermin at a salary of a guinea a year (Strekopytov 2021). Whether he was a genuine naturalist or just a copywriter and pamphleteer is not clear; his booklet has some scientific information, but also much myth and misunderstanding, and Weiss (1931) concluded that it was just a clever advertisement for pest control. Up until that point *Cimex* bedbugs had mostly been called wall-lice, for their habit of secreting themselves away in cracks in the walls, ready to emerge to suck the blood of their unfortunate victims during the night. The modern German word for almost any hemipteran true bug, *Wanze*, is a corruption of *Wandlaus*, meaning wall-louse, while the French *punaise* is a word also used for drawing pin or thumb tack – an item you might use to pin something to the wall. Like 'bug', 'louse' was another one of those vague old words used to mean almost any small crawling critter from head louse to woodlouse. *Cimex lectularius* is now classed as a genuine 'true bug' in the order Hemiptera, family Cimicidae.

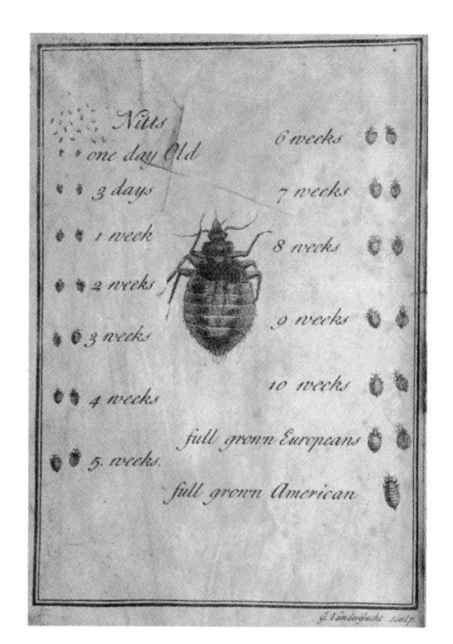

FIG 2. Frontispiece from Southall's *Treatise of Buggs* (1730), depicting the well-known Bedbug (*Cimex lectularius*), although he was of the opinion that European and American bedbugs were different species.

Uptake of the word 'bug' was slow to start; nevertheless, it soon crawled decisively out onto the entomological world stage. *Cimex* the wall-louse (also wall-lowse, or punie) is included in one of the most important early English books on insects, physician Thomas Mouffet's *Theater of Insects* (1658), along with three obvious shieldbugs, but he calls these 'wood wall-lice of the sheath-winged kinde', not bugs – a word he uses only once in his entire 400-page tract as an alternative name (along with 'klock') for beetle. Later, John Ray (1710) uses the Latin term *cimices sylvestres* (wood cimexes, i.e. wood wall-lice) for all of his shieldbugs too. After Southall's 1730 booklet, though, 'bug' increasingly meant not just *Cimex*, the bedbug/wall-louse, but gradually, by association, other hemipteran relatives.

WHAT'S IN A NAME?

So – **shield bug**, **shield-bug** or **shieldbug**? As is clear from the title of this book, I have already decided which one I will be using. But this is not without some trepidation, and I realise that I have exposed myself to potential argument. My reasoning for going down this route is as follows.

There was once supposed to be a tradition amongst naturalists that composite insect names were two words if they were scientifically logical, but one word if they were not. Thus: hover flies, house flies, bot flies and dung flies are all actual flies so two words each, but butterflies, dragonflies, mayflies and greenflies are patently not. By this same reckoning we should have lace bugs, spittle bugs, leaf bugs and shield bugs, which are all bugs, but Maybugs, which are beetles. But, as is all too common with English common names, they wash with the times and with the fashions of both popular and scientific publishing. Supposedly authoritative dictionaries and encyclopaedias are just as fickle, and the several I consulted from my extensive etymological library, published between 1895 and 1987, offer between them all three variants of shield bug, shield-bug and shieldbug, often as alternatives, and it is clear that no clear consensus has ever really existed.

Fly/flye/flie and bug/bugg/bugge are ancient words anyway, coined long before any real understanding of insect phylogeny or taxonomy was either necessary or available. There have long been transgressors and intermediates to the 'rule'. Maybugs, sometimes May-bugs but frequently May bugs, are beetles (*Melolontha* species) that just happen to fly about in May. They are also called cockchafers or cock-chafers (no hen-chafers known though) and are obviously chafers, from the Old English word *ceafor*, meaning beetle. However, the Black Beetle isn't a beetle, but a cockroach, *Blatta orientalis*; and cockroaches are nothing to do with cockchafers. Nor are they types of roach, which is a very modern contraction. And there are no hen-roaches. Fire-flies (family Lampyridae) and the Spanish Fly (*Lytta vesicatoria*) are not flies – they are beetles. Neither woodworms nor glow-worms are worms. Bumblebees and honeybees are certainly bees, and though they are sometimes two words, occasionally hyphenated, they too are frequently and inconsistently one. It turns out there isn't really a rule at all.

To conflate this muddle, there has been a general (by which I also mean rather haphazard) pattern in the entomological literature of the last 100 years or so, to first hyphenate some of these double word names, then unify them. My old A-level chemistry teacher Mr Warren maintained that this reflected the assimilation of large numbers of German scientists (particularly chemists) into British and American research programmes after the Second World War, bringing with them a long-standing German-language tradition of combining short words wholesale to make new longer ones. As hover flies have become more popular subjects for study, they first became hover-fly (Coe 1953), and now hoverfly (Stubbs & Falk 1983, 2002). Likewise we also regularly have housefly, and botfly, but strangely dung fly remains two words – for now. Bee-fly also has had to remain in limbo, hyphenated (Stubbs *et al.* 2001), since beefly looks like a brand of bovine stock cube.

Consequently I have partly followed this supposedly modern trend from the shield bugs of the nineteenth century, through a vague transition into shield-bugs in the mid-twentieth, to give us the shieldbugs we have today. This form was already well established, and although the more technical literature still often hangs on to two separate words, every popular British treatment I could find during the last 25 years now gives shieldbugs on a unified front. There are also leatherbugs, lacebugs, flatbugs and bedbugs, although contrariwise there remain stink bugs, ground bugs, squash bugs, tortoise bugs, Dock Bug and Parent Bug. No, this is not completely satisfactory, but it's what has evolved through the piecemeal jumble that is the English language, and it does just go to show how unhelpful, confusing and illogical English names can be.

In truth any logic in these common English names has long been broken. This is why scientific names were invented. Throughout this book the scientific names are the ones that hold authority, and any English names, if they are available, are given only subsidiary status. Nevertheless, certain groups of insects are currently enjoying increased popularity, particularly the relatively large and photogenic bug groups covered here. And many have acquired or have been newly christened with English names. The current style of the New Naturalist Library is to capitalise English common names. This works well enough for the well-established name pedigrees that support plants, fungi, birds, mammals, butterflies and moths, but can be counterproductive for the more obscure groups of insects. When I wrote the *Beetles* volume in 2018 I argued that not enough of the over 4,000 British beetle species had good enough common names, and the names that did exist (violet ground beetle, stag beetle, black oil beetle, red soldier beetle) were still too vague and nebulous to be conferred with capital letter importance. The editors of the series agreed, so in that case scientific clarity was upheld at the lesser expense of literary consistency.

Shieldbugs are on the cusp. They are now part of a burgeoning array of stepping-stone insect groups like dragonflies, grasshoppers, bees and hoverflies – slightly more difficult than butterflies and moths, but not as tricky as fleas or flea beetles. Here, English names can be a genuine help in understanding and do not get in the way of identification. Some are long-standing, some are inventive new coinings. In this case, with the scientific names to lead the way, I am happy to wash along with the publishing fashions of the day, and although I normally try to avoid too many extraneous capitals in my text, I do here meekly, if slightly grumpily, follow the prescribed style of the series.

Just over 30 years later 'bug' is already the default word used by Brookes (1763) for all manner of hemipterans, at least 40 species, from the original *Cimex* ('only a common bug') to *Notonecta* the 'water bug or boat-flie' and several probable shieldbugs, although he also includes May-bug (and May Bug) for the cockchafer.

Today the 'true bugs', the Hemiptera, are thought to comprise in excess of 100,000 species worldwide (that's an impressive 10% of all known insect species), and about 1,850 Hemiptera species occur in Britain and Ireland. Like most insects, many of these are tiny and never acquired useful common names, though very general terms like aphid, plant-louse, scale insect or leafhopper cover vast swathes of them. Shieldbugs represent a small group of some of the largest, brightest, most obvious and best-known amongst this broad and varied classification.

WHAT MAKES IT A SHIELDBUG?

It's not simply a case of claiming they are bugs which are shaped a bit like shields. What something looks like at first glance is not often a sound judgement on which to base any classification. There are now many hundreds of years of studious observation by which the organisms of this planet can be counted, accounted and arranged. In order to define shieldbugs properly, the other bugs in the insect order Hemiptera have to be identified and cut loose.

The Hemiptera in general are characterised by having four wings, sucking tubular mouthparts, prominent compound eyes, often also ocelli (simple eyes), and long to very short antennae, and they go through hemimetabolous development from egg to adult.[2] This 'incomplete metamorphosis' means that the tiny nymphs which hatch from the eggs are merely miniature wingless versions of the adults, already showing the correct body segmentation, full leg articulation, properly developed eyes and antennae. As they grow they moult their skins several times, and at each point the wing buds increase until the final winged adult form emerges. This is a very different process from the holometabolous development ('complete metamorphosis') of flies, beetles, bees and butterflies, which pass through a feeding and growing maggot, grub or caterpillar stage before the entire

2 The term hemimetabolous is sometimes only used for orders like Odonata (dragonflies) and Plecoptera (stoneflies) where an aquatic nymph or naiad then changes rather dramatically into a terrestrial/airborne adult. I use it here for the Hemiptera because it is a well-known umbrella term often used to cover *all* orders lacking the chrysalis stage. If we're getting technical, though, the gradually increasing wing-bud scheme of the Hemiptera is called paurometabolous growth, whilst that of silverfish and the like where there are no wings, or indeed wing buds, is ametabolous growth.

body plan is reshuffled (liquidised might be more accurate) inside the chrysalis to produce a winged adult insect wholly different in form and structure from the larva. The importance of the sucking mouthparts of the Hemiptera has long been appreciated, and indeed they are what characterise all true bugs. The order has also variously been designated as Rhyngota, Rhynchota or Rhyngotorum over the years, all derived from the Greek ρινο *rhino*, meaning snout or nose. The fine structure of the rostrum, as the beak-like tube mouthparts are called, is unique, and defines the Hemiptera. More details are given in Chapter 2.

Today the order Hemiptera (Greek ήμι *hemi* = half, and plural of πτερόν *pteron* = wing) is divided into 2–5 suborders, depending upon whose scheme you follow (Fig. 3). The suborder *Homoptera* (Greek όμο *homo* = same, and plural of πτερόν *pteron* = wing) have front and back wings very similar in shape, size and texture; at rest they are folded tent-like along the abdomen, and they are most often transparent and clear, but a few are coloured or patterned. This is a large group including leafhoppers, cicadas, aphids and psyllids, though it is often considered defunct because its evolutionary origins are unclear. Instead this slightly artificial grouping is usually subdivided along evolutionary lines into the suborders Coleorrhyncha (moss bugs, only 30 world species in a single family, Peloridiidae), Auchenorrhyncha Cicadamorpha (cicadas, leafhoppers, spittle bugs etc.), Auchenorrhyncha Fulgoromorpha (planthoppers) and Sternorrhyncha (aphids, scales, whitefly etc.). Little else needs to be said about these insects here except that they all share that hemimetabolous upbringing and tubular piercing sucking mouthparts.

The shieldbugs belong to the remaining suborder, Heteroptera (Greek έτερο *hetero* = different, and plural of πτερόν *pteron* = wing), which has very different front and back wings, usually resting more or less flat over the abdomen. The front wings are normally long and slim, partially coloured and sclerotised (hardened) in the basal section, or at least coriaceous (leathery) and opaque, whilst the hind wings are broadly triangular, membranous, clearly transparent and partially folded concertina- or origami-style. At rest the more delicate hind wings are furled away and covered by the tougher protective layer of the partly overlapping front wings. Together with the large toughened upper front plate of the thorax (the pronotum), this gives the 'het' bugs (as they are sometimes known) a more solid, robust form than many of the more flimsy Homoptera. Their under-bodies also tend to be tougher and harder.

This hard toughness comes from the chemical structure of the insect body. Insect cuticle is mostly made up of chitin, a long-chain polymer of *N*-acetylglucosamine; it is a polysaccharide analogous to the cellulose found in plants. Chitin is laid down in a series of parallel fibres to form thin flat sheets;

FIG 3. Examples of the Hemiptera. (a) Moss bug (*Xenophyes rhachilophus*, suborder Coleorrhyncha). (b) Rhododendron Leafhopper (*Graphocephala fennahi*, suborder Auchenorrhyncha Cicadamorpha). (c) Spotted Lanternfly (*Lycorma delicatula*, suborder Auchenorrhyncha Fulgoromorpha). (d) Cottony Scale (*Icerya purchasi*, suborder Sternorrhyncha). (e) Back-swimmer (*Notonecta* species, Hemiptera Heteroptera Nepomorpha). (f) Green Shieldbug (*Palomena prasina*, Hemiptera Heteroptera Pentatomomorpha), perhaps the most shieldish of shieldbugs.

a series of sheets, layered one on top of the other, with the grain of the fibres of each sheet at slightly different angles, gives a superior strength and rigidity much as do the laminated layers of plywood. The chitin fibres and layers are impregnated, interlinked and reinforced with proteins, and it is this combination of chitin sheets and protein glue that gives insects their tough yet flexible bodies. All insects use chitin, but in some hard and tough insects (het bugs are a very good example) the chitin layers are laid down so thickly that they achieve real armour-plating strength.

With their hardened wing structure, stout head and pronotum, and toughened underbellies, the Heteroptera slightly resemble beetles (Coleoptera), another heavily sclerotised group of insects. Beetles are one of the most diverse and species-rich groups of insects on the planet, a status they are thought to have achieved by the development of their hard shell-like bodies and tough protective wing cases (called elytra, singular elytron), allowing them the joint advantages of flying and of also pressing hard into tight spaces to hide without damaging their fragile flight wings. The convergent evolution that also toughened the bodies and hardened the front wings of the Heteroptera has conveyed a similar protective advantage to these bugs, and this large grouping has many species that regularly push down into flowers, the general foliage, plant root thatch or leaf litter, burrow into loose soil, live under rocks and stones, push between the scales of pine cones, under rotten bark or into fungi, occur as parasites on birds and mammals, live in or on water, including inside rocks on the seashore, or ride on the open ocean. Arguably the Heteroptera occupy a more diverse array of habitat types than any other insect group. No Hemiptera burrow through wood, or are internal parasites of other animals, but this is more a reflection of their non-chewing mouthparts, as discussed later.

The hemi 'half' in the name Hemiptera refers to the fact that most large bugs have the basal half of the front wings hardened and coloured, whilst the terminal half (called rather unimaginatively the 'membrane') remains transparent and membranous. This reiterates that comparison with beetle elytra; thus the hemipteran forewing is usually referred to as a hemelytron (occasionally hemielytron) – half a wing case.

A key body part that helps define a shieldbug is the large flat triangular plate, behind the pronotum, right in the middle of the back. This is the scutellum – quite literally Latin for 'little shield', being the diminutive form of *scutum*, the large, rectangular, slightly curved shield carried by Roman legionary soldiers. All hemipteran bugs have a scutellum, but it is particularly large and obvious in many of the shieldbugs, and this shield-plate is often given as the reason behind the common name, rather than these insects simply being generally shield-shaped. John Ray (1710) used the term *interscapulum* (Latin, 'between the shoulders'), but this word has almost completely fallen out of use except in medical textbooks referring to the upper back, or some rather technical bird descriptions to denote the area between the wings just behind the head.

Traditionally, true shieldbugs comprised the family Pentatomidae, so named for the five (*penta-*) rather than four segments that made up the antennae of most other bugs. But as I bemoaned above, English names for insects can be rather vague and amorphous, and this narrow classification rather unfairly excluded many groups with four-segmented antennae which had very obviously similar body form and which were also regularly called shieldbugs. It is at this point that we have to admit that 'shieldbug' is a somewhat arbitrary conglomeration of groups, of which most (but not all) are shield-shaped, most (but not all) have a prominent scutellum, and some (but not all) have five-segmented antennae.

Conveniently, modern scientific treatments of the suborder Heteroptera recognise an intermediate subdivision – the infraorder Pentatomomorpha, one of seven infraorders into which all het bugs are grouped. This will form the outer limits of what might reasonably be considered shield-type bugs. It is defined by numerous very abstruse technical characters including antennae with 3–5 segments, labium mostly four-segmented, large scutellum, costal fracture of forewings never present, pretarsus with claws equally developed, and pulvilli well developed, divided into basipulvillus and lamellate distipulvillus, and absence of a true egg operculum. This doesn't really help with fieldwork, where these awkward characters are often difficult to understand or impossible to see. And covering all of these groups would make this book unwieldy, and it would be overwhelmed by the very numerous much smaller and less shield-shaped members. Instead, I have chosen a personal pragmatic selection. For the basis of this book, I am including all the British families of the Pentatomomorpha, except the superfamilies Aradoidea and Lygaeoidea. This was, I admit, a bit of a wrench, but there is barely enough time and space so

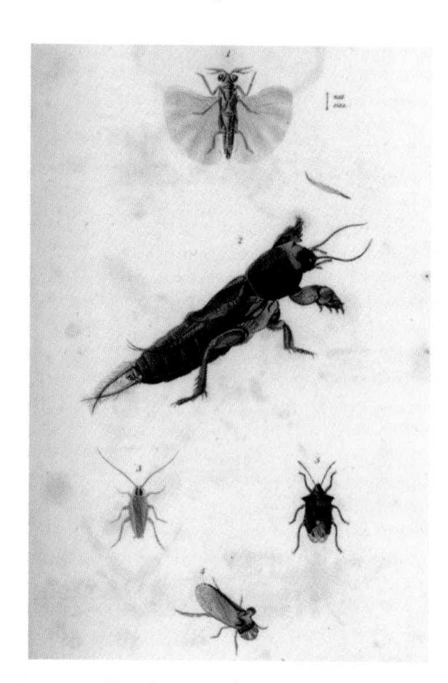

FIG 4. Keeping good company, *Pentatoma rufipes* (number 5) represents the Hemiptera Heteroptera, along with various 'other orders' (i.e. not the big four groups Coleoptera, Diptera, Hymenoptera, Lepidoptera) in this illustration from Kirby & Spence's *Introduction to Entomology* (1815–26). Also shown are: (1) *Xenos peckii* (Strepsiptera), (2) *Gryllotalpa gryllotalpa* (Orthoptera), (3) *Blatella germanica* (Blattodea) and (4) *Ledra aurita* (Hemiptera Homoptera).

FIG 5. Two extremes of shieldbug form: (left) typical shield-shaped *Palomena prasina* (Pentatomidae) and (right) lacy and spined southern European *Phyllomorpha lacerata* (Coreidae).

something had to give. But I can't quite bring myself to get rid of them entirely and have sneaked in, under the editorial radar, a paragraph on each (see box overleaf). This finally gives a workable 79 British 'shieldbug' species across 11 families, and although some are not particularly shieldish, there is a genuine biological cohesiveness in that they share a close evolutionary descent, have many similar body structures and forms, are likely to be found by the field naturalist on or under plants in the general landscape, without recourse to specialist searching techniques; and they are mostly large enough to appreciate and identify using a hand lens, often with just the naked eye (Fig. 5).

These, then, are shieldbugs, or at least my broad interpretation, and this is a permutation on the same group of families that have been covered in many other recent British and European publications including those by Hawkins (2003), Evans & Edmondson (2005), Boardman (2014), and Nielsen & Skipper (2015). I hope specialist entomologists will not consider my choices too jarring, or less specialist naturalists find it too confusing. I can see a logic here, and I hope I can justify it to you, dear reader. A summary of world and British shieldbug families, genera and species is given in Table 1, at the end of this chapter.

SOME NON-BRITISH SHIELDBUG GROUPS

As in most insect groups, shieldbugs are seriously under-represented in Britain when compared to the rest of the world. They are especially diverse in the tropics and subtropics. A limited number of genera and families occur here, but a much

EXCLUDED GROUPS – BRITISH PENTATOMOMORPHA DELIBERATELY LEFT OUT

FLAT, AND NOT REALLY SHIELD-SHAPED – ARADOIDEA

Commonly called flatbugs, or bark bugs, there are only seven British species anyway, in the family Aradidae; they are all small (less than 7 mm) and secretive, strongly flattened bugs that only occur under fungoid tree bark (Fig. 6). The commonest are *Aneuris laevis* and *Aradus depressus*. Worldwide, though, this is a huge group (2,050 species at least) and many are large (up to 20 mm) and attractive – not necessarily brightly coloured, but interestingly granulate or sculptured, and variously shaped oval, rectangular or triangular. Because of their secretive life under the bark, many tropical species no longer have winged forms. The mouthparts are much longer than those of other bugs, with the piercing stylets coiled around inside the head. These are used, under the bark, to penetrate deep into the hyphae of the fungal strands, and this mode of feeding may be ancestral for all of the Pentatomomorpha. Some exotic species live solely in termite nests, as do the eight known species of the related Termitaphididae. All British aradids are adequately keyed out by Southwood & Leston (1959).

FIG 6. Flatbugs occur only under fungoid bark. *Aneuris laevis*, adults (left) and pale nymphs (top left); and *Aradus depressus* mating pairs (right).

FAST, AND NOT REALLY SHIELD-SHAPED – LYGAEOIDEA

Usually called ground bugs for their habit of running about on the ground, or seed bugs because many feed on seeds, lygaeids are attractive under the microscope, but most are small or very small, and rather 'other' compared to shieldbugs (Fig. 7). And with over 100 British species, they would swamp this book if they were included. Most British species belong to the family Lygaeidae (although this has been subdivided by some authors), and typical species include the common *Heterogaster urticae*, which often occurs in large aggregations on Stinging Nettles (*Urtica dioica*), *Kleidocerys resedae*, a small but very smelly bug infesting birch trees, and *Rhyparochromus vulgaris*, a handsome well-marked and long-legged denizen of rough grassland where it rushes about on the soil at top speed if disturbed. The related Berytinidae are slim and very long-legged and despite their fragile forms some are thought to be at least partially predatory. Most British species can be identified using Southwood & Leston (1959), but the most up-to-date European treatment of the Lygaeidae is provided in the three volumes by Péricart (1998), PDFs of which are now freely available to download from the faunedefrance.org website.

FIG 7. *Rhyparochromus vulgaris* (left) is a typical ground bug, with its mottled earthy palette and large shield-like scutellum. *Lygaeus equestris* (right) could easily be mistaken for *Pyrrhocoris* or *Corizus*; it is widespread in mainland Europe.

broader diversity occurs elsewhere. Before going into more detail on our native species, it is worth a brief diversion across the globe to consider some non-British shieldbug groups.

Despite the fact that the Australian family Henicocoridae has just a single species and the Australian/South American Idiostolidae just five species, these families are interesting because they have several characters intermediate between true shieldbugs and ground bugs in the superfamily Lygaeoidea. These are mainly to do with the distribution of trichobothria, long sensory bristles that detect air movements or vibrations, on the underside of the abdomen. Tiny and insignificant as these hairs might seem, they are remarkably persistent across shieldbug groups, even in fossils, and they confirm the close relatedness of the various families. This deep southern-hemisphere distribution is typical of truly ancient lineages (like marsupials, kauri conifer trees, southern beech *Nothofagus* forests), and were this a 1950s science fiction book I might be describing them as missing links in the classification. They superficially resemble and were originally described as Lygaeidae, but are now given their own separate family statuses. Little is known of their life histories, but they inhabit moss and leaf litter in humid temperate forests and are thought to be plant-feeding.

FIG 8. Unnamed dinidorid shieldbug, its broad stout form showing its clear relatedness to the Pentatomidae.

There are only nine species in a single genus in the family Canopidae – all from Central and South America. These strange small hemispherical shining black, greenish or purplish bugs are nearly all scutellum, and both front and back wings have particular folding mechanisms to furl them away. They feed on the fruiting bodies of bracket fungi in tropical forests, and fungal spores have been confirmed in their guts. They are often found in conjunction with similarly globose shining metallic fungus beetles in the family Erotylidae. No British Hemiptera are known to feed on fungal fruiting bodies.

The Dinidoridae are large, heavily built insects mainly from Asia and Africa (Fig. 8). They are obviously shieldbugs, but separated into their own family on minor details of head structure and sensory hairs under the abdomen. Some have two tarsal segments like the Acanthosomatidae, some three like the Pentatomidae.

Two Australian species make up the family Lestoniidae. Though hemispherical and subglobular, with a huge scutellum and massive rounded pronotum, these lentil-shaped insects share many evolutionary features with the Acanthosomatidae, including two-segmented tarsi together with the presence and position of various abdominal glands. They feed on the growing tips of Australian cypress trees in the genus *Callitris*.

The Phloeidae are large broadly flattened bugs, up to 30 mm long, with wide leafy flanged body edges adapted to clamping down to hide on South American or Bornean tree trunks, where they vanish in the rough bark and lichen layers (Fig. 9). They are renowned for their maternal habit of protecting the young nymphs by carrying them about tucked in on the underside of the body, concealed by the skirt-like flange of their outline. Their eyes are separated into one pair of lobes on the top edge of the head looking upwards, and another pair just under the flange looking down at the tree bark. They also have the ability to squirt a jet of liquid a considerable distance from their bodies, though this is probably excretory rather than defensive.

The Tessaratomidae are a large group of large (to 45 mm long), mostly tropical shieldbugs which all have remarkably small heads. The nymphs of many species

FIG 9. Portrait of *Phloea corticata* (Phloeidae) showing the flat leaf-like edges to its body that help disguise it on tree trunks.

are broad and flat, often rectangular, frequently brightly coloured and regularly pictured in books for their attractive appearance. Specimens of the 'Lychee Beetle' (*Tessaratoma papillosa*; Fig. 10) are often sold embedded in resin blocks as collectibles; again, these are usually the brightly coloured red, orange, white and blue nymphs, rather than the brownish adults. They are reputed to be able to squirt a jet of chemicals 10–15 cm, presumably as a defence against predators. A green edible species, *Encosternum delegorguei*, is collected by the bucket-load in Zimbabwe and South Africa. They are first doused in water and agitated to get them to release the noxious chemicals (which can stain the skin brown). They

FIG 10. Adult and nymph of the 'Lychee Beetle' (*Tessaratoma papillosa*, Tessaratomidae).

FIG 11. A bucket of harvested *Encosternum delegorguei* (Tessaratomidae) ready to be cooked and eaten.

are then sorted to remove any dead bugs – these still contain the bitter defensive secretions since only live bugs can eject the contents of the glands. They are then cooked, dried and eaten, after discarding the wings (Fig. 11). With a little salt they are consumed as snacks, but are also claimed to aid digestion, or cure hangovers.

DON'T GET MUDDLED – OTHER INSECTS A BIT LIKE SHIELDBUGS

British shieldbugs are a fairly neatly precise group of insects, despite my apparent difficulty in pinning them down for this book, but to the novice entomologist they can be muddled up with several other insect types. A common confusion is with beetles, which are perhaps more familiar to a general naturalist. Indeed, bugs were sometimes classified as beetles in many pre-modern schemes, as demonstrated by Thomas Mouffet (1658), who includes a probable lygaeid ground bug amongst his 'lesser beetles'. Pentatomid nymphs in particular are often mistaken for ladybirds – being small, round, domed and shiny, frequently brightly coloured red or yellow, with black spot-like markings. Conversely, tortoise beetles (*Cassida* species) just don't seem domed enough to be real beetles and are often mistaken for shieldbugs, particularly as many of them are the same leaf-green as the well-known *Palomena prasina*. The easy distinction is the four- or five-segmented antennae of the bugs, compared to the normally 11-segmented antennae of beetles (Fig. 12). And even the smallest of shieldbug nymphs has a prominent segmented rostrum (snout) underneath its body where beetles all have small biting jaw mouthparts.

It's still easy to make mistakes, though. I am reminded of a meeting of the British Entomological Society several years ago, where I exhibited what I thought was a curious large and furry flower beetle (family Nitidulidae) that I had found on holiday in Greece (Jones 1991). I had hastily (= foolishly) relied on general form and shape to make my tentative appraisal. I was slightly embarrassed when at the next meeting of the society it was revealed by Roger Hawkins, who had taken the specimen away with him, to be a shieldbug of the family Scutelleridae – the circum-Mediterranean *Irochrotus maculiventris*.

WHAT'S IN A PLACE NAME?

The blurb for the New Naturalist Library claims that the series is aimed at the general naturalist interested in the wildlife of Britain. That's all very well, but wildlife does not respect national or political boundaries, and even after Brexit the UK remains, effectively, just a bit of north-west Europe as far as plants and animals are concerned. I've always lived in England and the English are frequently accused of being too Anglocentric in their outlook, often with embarrassing overtones of imperial pomposity and entitlement. I really have done my best to step outside this stereotype.

Throughout the text of this book I've tried to be as accurate in my geographical descriptions as I can. England really does mean that single country, which together with Wales and Scotland make up the island of Britain, sometimes considered Great, sometimes not. Likewise Northern Ireland is that part of the island of Ireland still politically, royally and governmentally aligned with Great Britain to form the United Kingdom of etc., etc., whilst the Republic of Ireland is the rest of it. These, together with various outlier islands, form the British Isles. And although I sometimes use 'British and Irish' and at other times just 'British', the two descriptors are often interchangeable. I apologise if I have sometimes fudged my words a bit: this is in the interests of readability, and the avoidance of repetition or clumsy word constructions. The Channel Islands are also often included as part of the British Isles, though they are not part of the UK, and biogeographically they are merely a few large rocks offshore from mainland Europe. But because of a tradition of shared language and monarch, they are often included in books on 'British' natural history, and I mention them often enough here. The anomalous Isle of Man is 'British' in geographical terms rather than political. Unfortunately it barely gets a mention here, mainly because its shieldbug fauna is little reported and poorly understood – although by my admittedly lazy intuition it ought to be intermediate between those of Britain and Ireland. Perhaps Manx naturalists will take this as a challenge rather than an insult.

Although I regard myself as variously English or British, depending on the context, I have friends, family and colleagues in Europe and so I also consider myself very European. Shieldbugs hold no national, political or social allegiance, they go where they will, or where they are accidentally taken by human actions. Any notional British fauna is constantly fluctuating and changing, and the rest of Europe plays a large part in what is happening now and what happens next. This book is therefore *mainly* about the shieldbugs that occur in the British Isles (in that very broadest sense), but constantly makes reference to species beyond our shores, for their potential imminent arrival here, for the input they make to a broader understanding of ecology, physiology and evolution, or by way of an interesting exotic contrast to our parochial British fauna. For these jaunts I make no apology.

FIG 12. Mating pair of Fleabane Tortoise Beetles (*Cassida murraea*). Like shieldbugs, beetles can be broad, round and domed, but the scutellum is usually small, tiny or invisible, and the antennae have about 11 segments rather than four or five.

It is possible that cockroaches, four British *Ectobius* species, might be confused. They are uncommon rough-grassland insects, pale straw-coloured like *Aelia*, but longer, flatter and more smoothly oval than shieldbugs. They also run like the wind, far faster than any sedate pentatomid, and they fly readily. The occasional large introduced domestic pest cockroach like *Blatta orientalis*, which has shortened wing cases, is also a bit shield-like. Actually, shieldbugs are probably more likely to be mistaken for cockroaches by the non-expert, and unfairly persecuted accordingly. All cockroaches are easily distinguished by their long thin hair-like antennae and short multi-segmented feeler-like tail appendages called cerci, which are completely absent in the Heteroptera. These same features will also distinguish the Field Cricket (*Gryllus campestris*): although it has jumping hind legs like a grasshopper, the legs are shorter and the whole body form is shorter, broader and flatter, perhaps very slightly shieldbug-like ... in some lights ... if you squint. *Gryllus* is highly unlikely to cause problems, though, because it is very rare here. Other possibilities are the widespread but uncommon House Cricket (*Acheta domesticus*), or any of the similar species bred by commercial suppliers as food for pet tarantulas, lizards or the like; these crickets are fed live to the pets, so often escape into urban gardens or courtyards.

FIG 13. Named for its obvious resemblance to het bugs, the parasitoid fly *Phasia hemiptera* has broad darkened hemelytra-like wings, especially in the male, as seen here.

One of the strangest bug-like insects is the fly *Phasia* (formerly *Alophora*) *hemiptera*, named for its obvious and close resemblance to shieldbugs (Fig. 13). It has broad coloured black and brown wings that look very like hemelytra, especially in the male where they are broader, more triangular and very un-fly-like. The similarity, first officially noted by Danish naturalist Johan Christian Fabricius (1745–1808) when he named it in 1794, is more than coincidental or decorative. The fly, one of a large group of insect parasitoids in the family Tachinidae, lays its eggs on shieldbugs, and the developing grubs eat their poor hosts alive from the inside. The fly is widespread throughout the British Isles, a frequent denizen of damp woodlands where it visits hogweed and other flowers. There are several other shieldbug parasitoids in this fly group (see page 116–19), but no other so closely matches its victims in physical appearance.

The insects most likely to be confused with shieldbugs are actually other bugs. The Saucer Bug (*Ilyocoris cimicoides*) is very shieldbug-shaped, and has a large scutellum to match (Fig. 14). It is an aquatic insect (no shieldbugs live in water), but some individuals fly and it has turned up in moth light traps to the confusion of the watchers. It is easily distinguished from shieldbugs by its tiny

FIG 14. Out of water the Saucer Bug (*Ilyocoris cimicoides*) is rather shield-shaped, but it can easily be distinguished by its feathered swimming back legs and its tiny, almost invisible, antennae.

FIG 15. The cone bug *Gastrodes abietum* is one of the broadest of the seed bugs (family Lygaeidae) and rather shieldbug-like at first glance.

antennae, almost invisible beneath the eyes. And it can give a painful bite with its sharp predatory rostrum; no shieldbug would ever do such a thing. Some of the larger ground bugs (family Lygaeidae), deliberately excluded from this volume, share the shieldbug look of having a large scutellum. They are a valid part of that overarching infraorder Pentatomomorpha, but generally have longer legs and are more active, running about under the herbage, rather than over it. Perhaps the most likely of them to be muddled are the cone bugs (*Gastrodes* species), which are brown, broad, oval and flat for squeezing between the scales of pine cones after the seeds (Fig. 15), and some of the oval *Graptopeltus lynceus* or *Emblethis* species that run about on the soil. Several large lygaeids, broad, bright black and red species like *Horvathiolus superbus, Melanocopryphus albomaculatus* and *Spilostethus pandurus*, are currently scarce, possibly vagrants here at the moment, but are perhaps on the verge of colonising the British Isles; they superficially resemble *Pyrrhocoris apterus* and *Corizus hyoscyami*. The large, broad, and squat leafhopper *Issus coleoptratus* is also remarkably shield-shaped, but like *Ilyocoris* it has tiny antennae, and all doubt is dispelled when it pings suddenly away using its large, spring-powered hind legs – no shieldbug can hop, skip or jump (Fig. 16).

And at the edges of my broad shieldbug selection, several rather non-shield-shaped bugs start to merge with other bug families in shape and habits – always a good opportunity for confusion (Fig. 17). The rare migrant *Mecidea lindbergi* and narrow rhopalids like *Myrmus miriformis* and *Chorosoma schillingi* look more like some of the grass bugs *Stenodema* or *Leptopterna* (family Miridae), *Alydus calcaratus* rather resembles some of the *Nabis* or *Himacerus* damsel bugs (family Nabidae), and the striking *Corizus hyoscyami* is easily muddled with many of the large bright ground bugs mentioned earlier, or *Lygaeus* that occur in Europe though not quite yet in the British Isles.

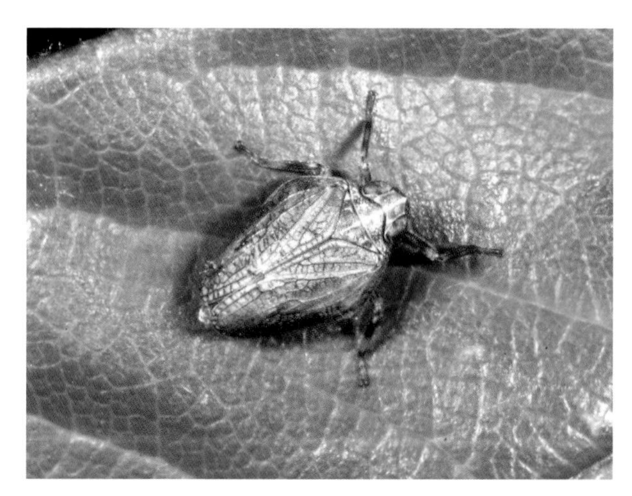

FIG 16. *Issus coleoptratus*, Britain's broadest and squattest leafhopper. It might be mistaken for a small shieldbug – until the moment it suddenly jumps away on its powerful back legs.

FIG 17. The shieldbug *Chorosoma schillingi* (left) looks very like one of the grass bugs, *Stenodema*, family Miridae (below).

FIG 18. A mirid plant bug, *Plagiognathus arbustorum* (left), looking slight and slim beside a typical broad and stout shieldbug, *Eysarcoris venustissimus* (right).

On the whole, though, once you get your eye in, shieldbugs have a clear look to them (Fig. 18). Now the question becomes not 'is it a shieldbug?' but 'what type of shieldbug is it?' Table 1 lists the shieldbug families and indicates the number of genera and species in each. Chapter 7 is an illustrated key to British and Irish (and nearby) shieldbugs, and Chapter 8 is an illustrated catalogue of them. Many British species are large and distinctive and can readily be identified from photographs, but some are easily confused and need to be examined under a lens, or at least with an understanding of their form and structure. So rather than jump straight into an identification guide, the following chapters will look first at the ground-plan layout of the insects, their strange and sometimes unique structures, which themselves are tied intricately into their ecologies, life histories or behaviours.

TABLE 1. World and British Pentatomomorpha families, genera and species – with numbers derived from Schuh & Weirauch (2020) and British numbers mostly taken from the latest checklist published on the britishbugs.org.uk website in June 2022. Genera and species covered in this book are marked in bold.

Family	*Description*	*World*		*British Isles*	
		Genera	*Species*	*Genera*	*Species*
Aradidae[1]	Small, flat, under tree bark, feeding on fungal hyphae	≈285	≈2,050	2	7
Termitaphididae	Tiny, flat, in termite nests, maybe feeding on fungi	2	8	–	–

Family	Description	World		British Isles	
		Genera	Species	Genera	Species
Henicoridae	Small, oval, in Australian leaf litter	1	1	–	–
Idiostolidae	Small, oval, in southern-hemisphere forest leaf litter	3	5	–	–
Acanthosomatidae	Small to large elongate/oval, or kite-shaped, tarsi 2-segmented	57	>287	4	5
Canopidae	Small, subglobose, shining black, in the Neotropics	1	9	–	–
Cydnidae	Small, dark, often shining, legs with spines for digging	>113	>965	7	10
Thyreocoridae	Small, dark, domed, shining, legs with spines for digging	31	220	1	1
Dinidoridae	Large, subquadrate, heavy, from Africa, Asia, Australasia	>16	>107	–	–
Lestoniidae	Tiny, convex, circular, heavily punctured, Australian	1	2	–	–
Megaridae	Tiny, ovoid, globose, with massive scutellum, Neotropical	1	16	–	–
Pentatomidae	Medium to large, mostly shield-shaped, but also broad to long	≈940	≈4,948	25	31
Phloeidae	Large, flattened, edges flanged, foliate, on tree trunks	3	4	–	–
Plataspidae	Tiny, globose, shining, beetle-like, scutellum huge	3	4	1	1
Saileriolidae	Small, oval, head protuberant, South-East Asia	2	3	–	–
Scutelleridae	Medium to large, broad oval, with scutellum covering body	≈100	>531	2	5
Tessaratomidae	Large, broad oval, shield-shaped, head very small	≈62	≈252	–	–
Urostylidae	Medium-sized, elongate, shield-shaped, south and east Asia	4	80	–	–
Largidae	Medium-sized, narrow, head broad, eyes bulbous	24	≈222	–	–
Pyrrhocoridae	Medium-sized to large, narrow, often black and red	46	453	1	1

continued overleaf

TABLE 1. *continued*

Family	*Description*	*World*		*British Isles*	
		Genera	*Species*	*Genera*	*Species*
Alydidae	Medium-sized, long and narrow, head broad, legs long	54	282	1	1
Coreidae	Medium to large, parallel-sided or lobed like dead leaf, often brown	436	2,567	10	11
Hyocephalidae	Large, narrow, head long and tapered, Australian	2	3	–	–
Rhopalidae	Medium, long oval or narrow, head broad, mostly dull brown	23	224	7	11
Stenocephalidae	Large, narrow, parallel-sided, mostly on spurges	1	30	1	2
Lygaeoidea[2] (16 families)	Tiny to large, mostly narrow ovoid, huge diversity, mostly seed-feeders	752	4,424	51	104
Totals[3]		**≈3,000**	**≈17,700**	**113**	**190**

[1] The flatbugs (Aradidae) are not included in this book. Worldwide this is a large and important group, but there are only seven British species and they are behaviourally completely separate from shieldbugs.

[2] The ground bugs (Lygaeidae and other families) are also excluded from this volume. They are mostly ground-dwelling seed-feeders, little like 'true' shieldbugs, and would take another monograph (perhaps several) to cover anyway.

[3] Consequently, useful totals for 'British' shieldbugs and related groups covered in this book are **60 genera** and **79 species**.

CHAPTER 2

Shieldbug Structure

Like most insects, shieldbugs have a head at one end and genitals at the other, here described in traditional travelogue style from start to finish. Unlike, say, ants, earwigs or butterflies, shieldbugs have a squat, generally rounded or oval form, and the usual insect body-part distinctions – head, thorax, abdomen – are less immediately obvious. But they are there. Shieldbugs are like beetles in having those hardened and coloured hemelytra snugly obscuring the abdomen, and these fit tight up against the thorax to give, along with the head often recessed into a notch at the front, a unified single-outline silhouette. The true tripartite insect form is only really visible as the shieldbug takes flight – the wings are opened and the abdominal segments are revealed. The accompanying diagrams, showing whole and part bugs, wings closed and wings open, should be self-explanatory (Figs. 19–22). As ever with particular insect groups there are always specialist technical jargon terms, but these are kept to a minimum here. They are shown on the anatomical diagrams, and many are further explained in the glossary on pages 401–8.

FIRST – THE HEAD

Although it is sometimes difficult to see, the most distinctive part of the shieldbug head is the rostrum, the feeding tube that so clearly defines the Hemiptera. It was often called the 'beak', even in specialist technical monographs. This is normally held underneath the body, and is usually only easily visible when the bug is feeding. It is a complex cylindrical structure made up of a close-fitting bundle of elongate structures that originally corresponded

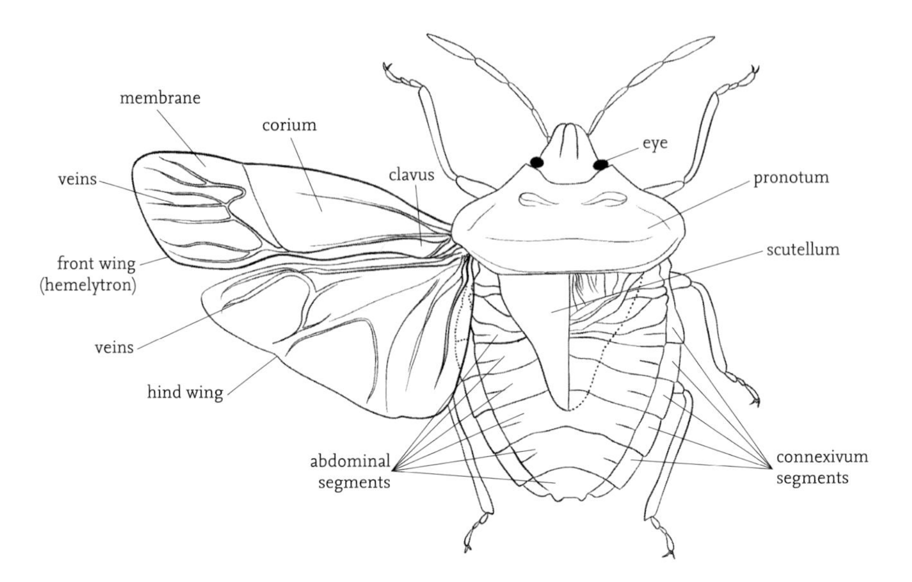

FIG 19. Generalised shieldbug structure: from above.

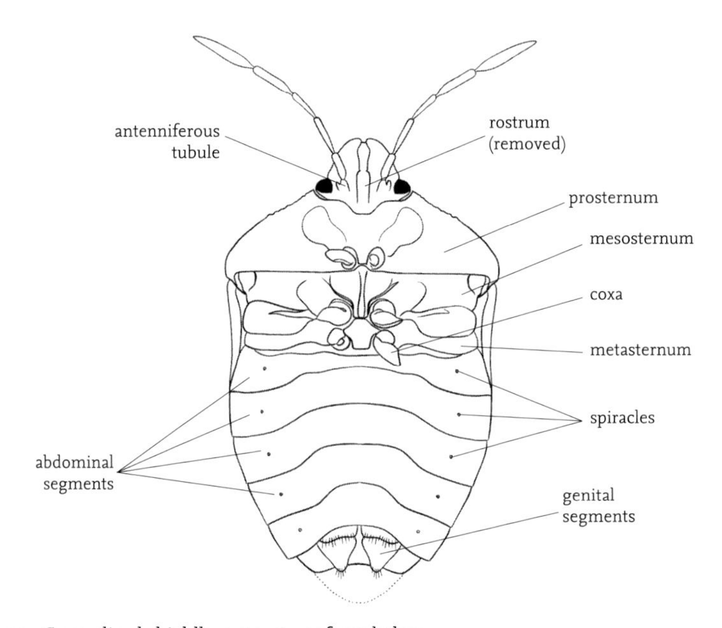

FIG 20. Generalised shieldbug structure: from below.

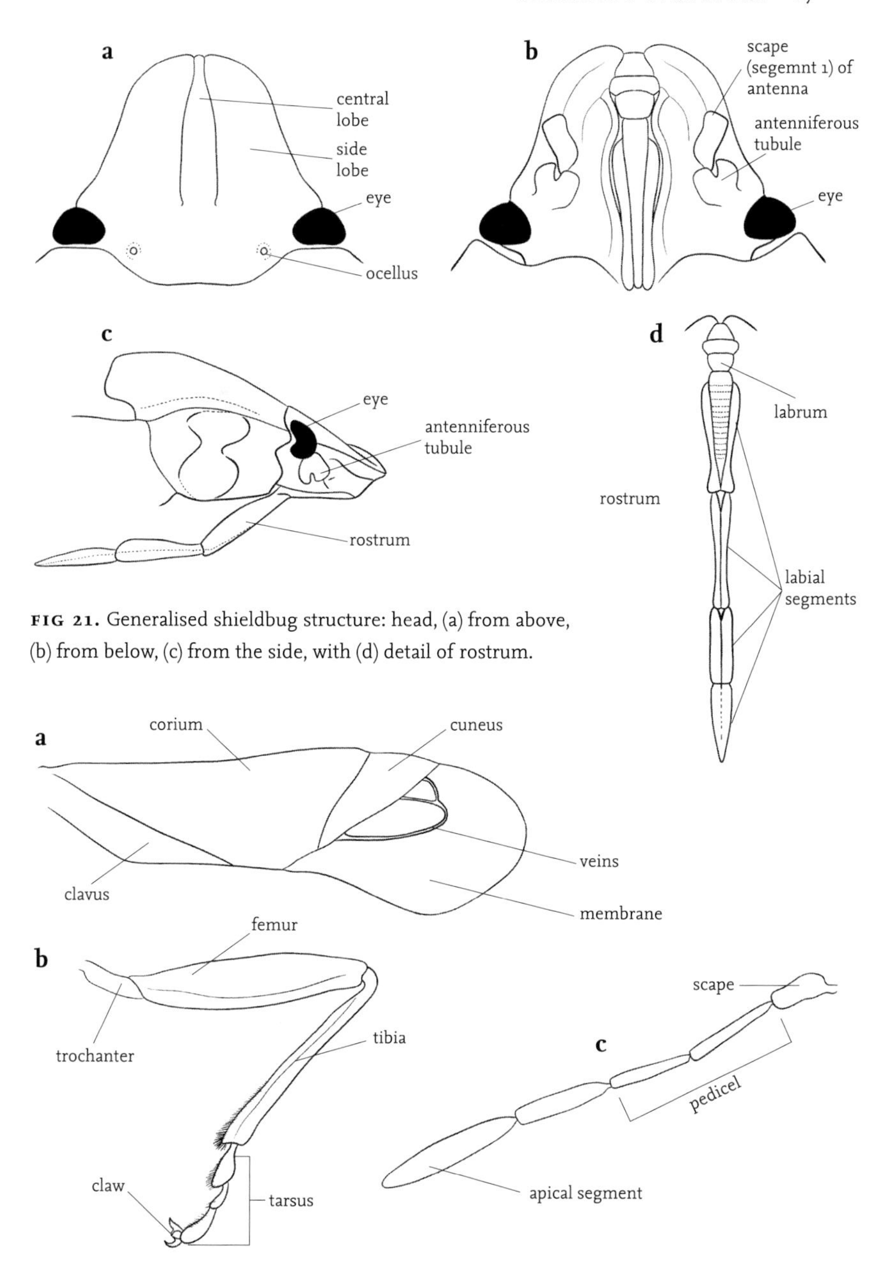

FIG 21. Generalised shieldbug structure: head, (a) from above, (b) from below, (c) from the side, with (d) detail of rostrum.

FIG 22. Generalised bug structure: (a) hemelytra (forewing) of a leaf bug (Miridae) – shieldbug families lack the fold marking off the cuneus as a distinct portion; (b) leg; (c) typical five-segmented antenna of a pentatomid bug.

to segmented limbs in some long-extinct primordial multi-legged ancestor of insects. Today the mouthparts of some insects are still identifiable as being the evolutionary remains of three separate body segments, each of which had limb-like articulations, but which are now reduced to: mandibles (jaws) of just a single scissor-blade segment; maxillae, with a basal food-manipulating lobe and short antenna-like maxillary palpi of up to about five segments; and the labium, with a lip-like shield and very short antenna-like labial palpi of two or three segments. Some of these structures are still visible in modern beetles, earwigs, grasshoppers, ants, wasps etc. In the Hemiptera they have evolved beyond easy recognition; many structures are missing, but some can still just about be interpreted under very high-power microscopic dissection.

In shieldbugs (indeed all Hemiptera) the labium (the lower lip) forms a three- or four-segmented outer sleeve that almost completely envelops the pair of long sharp piercing mandibular stylets (Fig. 23); these are evolved from the same structures that give two biting mandibles (jaws) in other insects. These mandibular stylets run up and down grooves alongside the pair of maxillary stylets (evolved from two maxillae). These four stylets form an inner core with two internal cylindrical tube-like channels running their lengths – the salivary canal takes salivary juices down into the food, and the larger food canal allows the partially digested liquid food to be sucked up. There are no palpi, the segmented feeler-like appendages that many insects use to fine-manipulate food particles near the mouth.

The outer labium sheath is narrowly but flexibly open along its front edge, a bit like the cylindrical foam insulating sheaths you can get to lag water pipes

FIG 23. Close-up of *Gonocerus* head, showing the split sheath of the labium wrapped around the rostrum mouthparts.

in the loft. This enables the whole thing to disarticulate from the stylet bundle inside, folding back on itself elbow-like as the long straight piercing skewer spike of the stylets is pressed down into the food source. This can sometimes be seen in side view if the bug is found engrossed in feeding. A very few notorious shieldbugs, but a broader swathe of heteropteran bugs in general, are well-known plant pests, so more details of feeding are given in Chapter 3.

In addition, the labrum (upper lip), which in many other insect groups forms a protective cover across the top of the jaw articulations, is a long narrow plate which acts like a reinforcing hem at the end of the labial split at the base of the rostrum. When the rostrum is held under the body, as the bug is walking about, the base of the labium sits neatly in a shallow gully underneath the head. This groove is ridged each side by two raised flanges, the bucculae (singular buccula), which are prominent enough to be visible in side view in some species. Behind and above the labium is the clypeus, a flattish protective plate at the front of the top edge of the head. In some insects (beetles and ants, for example) this is clearly visible as a smooth extension of the forehead or frons, but in shieldbugs it is divided into three – a central clypeal lobe sometimes called the tylus, and two outer finger- or sausage-like projections called variously juga, lora, lamina mandibularis, paraclypeus, or more prosaically simply side lobes. The relationship between these three clypeal segments is sometimes a very useful character in species identification. In the South American pentatomid *Ceratozygum horridum* the outer clypeal lobes of some males are enlarged into long, flat, toothed antlers which project forwards far beyond the front of the head. The function of these strange structures is unknown. Other males show 'normal' short-lobed forms, much like the females. Both sexes have the abdominal segments and pronotal margins produced into similar toothed leaf-like flanges, which presumably break up the insect's outline and help it blend in to the rough foliage on which it feeds.

On the upper side of the head, the compound eyes of shieldbugs are relatively small compared to some insects, but are nevertheless obvious, often prominent and bulbous (Fig. 24). They work in the conventional manner as in other insect groups. Each is made up of many narrow cylindrical light-sensitive slightly tapered or cone-shaped columns, which align together to give a curved outer surface. Each column, technically called a rhabdom, is topped with a tiny light-focusing lens, and these are visible under the microscope as the adjacent facets, approximately hexagonal, that make up the eye. Vision in shieldbugs is probably rather pixelated, giving a highly granular view of the world, and nothing like the complex photographic detail we as humans can visualise. It is probably barely more detailed than areas of light and dark, but this is sufficient for an insect that has little need to identify complicated patterns. As in many insects, however, the

FIG 24. Close-up of *Nezara viridula* head, showing the compound eyes, ocelli and side lobes of the clypeus enclosing the central lobe.

FIG 25. Head and shoulders of *Rhaphigaster nebulosa*. The armour-like quality of the head, pronotum and scutellum is clearly evident.

eyes are highly sensitive to movement, and there is something disconcerting about a Dock Bug (*Coreus marginatus*) angling its body towards you as you approach on a sunny day, as it eyes you up, deciding whether you might be an enemy to flee from.

In addition shieldbugs have two ocelli, sometimes called simple eyes, towards the rear of the head. These very small hemispherical structures are present in many insects like dragonflies, wasps and ants which all have three, but are absent from others (most beetles, for example, and many bug groups like leafbugs in the family Miridae, lacebugs in the family Tingidae, and all water-dwelling bugs), suggesting that their function is not universally important. They seem to offer little other than a simple measure of light intensity, and are thought to aid horizon-monitoring during flight, to maintain a level trajectory. This link to aerial movement is supported by the fact that the ocelli are absent from wingless (and therefore non-flying) shieldbug nymphs, and are often missing in the short-winged (also non-flying) morphs in some other bug groups.

The main sensory organs of shieldbugs, again similar to most other insects, are the antennae. In most groups they comprise four narrowly cylindrical segments, few enough to be individually named rather than merely numbered – thus, from base to tip: scape, pedicel, basiflagellomere and distiflagellomere. The five, which give the pentatomids their name, are achieved by a division of the pedicel into two, although oddly the resulting two pedicel segments do not appear to have been given formal names – I'm tempted to suggest basipedicellomere and distipedicellomere to keep up with the multisyllabic fashion of the day. In a few other obscure heteropteran groups the antennal segments become fused to give one, two or three segments. The process that gives rise to more or fewer segments occurs in the embryonic stage, but is obviously flexible enough to occur at almost any time, and odd specimens of bugs across many different families frequently occur where left and right antennal segment numbers differ asymmetrically.

The antennae arise from the front of the head, usually just in front of the eye, and the ball-and-socket joint at the base may be recessed under the front or side of the head or from a slightly projecting lobe – the antenniferous tubule or antennifer. In shieldbugs the antennal segments are long and sausage-shaped; sadly, allantoid, a technical term meaning 'sausage-shaped', has not yet been embraced by hemipterists. The antennae of different species vary in their degrees of hairiness; some are smooth, others granular or bristly.

As in most insects, the antennae (wrongly called feelers in some introductory texts) are the 'smell' organs, and although a detailed analysis of the chemoreceptors is beyond the scope of this book, it is clear that the antennae of shieldbugs are attuned to finding food, or each other. The apical segment (the distiflagellomere) in particular is covered with sensilla – tiny spines, pits, slits, or knob-shaped growths on the integument, only visible under very high-power scanning electron microscopy. These react with airborne molecules to create nervous impulses which the bug's brain interprets as 'food' or 'mate'. For the more scientifically adventurous readers, numerous highly technical research articles report on the sensilla morphology and chemoreception of various shieldbug species (e.g. Silva *et al.* 2010, Zhang *et al.* 2014, Ahmad *et al.* 2016), and for a broad review of insect chemoreception see the paper by Dahanukar *et al.* (2005).

The general consensus on insect smell is that individual chemical molecules in the air[3] land on specific sites on or inside a sensillum where they react to depolarise a nerve membrane causing a nerve signal to be sent. These airborne molecules fit like lock-and-key into the chemoreceptors so that each receptor only reacts to one particular type of chemical. Chemical sensitivity is greatest where large complex-shaped molecules, such as species-specific sex pheromone scents, are the only molecules that can physically fit into the highly species-specific shape of the correct chemoreceptor. This allows an insect to find a mate of the right species, and not get side-tracked by similar-smelling aliens in other genera or even families.

Smaller and simpler biological molecules like sugars, alcohols, and esters fit roughly into a much broader array of chemoreceptor types. This gives less sensitivity or olfactory acuity, but it does mean that plants can entice all manner

---

3 This is emphasised by the fact that fully aquatic bugs like *Notonecta* and *Corixa* boatmen, *Nepa* water scorpions and *Ilyocoris* saucer bugs have tiny, barely visible, and therefore probably hardly functional, antennae.

of pollinators by producing a few straightforward scent compounds, without having to create a unique chemical for each type of insect. Somewhere in the middle of this sensitivity spectrum are chemicals detected by the bugs which draw them to their own particular foodplant choices. As with sex pheromones, sensitivities of a few parts per billion allow an insect to detect just a few molecules on the wind to find a single plant of its chosen food in a large field of non-edibles. Some attempts have been made to synthesise chemicals that mimic specific shieldbug sexual pheromones, for example that produced by *Piezodorus hybneri*, a major pest of soya beans in Japan (Endo *et al.* 2010), and recently sex pheromones of *Halyomorpha halys* have been used in monitoring traps to check for its arrival in the British Isles.

A serendipitous olfactory discovery was made in studies of the Central American mesquite bug *Thasus acutangulus* (family Coreidae) (Aldrich & Blum 1978). The adults of this species are large, dark brown and solitary, but the sumptuously coloured harlequin nymphs huddle together on the foodplant trees and probably get protection in numbers, their bright colours advertising their foul taste to would-be predators. Displacing the nymphs into varyingly sized small groups on the ground beneath several trees, and allowing them to gather back together, showed that they mostly (84%) migrated to the tree with the largest start-up group. This implies that there was probably a snowball effect with increasing huddlers attracting increasing incomers. An aggregation pheromone was suggested, and this is partly borne out by the observation that surgical removal of the fourth antennal segment resulted in near complete inability of the nymphs to reaggregate.

MOSTLY THE THORAX

Most of what appears to be the upper surface of the shieldbug is actually the thorax, since the wings, which are most decidedly attached to the thorax, cover and conceal almost all of the abdomen. The thorax is technically made up of three segments, each with a pair of legs attached underneath, but it is the first and second of these that dominate the top of a bug. What might in ordinary conversation with a non-specialist be called the thorax of a shieldbug is actually just the upper plate of the first thoracic segment. Back in way-distant evolutionary time, when those imagined multi-segmented, multi-legged ancestors of insects were roving about, each body segment might usefully be considered to be a bit like a box kite, with a top plate (the notum), a bottom plate (sternum) and side plates (pleura, singular pleuron). During the last however-many-million years (see Chapter 5 for more on shieldbug evolution), these segments have evolved and morphed into different shapes and sizes, just as the head segments have fused and combined to form the elaborate rostrum construction. However, in contrast to the head, the thorax of all insects retains its three original segments (termed pro-, meso- and meta-thoracic), and these are clearly distinguished from each other by visualising the legs, two of which sprout from each segment below. In side view, too, the three separate pleura can still be visualised in many shieldbugs, and these are sometimes important diagnostic and identification characters. On the upper side, though, the upper plate of the first segment has become super-enlarged and almost completely covers the front half of the shieldbug to become the dominant structure of the thorax – the pronotum.

Just as the shieldbug hemelytra have the same protective function as beetle elytra, so too the shieldbug pronotum has the same purpose as the beetle pronotum, a broad flat shield to protect the front portion of the insect's body as it pushes through herbage, roots or soil. The size and form of the pronotum vary tremendously across different groups, and its shape, corner prominence, granularity, polish, texture, sculpture and hairiness are important identification points. The collar of the pronotum fits tight around the articulation with the head, which sits snug up against it in shieldbugs (in other bug families the collar may be raised or crisply defined around a narrow neck). The pronotum is normally trapezoidal, with front corners rounded or projecting; likewise the rear corners, which in some species are extended into spines or hooked prominences. Sometimes the rear corners are extruded into sharp spikes (notably in *Picromerus bidens* and *Elasmucha ferrugata*, and taken to even further extremes in some exotics) – possibly a defence against being eaten, or at least being swallowed whole (Fig. 26). Across the front centre of the pronotum there is often an indented ridge or valley seemingly dividing the plate into front and back portions. The front half may have small slightly domed mounds, the outward expression of the front leg muscle attachments inside the thorax.

Behind the pronotum is the upper plate of the second (meso-) thoracic segment, the large shieldbug mesoscutellum – usually just referred to as the 'scutellum', as in Chapter 1. Beetles also have a scutellum, but it is tiny in comparison to their large pronotum, often reduced to a small dimple between the bases of the elytra, and frequently invisible. The shieldbug scutellum varies from large to enormous, sometimes completely covering the entire abdomen, a continuation of the protective quality of the upper carapace of the insect. In some groups, like the Scutelleridae (well named), Plataspidae and Thyreocoridae, the bug is almost all scutellum. A similarly huge scutellum has also evolved in a few

FIG 26. The dangerous-looking pronotal spines of *Elasmucha ferrugata* may make this an unpalatable morsel for any would-be predator. Or they may help to break up the silhouette outline to confuse the search-image-based hunting mode used by most birds.

FIG 27. The Striped Shieldbug (*Graphosoma italicum*) is practically all scutellum.

Pentatomidae – *Podops* and *Graphosoma*, for example (Fig. 27). In North America these few giant-scutellum groups are sometimes called shieldbugs rather than stink bugs, whilst in Britain names like turtle bug and tortoise bug are used. In the Neotropical pentatomid subfamily Cyrtocorinae, the scutellum is produced

FIG 28. *Acanthosoma haemorrhoidale* seen in side view, showing the orifice to the metathoracic gland between the middle and hind legs. The dark spots along the abdominal segments are the spiracles.

into a humped protruding lobe or large blunt spine; combined with a wrinkled surface texture and expanded body edges, this helps them blend in to the plant stems on which they feed (Packauskas & Schaefer 1998).

The underside of the thorax is more clearly three-segmented, and the side plates (pleura) are usually identifiable; each pleuron is also variously divided completely or nearly into front (episternum) and back (epimeron) portions. Apart from their use in identification keys, most of the underside characters of shieldbugs are rarely examined, but it is worth looking at the occasional shieldbug under a hand lens to see the metathoracic scent glands (Fig. 28). These are the major glands of the adult shieldbug body, and are responsible for ejecting either that delicate almond scent I like so much, or the startling skin-staining brown liquor. The oval, ear-shaped or slit-like gland opening is on the metepisternum, and is marked out by a surrounding area of specialist cuticle which appears mat or furry, called the evaporatorium, even though its function may be nothing to do with evaporation; see Chapter 4 for more on shieldbug defensive secretions and glandular structures. Between the legs, on the underside of the thorax, the meso- and/or metasternum may have long spines pointing forwards or backwards, the function of which is unclear.

HEMELYTRA AND FLIGHT WINGS

Shieldbug wings (or 'alary organs' in the quaint jargon of several bug textbooks), like those of other insects, arise from the second (meso-) and third (meta-) thoracic segments. At the front, the forewings, the hemelytra, are easy to examine, forming the upper surface of the bug when it is not airborne. The hardening (technically sclerotisation) varies from full colour across the major portion (most pentatomids and coreids) to merely speckled veins around clear windows in some of the Rhopalidae. The membrane at the tip of the wing can vary from clearly transparent to pale smoky, spotted or completely darkened, but the texture remains fine and membranous, if sometimes wrinkled, or with visible veins. The membranes of left and right wings overlap each other at rest, but there seems no hard-and-fast rule, and left- and right-wingedness occur throughout, even across a single species. As is shown in the diagrams (pages 26–27), areas of the forewings form discrete zones, and these are variously named with technical terms. The two largest and most significant areas are the corium, the broad stiff long-triangular front leading edge of the wing when it is opened in flight, and the clavus, the narrower trapezoidal flap that rests hard up against the scutellum when the wings are closed. The narrow outer rim along the leading fore-edge of the hemelytra is sometimes called the embolium. In many other het bugs, the cuneus, the small triangular tip of the corium where it meets the membrane, appears as a separate zone, but this differentiation does not occur in shieldbugs, which lack the groove – the costal fracture – that demarcates it in other groups.

In some bug groups, wing morphology varies within species, and appears to be under some as yet uncertain genetic and environmental control. This affects the length of the membrane portion, and of the underlying hind wings, and gives rise to individuals with longer or shorter wings variously described as macropters (long wings), brachypters (short) and micropters (very short). Most adult shieldbugs are constant macropters and fly well; Westwood (1839–40) suggests that in flight *Coreus marginatus* makes 'a humming noise as loud as a hive bee'. A few outlier groups (*Myrmus miriformis, Micrelytra fossularum* and *Pyrrhocoris apterus*) show frequent brachyptery, where flight is reduced or absent. In these cases the short wings may nevertheless give protection to the delicate junction between thorax and abdomen. Generally the hind wings of shieldbugs are broadly triangular and resemble the flight wings of flies and bees in being clear and membranous, but supported by narrow vein wing-struts.

The original precursor of all insects was thought to have had eight radiating veins in each of its four wings, but these have variously merged or become vestigial through the process of evolution. Betts (1986) gives a good review of the comparative morphology of the veins and associated thoracic hinges in several heteropteran bugs, including *Palomena prasina* and *Coreus marginatus*.

Shieldbug wings work on the principle of functional deformable aerofoils. They are not hard fixed struts like the wing of an aeroplane or the blades of a helicopter, and the bugs do not glide through the air as might some large birds.

Instead the wings work by folding and billowing as they move through upstroke and downstroke, effectively producing thrust by pushing the air out of the way downwards and backwards as if the bug were swimming through the atmosphere. This is, indeed, how all insect wings work. A good analogy is to imagine insect wings as being umbrella-like, with a taught but flexible membrane supported by the reinforcing spoke struts of the veins. Stretching this a little, insect flight is like using, say, a bit of an umbrella on each foot instead of flippers when you go snorkelling and are treading water at the side of the boat. The struts of the umbrella portions give supporting stiffness to the soft membrane, but they are not completely rigid, and allow that flexibility to flip-flop to and fro through the water, pushing it down with both forward and backward kicks to keep yourself afloat. The veins of insect wings give a similar fan/strut support to the flexible aerofoil membrane.

The evolutionary remains of the original proto-insect eight veins can still be visualised in dissected shieldbug wings, although precise interpretation is difficult and over the years various researchers have contradicted each other or come to different conclusions as to exactly which vein is what. One of the earliest studies, still very interesting and readable, was by Sara Hoke (1926), who dissected out the more easily accessible membranous hind wings, including those of several shieldbugs (Fig. 29). She found that the veins across the rear wing seemed to coalesce into two distinct trunks at the wing base and that the gap between them was an area of flexibility, now termed the median flexion line (MFL). A similar flexion line appears in the front wings (although it is often obscured by their leathery texture), and in some books is used to divide the corium into the exocorium towards the front edge of the wing and the endocorium behind. Hoke intended to write a subsequent paper comparing Hemiptera front wings, but sadly it was never completed. Nearly a century later, this is perhaps an area of study still amenable to field entomologists today.

Wing-vein terminology can be complex and confusing. In bugs this is partly because of the development of the front wings into hardened hemelytra and the need for the back wings to fold for stowage. In bees, wasps and flies, the wings really only have one function – to lift the insect into the air. To this end they have evolved various subtly different shapes and colours, but retain the vein structure rather unconstrained by other survival pressures. Wing maps showing the different arrangements of the veins play a prominent role in identification books on flies, bees, lacewings, dragonflies and the like. In bugs (and similarly in beetles) the wings have had to adapt to new, competing evolutionary roles, so the front wings have taken on additional protective function, whilst the back wings have had to evolve complex folding and counter-folding mechanisms without

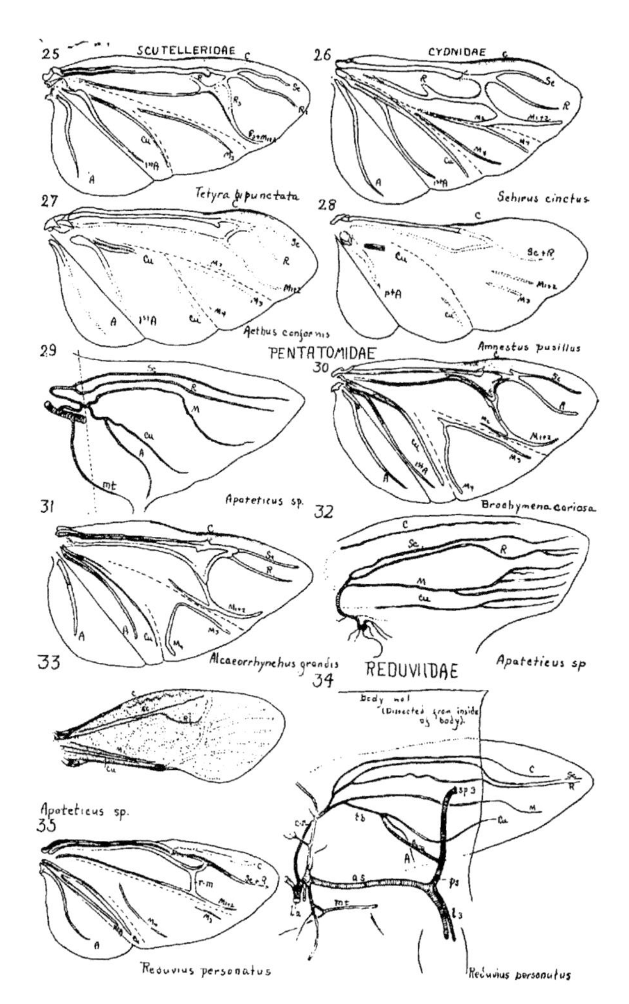

FIG 29. Hind wings of various shieldbugs (and other Heteroptera), drawn by Hoke (1926).

too much loss of aerobatic strength and ability. Revisiting my umbrella analogy, I can't help but think that some clever designer ought to be able to come up with a brilliantly organic rain-protection device based on shieldbug wings, part-hardened sail-plates, flexion lines, and a variable supporting internal struttage.

There are useful reviews relevant to shieldbug wings by Betts (1986) and Wootton & Betts (1986), and others on general insect wings by Dudley (2000) and Alexander (2015). Basically, the wing veins are thought to have evolved from tubes used in gas exchange in gill-like flaps of some imagined primordial ancestor, and these now come in various sizes, thicknesses, cross-sections and stiffnesses to best support the wing just as in my umbrella analogy. Near the wing base, and along the leading edge of the wings, the veins are large, stiff, thickened, and more or less circular in cross-section. However, out across the wing they are curved and bent, merging into and out of each other to give a mesh-like network for added strength and support in a highly architectural design.

The veins are not, though, simply hollow tubes. It has long been known that haemolymph, the insect equivalent of blood, can be seen flowing along them. Baker (1744) observed this in grasshopper wings. Particles in the haemolymph can be seen passing out through the front veins closest to the wing edge, and back through the rear veins, or sometimes showing a tidal in–out flow. There is also leakage into the surrounding wing membrane. Arnold (1964) reports on the wings of many insects including several Heteroptera (but sadly no shieldbugs); he comments on how difficult it is to see the haemolymph passing through a bug's hemelytron even in the flimsy mirid leaf bugs he studied. This haemolymph circulation is partly a hangover from when the adult wings were first pumped up after the final instar moult, where their convoluted form inside the wing buds was at last expanded and inflated into their full adult aerodynamic form. This happened a bit like a swimming-pool inflatable lilo being pumped up after the compact folded and rolled item was first tipped from its cardboard box wrapper. But it is also because the wing is not an inert structure. As well as haemolymph, the veins sometimes also contain gas-exchange tubes (tracheae) and nerves to supply sensory cells. They hydrate and maintain the function of living tissues in the membrane. A flexible rubbery compound, resilin, an elastic protein, occurs throughout insects and is especially important in wing suppleness and strength. Described as a 'disordered' protein, its complex amino acid chain crumples to resemble a tangled ball of wool; as it is stretched, sections become untangled into long hairpin loops, but on relaxation it scrunches back into disorder with 97% transfer of energy, making it the most efficient elastic substance known. It is amazingly resilient and must undergo hundreds of millions of extensions and contractions during the lifetime of the insect. Without haemolymph circulation in the veins, the wings desiccate and deteriorate – something that entomologists will recognise from handling long-dead museum specimens in which the wings have become dry and brittle. For anyone wanting to take this study further, Pass (2018) and Salcedo & Socha (2020) give thorough reviews of wing-vein tracheation and circulation.

In the majority of insects, most of the supporting gantry of veins occurs in the stress-bearing leading half of the wing, an area usually referred to as the remigium. In shieldbugs the remigium corresponds to the corium in the forewings. Behind this, a vein-free rather flap- or sail-like area is termed the clavus of the forewing, while the rear portion of the shieldbug's membranous hind wing is termed the vannus. Even here, though, the wing is not just flat. Towards the wings' edges, in the outer 'membrane' section of the forewings and in the large apparently clear margins and hind edges of the hind wings, the wings are still reinforced, but this time the veins, often visible as mere shadows, are simply convex or concave thickenings in the wing cuticle, sickle-shaped in cross-section, creating pleat-like structures that offer support but at a highly efficient weight-to-strength ratio. A similar offsetting of intrinsic strength to gain flexibility and in-flight manoeuvrability occurs where the major veins need to cross the flexion lines; at these narrow points the hardening of the veins may be relaxed, hinge-like, the cross-section changing from round to flat, or the vein may be completely interrupted at a brief flexible gap. Shieldbug front wings, being hardened and stiffened into hemelytra, nevertheless retain these crease-like lines along the median flexion line and claval fold, and these allow the wing to twist and distort in flight and also when the hemelytra are clicked back together when the bug lands. In the rear wings these folds are clearly discernible and are actual creases, allowing the relatively deep back wing to fold up for compact storage when the bug comes to rest.

For added efficiency, fore and hind wings work together in flight to give a unified aerofoil, unlike the situation in, say, dragonflies, where front and back wings flap independently of each other. Most aerial insects achieve this unification by securing front and back wings together – by means of a series of hooks (the hamulus) in bees and wasps, or a tuft of bristles (the frenulum) in butterflies and moths. Conversely, beetles do not have a unifying connector; most hold out their stiff hard elytra sideways, and power only the membranous hind wings to give lift.

FIG 30. *Cyphostethus tristriatus* takes wing. The gap between scutellum and pronotum opens up when the wings are unfurled, but when they are folded back into place on landing, the scutellum is clicked back into place to anchor the wings closed.

In shieldbugs, with their less stiffened and partly membranous hemelytra, front and back wings are coupled together by a clip- or clamp-like arrangement. Two tiny (0.1–0.2 mm) overlapping wave-shaped growths or nodules on the underside of the rear edge of the hemelytron grip the ridge-like leading edge of the hind wing, which slots in between them. These two nodules are covered with microtrichia, minute hairs found across insect integument in all orders. The front node (termed the costal side wave in the literature) is still covered with an array of these long thin spines, 20–200 of them, and it looks a bit like a toothbrush – shaped long oval or round, ragged or neat, depending on the species. On the rear node (the anal side wave) the microtrichia have evolved into flat fish-scales, and this projection can be ear-shaped, mound-like or narrowly ridged according to the species. This front-wing clip works like a clasp, gripping the reinforced front edge of the back wing, which sits hard up against the rounded fish-scaled anal side wave whilst the flexible toothbrush microtrichia of the costal side wave hold it in place. The brush of microtrichia on the costal side wave is rigid enough to hold the rear wing in place, but flexible enough to allow the two wings to slide against each other slightly as they flex, flap and deform in flight. This wing-coupling mechanism appears to be a ground-plan feature for all the Heteroptera, including aquatic species, but for a detailed analysis of these structures in shieldbugs, complete with scanning electron microscope pictures of the mounds and toothbrushes, see Gorb & Perez Goodwyn (2004) and Czaja (2012).

One of the main functions of the clavus at the back of the front wing is to secure the hemelytron in its normal flat resting position when the bug runs about. This it does using a series of microscopic velcro-like hooks, also evolved from microtrichia, which anchor the rear edge of the forewing to the side margin of the scutellum. Here they interact with ridges either side of a groove (together called the frena) on the scutellar process, a narrow ridge-like fold of the scutellum which is normally invisible, covered by the wings, when the insect is not flying. The ridges either side of the groove are again covered with cuticle that looks hairy with microtrichia, or as if it is covered in downwardly directed fish-scales. The slightly thickened rear edge of the clavus tucks into this groove and sits snugly, the mechanism holding the wing aligned in place. It is likely that the claval fold flexion line across the hemelytra, as well as providing flexibility in flight, also helps hinge and click the stiff wing case into its frenal groove. The frenal groove runs about half or two-thirds the length of the scutellum.

Meanwhile a folded ridge-like process near the base on the underside of the front margin of each hemelytron is covered with a similar scale-like texture, and this clicks onto a flap-like knob on the mesepimeron (front side plate of the second thoracic segment) that is also covered with these tiles or fish-scales; in some pentatomids the scales around the edges of the lobe are slightly swollen and developed into mushroom-like processes with their tops directed forwards. This scaly structure is called the druckknopf, a term for popper fastener borrowed from the German fashion industry, and this perfectly describes the pop-and-click gripper mechanism by which the hemelytra are kept closed over the body as the bug goes about its normal business of climbing through the herbage, hiding in leaf litter or pushing down into the root-thatch. Frena and druckknopf wing-closing systems have been found in most heteropteran groups and are suggested as being ancestral traits in the order.

Despite the need for aerodynamic flow, the seemingly flat wings of shieldbugs can still have lumps, bumps, nodes, nodules, ridges, dimples and points across them. Many of the Cydnidae have a sound-producing mechanism on the underside of the hind wings – a series of tiny pegs (called variously the file, strigil or stridulitrum) on the post-cubital vein which is rubbed against a ridge (called the plectrum) on the abdomen to produce a faint rasping squeak. This behaviour, termed stridulation, is very well known in grasshoppers and crickets, but is remarkably widespread across many other orders, though the noise produced is much less obvious. Many insects have similar comb-and-scraper sound-making devices, and these are usually thought to be a defence against being eaten. Their function seems to be to startle a would-be predator by causing a sudden jarring buzz in the mouth or beak, hopefully sufficient to slacken the jaws just long enough for a quick wriggle escape. In shieldbugs this sound production may also be part of courtship. Even so, there is something slightly unnerving about feeling *Sehirus luctuosus* vibrating plaintively as you hold it carefully between forefinger and thumb. Other shieldbug groups are reported to have similar structures, and this is an area where field entomologists could make significant contributions to knowledge. Leston (1957) gives a good review after using a simple stethoscope to listen, and Lis & Heyna (2001) illustrate the stridulitra of 160 species of Cydnidae.

The tight-fittedness of the hemipteran hemelytra not only provides a hard sheath or cover for the flight wings, it also produces an insulating layer across the abdomen to help reduce water loss – an especially important danger to small insects. And it also accounts for the ability of many bugs to live underwater, where the effectively watertight hemelytra keep the membranous hind wings dry until they are needed for flight between waterbodies. Similarly encased beetles are the only other insects to have evolved truly aquatic adult stages.

Sadly, wing-opening mechanisms have not been well studied in the Hemiptera. In beetles, wing cases are flipped open by muscle action and flight wings expand by a combination of muscle, resilin elasticity, and haemolymph pressure. Something similar is quite likely in shieldbugs, but this remains a good area for potential scientific research in the future. The wing articulations are described well enough by Betts (1986) in terms of a complex interaction of hardened plates (axillae or axillary sclerites) and elastic ligament hinges made of that tough rubbery protein resilin. More aerodynamic (and perhaps acrobatic) flight in heteropteran bugs seems to have been made possible by a simplification of these hinges. In particular, shieldbugs (indeed all het bugs) lack a tegula where the main costal vein joins the thorax. This is a domed epaulette-like cover still clearly visible in many homopteran planthoppers.

THESE LEGS WERE MADE FOR WALKING

Beneath the three thoracic segments is the usual insect complement of three pairs of legs. Shieldbug legs are all fairly similar, and although other heteropteran bug groups show some modifications like flat paddles for swimming, long paddles for rowing, spines or claspers for grasping, enlarged muscular back legs for jumping, or broad shovel legs for burrowing in soil, most shieldbug legs are very clearly for walking. Having said this, the Cydnidae have very spiny legs, an adaptation thought to help with pushing through the root thatch or dense herbage near the soil surface. The legs of these bugs often show signs of wear and tear, caused by the constant scraping of sharp grit and sand particles on the soil. This is taken to extremes in members of the mostly New World genus *Scaptocoris*, which have become almost globular soil-burrowers closely resembling soil-dwelling chafer beetles; they lack tarsi on their broad shovelling hind legs, a feature shared with some large earth-digging dung beetles, especially those that roll and bury dung balls, which lack tarsi on their front legs. *Scaptocoris* has special spiny combs on the coxae that prevent soil particles getting into the coxa/trochanter joint. Individuals have been dug up from about 1.5 metres down in the earth, making them truly hypogean (Greek for subterranean). There is always the possibility that other soil-dwelling species might be found. *Stirotarsus abnormis* is a curious insect (Rider 2000), more obviously pentatomid-shaped, but with broad, flattened blade-like tibiae, inflated antennae and scabrous body texture; it is only known from a handful of specimens from South America. Might its apparent rarity be connected to a semi-subterranean existence? Meanwhile the strange flattened front tibiae of a *Stiretrus* species (Pentatomidae, Asopinae) I found on the Pacific coast of Costa Rica in 1991 are smooth and blade-like rather than spine-reinforced shovels, and combined with a prominent sharp tooth on each front femur look more like tools for courtship or mating rather than soil excavation. Many others in the genus have normal slim tibiae, suggesting that this is a trait that has evolved recently.

Closer to home, British shieldbug legs follow the standard insect leg format of segmentation, with five articulating sections: coxa (plural coxae), attached to the thorax; trochanter (small, barely articulating); femur (plural femora), the long thigh segment; tibia (plural tibiae), the long shin segment; and tarsus (plural tarsi), the foot. Within this formula there is little variation, although the Acanthosomatidae show only two tarsal segments (technically tarsomeres) where other shieldbugs have three (but two in nymphs of some species). As mentioned earlier, members of the family Cydnidae have prominent spines along the tibiae, and a newcomer to Britain, the Western Conifer Seed Bug, *Leptoglossus occidentalis*, has swollen hind tibiae that make it look as if it is wearing flared trousers. This leaf-like structure appears commonly on coreids around the world (often dubbed leaf-footed bugs), and may be a predator avoidance device, distracting or confusing any attacker, especially when the bugs are flying, trailing their long hind legs behind them in the air (Fig. 31). This is borne out by studies of exotic

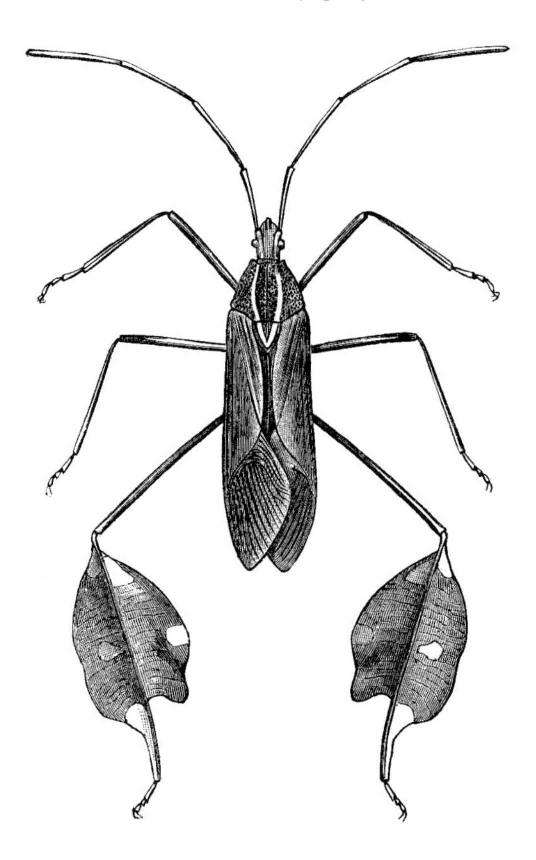

FIG 31. Extreme plate-like tibial development in the South American coreid *Diactor bilineatus*, from Sharp (1909). Such flattened legs are thought to confuse the eye of predators, either deflecting attack away from the body or causing momentary hesitation enough for the bug to escape.

coreids which frequently have missing legs. In their study of nine common North American species Emberts *et al.* (2016) recorded 7.9–21.5% of wild observed specimens lacking one or more legs, and of these the hind legs were missing disproportionately often. They also showed in laboratory studies that a bug would readily autotomise (jettison) a limb if it was gripped by an enemy. The break occurred at the junction of trochanter and femur and was typically accompanied by a rise then sudden drop of the abdomen, shearing the limb at this natural weak point. There is good evidence that these enlarged hind legs, often with swollen and spiny femora (Fig. 32), are used as weapons by males against each other to maintain territories (though not in our demure slim-legged British species). In their missing-leg survey Emberts *et al.* (2016) noted that males of the most swollen-thighed genera (*Acanthocephala*, *Narnia* and *Euthochtha*) were far less likely to have lost a limb than females of the species, suggesting that there was some innate reluctance on their part to give up such a valuable weapon; in less weaponised genera (*Leptoglossus*, *Chelinidea* and *Anasa*), males and females were equally likely to have the limbs missing.

It is often pointed out in insect monographs that leg structures concerned with burrowing, hunting, gripping, jumping, swimming or mating are often well researched for the curiosity value of their strange forms, but that simple walking legs are frequently glossed over. This is a shame, because most shieldbugs are obviously built for walking. Monitoring how insects walk can be quite tricky, since they are often small and move about with incredible rapidity, but shieldbugs are slow and cautious and readily lend themselves to close observation with the naked eye. Although it was written over 55 years ago, the review of insect walking by Wilson (1966) is still pertinent, relevant and understandable.

FIG 32. North American *Narnia* species, showing the large toothed hind femora thought to be used as weapons when competing for mates or maintaining territories.

A standard shieldbug gait is an alternating tripod where front and back left legs and middle right leg touch the substrate at the same time (the stance or retraction phase) whilst the other three legs are swung forwards (protraction phase) ready for the next step. This moves the bug onwards at a steady walking pace, with the centre of gravity always supported at the centre of the tripod. Using the leg nomenclature set out by Wilson, (R)ight and (L)eft legs front-to-back 1, 2 and 3, this alternating tripod is R1/L2/R3 followed by L1/R2/L3 and so on. However, shieldbugs are rarely in such a hurry, and at mosey speed they adopt another standard insect gait: R3/L1, then R2/L3, then R1/L2, then R3/L1 again, or sometimes its mirror image (L3/R1, L2/R3, L1/R2, L3/R1), effectively moving two legs up in the air at a time rather than three. Notice that the wave of leg movement starts at the back of the body – on the right-hand side R3 then R2, then R1 then back to R3 again, with the left-hand legs moving one segment out of phase: L1, then L3, then L2, then L1 again. This follows a 'rule' of arthropod limb movements, that a leg is not lifted and swung forward until the leg behind it has been swung and come to stand firmly on the substrate. In six-legged insects this is quite difficult to see, but anyone who has ever watched a millipede forward-ripple its legs like a miniature Mexican wave (technically a metachronal wave) will immediately recognise the similarity. In real life, however, these formalised gaits are seldom precisely choreographed, and as it walks over an uneven surface a shieldbug is likely to be out of rhythm as much as in. Even under experimental conditions the alternating tripod legs may move slightly out of phase with each other and rarely move simultaneously as a unified threesome.

This standardised walking scheme only works for movement on a broad flat surface like a tree trunk, fence panel or entomologist's open hand. In the normal course of events, shieldbugs spend a lot of time walking up thin plant stems and along narrow leaf edges, so studies of their natural movements are perhaps best understood by a comparison with tightrope walkers. This was exactly how Frantsevich *et al.* (1996) described their experiments with *Graphosoma lineatum* (*italicum*). The first thing to realise is that when a shieldbug 'walks' along a narrow stem it mostly hangs down rather than stands up. The angles of femora and tibiae are also changed, so that the body of the insect is further away from the stem than it would be from a flat tree trunk. Imagine the difference in your elbow angles when carrying a vertical light broom handle well away from you, compared to a large and rather awkward sheet of hardboard held close to your body as you grip the edges with outstretched arms. Obviously, it's slightly different for shieldbugs, but you get the picture. In *Graphosoma* the change from flat walking to stem walking amounts to about 35–40 degrees more depression of the femur. Previous workers, studying stick insects, had suggested that this change in stance was due to the body weight of the insect hanging below the twig. But *Graphosoma* only weighs about 80 mg, and the same changes in leg angles occur in an experimentally tethered and supported bug whether it is given a narrow (and light) ring to run through its feet or a broad ribbon hoop. This was subtly different from the walking style measured in some equally stout beetles. Similar measurements in *Leptinotarsa decemlineata* (the Colorado Potato Beetle) and *Coccinella* ladybirds showed that they maintained close body contact to varying substrates, narrow stem or broad leaf, because they constantly kept the maxillary palpi (short multi-segmented feelers on the mouthparts) touching the leaf or ground. Bugs do not have these palpi so do not operate this type of sensory distance control.

The two major shieldbug leg segments are femur and tibia, but during walking these act rather like hinged stilts, and the main contribution to body propulsion is provided by the coxae, particularly the front and middle. This is not really surprising, since the power to move a coxa is provided by large muscles inside the thorax. The rotational axis through each coxa is what controls the stride swing of the leg through its forward airborne protraction arc and powers the push backwards as the body of the insect is propelled forwards. The oblique angle of the coxal pivot in the thorax is what gives the shieldbug leg its ability to give long stride length even over variable substrates – flat leaf or narrow twig. Imagine one extreme where straight legs were held vertically downwards: the optimum pivot (allowing the legs their longest stride length) would be 90 degrees from vertical. This is the system adopted by human legs during military square-bashing parade marching, and best understood by visualising the line of the pin through the hips about which the leg of, say, a Barbie doll swings. Then imagine the other extreme where the legs are held straight out to the side of the body. Now the optimum pivot axis is 0 degrees (completely vertical), with the sweep of the limbs resembling the arc of human arms during the powered pull-back in breaststroke swimming. Neither of these extremes is any use on its own in a real live creature needing adaptability in an uncertain and uneven world. There must be some sort of optimal compromise – and there is. There is some complicated three-dimensional mathematics here, but in an earlier paper on a theoretical optimal hexapod invertebrate leg design Frantsevich (1995) calculated the coxal angle (that subcoxal axis) which gives most efficient movement as arctan $\pi/2$, or more conventionally 57.5 degrees from vertical. It is perhaps no surprise that the coxal angles of *Graphosoma* were measured as very similar: 60 degrees (front leg), 64 degrees (middle) and 64 degrees (rear). Shieldbugs really are perfect walking machines.

Again, here is a rich seam of entomological research available to the field entomologist, particularly in a time when slow motion is a fairly standard camera option on even the cheapest smartphones. The laboratory work of Frantsevich *et al.* (1996) offers a few starting parameters for shieldbug walking studies. Speeds of 3.0 steps per second were recorded on a horizontal plane, 2.2 on a horizontal rod, or in tethered insects 2.5 steps per second rolling a narrow ring and 2.2 rolling a broader hoop. Stride lengths were 5.5 ± 0.8 mm on a horizontal plane, 4.5 ± 0.3 on a rod. The pitch angle of the bodyline (front to back) relative to the substrate was 6–7 degrees positive on a flat plane, but negative on a narrow rod. Walking on a flat surface, middle legs (ends of tibiae) were 9.5–10.5 mm apart, front and back legs 2–3 mm closer together. My own non-laboratory observations of *Palomena prasina* in my garden suggest that they do not get to alternating tripod until they reach canter speed (about three steps per second) and then they are quite likely to fly away.

Finally, at the tip of each foot there are two tarsal claws by which the bug clasps the substrate. These are large enough to see under the hand lens or microscope, but small enough not to cause discomfort as the bug crawls over the back of your hand. In shieldbugs they are pretty uniform – slim and hook-shaped. Between the claws, bugs have a variety of tiny hair and plate-like structures used to sense the surface of the substrate or adhere to it (Fig. 33). Shieldbugs lack the arolia (singular arolium), tiny bristles or slim bladders, found in many other Heteroptera, but they do have pulvilli (singular pulvillus), small flat pad-like structures arising from the bases of the claws as a stalk (basipulvillus) leading to a membranous distipulvillus. These are thought to help them adhere to very smooth surfaces, like glass, by friction. As with many insect groups, research is often focused on potential pest species, and an interesting paper by Salerno *et al.* (2018) describes experiments on *Nezara viridula* walking on glass and various smooth, hairy or waxy leaf surfaces.

FIG 33. *Palomena prasina* walks easily up a pane of glass. In this view the spiracles are clearly visible at the side of each abdominal segment.

ABDOMEN

The abdomen of most British shieldbugs is hidden beneath wings and hemelytra, but the edges are visible from above, to varying degrees, as the skirt of the connexivum – the edges of the abdominal segments projecting beyond the corium. This is sometimes a contrasting colour, sometimes chequered or spotted. Each abdominal segment is made up of a lower plate (sternite) and upper plate (tergite): seven in females and eight in males, plus the genital capsule. The remains of two further segments can sometimes be seen on dissection, termed the proctiger or anal tube. The upper surface of the abdomen is usually only revealed when the bug flips open its hemelytra, unfurls its flight wings, and takes to the air (Fig. 34). In some species, notably *Coreus marginatus* and *Alydus calcaratus*, the upper abdomen is brightly coloured and slightly startling to the eye of any would-be predator, or indeed entomologist. The upper abdomen of a nymph, however, is easily viewed, because they are wingless.

Nymphs of Pentatomidae are often more strongly coloured than the adults, and this colour is dominated by a bright abdomen, marked with darkened spots,

FIG 34. With its wings open, the red flash of the abdomen is obvious in *Coreus marginatus*.

blobs or streaks, especially down the centre line (coreids, on the other hand, have rather camouflaged nymphs with muted mottled colours, bristly or knobbly outlines). For example, although the adult of *Nezara viridula* closely resembles the common green *Palomena prasina* in being uniformly green, it is unmistakably harlequined in the nymphal stages, with bright black, white and pink spots and blotches to startling effect. It is a common error for the novice hemipterist to mistake the domed and brightly spotted nymphs of many shieldbug species for ladybirds (Fig. 35). Against the paler ground colour of the abdominal segments, the dark areas on the back often mark the openings of the dorsal abdominal scent glands. There are usually three glands, near the front margins of segments 4, 5 and 6, but they vary across the world and across other bug families, from zero to four. The glands are paired, with a common reservoir opening onto the back of the nymph, and, like the evaporatorium around the metathoracic gland, the hardened coloured cuticle may act to disperse the exuded chemicals. There is more on their structure and biochemistry in Chapter 4.

The underside of the abdomen is fairly easily seen in side view in most British shieldbugs. The segmentation is clear and the arrangement of the spiracles (breathing pores) can be observed. Many female members of the Acanthosomatidae have roughly ovoid areas of slightly hairy-looking cuticle on either side of the sixth and seventh (sometimes also fifth) sternites; these are the Pendergrast's organs, used during egg-laying. The female rubs her hind tarsi across them, then smears as yet unidentified secretions onto the eggs as they are attached to the leaf surface. The secretions are thought to have some protective role for the eggs, and the organs are lacking in *Elasmucha grisea* and other acanthosomatids that stand guard over their egg batch (Fischer 2006).

Internally, the abdomen encloses the breathing tubes (tracheae), the digestive tract, the rudimentary blood-pump and the genital paraphernalia. A forensic knowledge of shieldbug innards is not essential to the field entomologist, so what follows is just a brief synopsis. Detailed descriptions and dissections can be found in standard textbooks such as Schuh & Weirauch (2020).

The digestive canal is essentially a flexible tube or series of pouches, through which the liquid food sucked up through the rostrum passes to be digested. Digestion starts inside the food source itself, with the injection of saliva down the salivary canal in the rostrum. The saliva also acts as a lubricant, allowing the paired mandibular stylets to rub past each other within the maxillary housing. A pump just inside the head sucks up the food and then pushes it down into the gut. The rostrum and oesophagus are chitin-lined to give a stiff inflexible tube – analogous to using a straw to suck up liquids. The food passes through the midgut, which in shieldbugs has three, four or rarely five sections – the first,

FIG 35. Top: A huddle of newly hatched nymphs of *Palomena prasina*, all looking very ladybird-like. Bottom: A five-day-old nymph of *Coriomeris denticulatus*, on the other hand, is remarkably well camouflaged, with its muted mottled colours and knobbed form.

the largest, and often termed the stomach, is effectively a flexible food-holding bag. The penultimate section has 2–4 rows of blind bag-like invaginations called gastric caeca. These have long been known to harbour symbiotic bacteria that are reckoned to help with the breakdown of especially tough plant material, just as the large caecum of ruminant sheep, cattle, goats and deer uses a bucket-fermentation process to break down the copious amounts of cellulose that these animals eat. Tellingly in shieldbugs, these gastric caeca are missing in the predatory subfamily Anopinae, which includes *Picromerus bidens* and *Zicrona caerulea*. Presumably they do not need help from bacteria to digest their protein-rich insect victims (although oddly gastric caeca are present in other similarly predatory bugs and also those that feed by sucking vertebrate blood). The bacteria are thought to be first taken into the gut by the nymphs feeding from faecal smear deposits made by the mother when she lays the eggs. This midgut portion of the alimentary tract is where most nutrients are absorbed from the food for metabolic use in the body of the shieldbug. The end of the alimentary tract, the hindgut, is where water, salts and other substances are retrieved if needed, and terminates in a hardened pear-shaped container that functions as a rectum, storing waste materials until they need to be ejected.

The genitalia are also located at the tail end of the shieldbug. The shape of the hardened (sclerotised) components can provide useful identification characters in some difficult groups or species pairs. Basically, the genitalia comprise the eighth or ninth segment onwards, but these have become so small and complex through evolutionary history that it is often very complicated to determine which is what. In males sternite 9 forms a hardened structure called a pygophore, containing the aedeagus (phallus or 'intromittent organ') which delivers the sperm, and parameres – paired claspers that hold the insects together during copulation. Male genitals are usually asymmetrical. Female genitals are complicated in other ways because they are made up of articulated ringed structures into which the aedeagus fits, onto which the claspers clasp, and through which eggs are subsequently laid. The sperm-receiving apparatus includes the spermatheca, a reservoir in which sperm is stored until it is needed for egg fertilisation. The ovaries are made up of several egg-producing tubes called ovarioles. Rather prosaically, mating, insemination and fertilisation in shieldbugs have been described as 'unremarkable'. A few personal observations are made in the next chapter. For more detailed descriptions of hemipteran genital structure see the general reviews in Genevcius & Schwertner (2017) and Schuh & Weirauch (2020).

Insects breathe through a system of tubes, the tracheae. These divide and subdivide into an intricate network that delivers air to virtually all tissues within the body. The orifices are called spiracles, and long ago there were eight down each side in an ancestral proto-bug which had eight major segments to make up the abdomen. Number 1 is absent in all shieldbugs, and the number in surviving species is sometimes reduced. The spiracles can be observed in side view in most species. In other groups, like the superfamily Lygaeoidea specifically omitted from this book, the spiracles have migrated, over evolutionary time, and some or all of them can be on the upper surface of the abdomen, more or less hidden by the wings and hemelytra.

With oxygen delivered via the tracheae and carbon dioxide removed by the same route, the blood of insects usually plays no part in gas exchange. Instead the fluid (technically haemolymph) functions primarily to move nutrients, metabolic waste materials and hormones around. Shieldbug circulation follows the standard insect norm of a free-flowing liquid bathing all the internal organs so that chemical transfer can occur. A rudimentary pump, variously termed heart to the rear and aorta to the fore in entomology textbooks, takes in the abdominal fluid through pores (ostia, 2–7 pairs corresponding to 2–7 abdominal segments) and pumps it forward through a narrow dorsal vessel to release it into the head capsule and thorax, where it enters the general body cavity, called the haemocoel. It then flows passively back through the body into the sump of the abdomen for the process to begin again. In addition, most bugs (indeed most insects) have extra pumps in the thorax, to supply pressurised haemolymph to the wing veins. In the Hemiptera, these 'wing circulatory organs' (sometimes charmingly called wing hearts) are not connected to the main heart/aorta pump-vessel as they are in supposedly more 'primitive' insect orders of earwigs, dragonflies, cockroaches, grasshoppers and the like, but seem to work independently, taking free fluid from the thoracic cavity and pushing it into the wings by a pulsatile diaphragm attached under the scutellum (Krenn & Pass 1994).

The other major internal structure is the nervous system, which comprises a series of nerves and ganglia that control and orchestrate bodily activity and behaviour. As in other insects, this is less centralised than in, say, humans, although there is a large block of neural material in the head, and this is usually called the brain. This is vaguely divided into three parts, the first of which receives the optic nerves, the second input from the antennae, with the third section connected to the mouthparts. Smaller knots in other segments work semi-autonomously to synchronise leg or wing movements, regulate the alimentary tract, interact with mating and egg-laying behaviour, or control abdominal breathing. Most studies of shieldbugs require little detailed knowledge of nervous-system anatomy. Thankfully there appear to be fewer anecdotes about experiments on headless shieldbugs compared to those on headless cockroaches or headless ants.

This, then, is how a shieldbug is built. And although they vary from broad to narrow, flat to globular, squat to lanky, there is an underlying and unifying theme that defines shieldbug structure. But this is just the outward appearance. Traditionally an appreciation of natural history, particularly entomology, was based on the minutiae of the insects' morphology, and how one species might differ from another in the formal museum collection. Today scientific understanding extends well beyond drawers of pinned specimens and must address the bugs in their natural habitat, doing all the things that shieldbugs have done for millions of years.

CHAPTER 3

Life Histories: Breeding and Feeding

EXPERTS IN HEMIMETABOLY

Shieldbugs, like all members of the Hemiptera, go through the 'incomplete' hemimetabolous developmental route mentioned earlier (Fig. 36). This is the path from egg to adult that does *not* have the chrysalis/pupa stage which is so characteristic of the 'complete' holometabolous development from, say, caterpillar to butterfly. Hemimetaboly is often regarded as the more primitive state in the evolution of insects, because it is more ancient than holometaboly. It is also an extension of the type of growth shown by all other arthropod groups – spiders, scorpions, centipedes, millipedes, woodlice and crabs – and was probably similar to the growth of extinct trilobites and giant sea scorpions, as well as the first multi-legged proto-insect precursor that hauled its way out of the primordial murk half a billion years ago. The success of the more 'advanced' holometaboly is clear enough though. The larval form, completely different from the adult in appearance and behaviour, can develop without having to compete with its usually larger and tougher adult form. Living in a different habitat from the adults, feeding on different foods, often at a different time of year, enables these larvae to utilise a huge diversity of extremely narrow, very specialist, frequently rather hostile ecological niches. This very different-looking feeding and growing stage of the life cycle ends with metamorphosis, and when an adult insect emerges from a chrysalis, that is it – it is fixed and complete, with no more changing or growing to be done. It is then up to the

FIG 36. Life cycle of *Troilus luridus*. Top row, left to right: eggs, first-instar nymphs, second-instar nymph. Centre row: third-, fourth- and fifth-instar nymphs. Lower row: fifth-instar nymph, and adults. Photomontage by Maria Justamond.

free-living usually winged adult stage to mate, disperse and colonise new areas. So successful is this strategy that about 90% of extant insect species are holometabolous – these are the well-known hyperdiverse orders Coleoptera (beetles), Diptera (flies), Lepidoptera (butterflies and moths) and Hymenoptera (bees, wasps, ants etc.) together with a few small orders like Neuroptera (lacewings) and Trichoptera (caddisflies).

This is not to disparage the very many and very diverse hemimetabolous groups – like Dermaptera (earwigs), Blattodea (cockroaches), Odonata (dragonflies), Orthoptera (grasshoppers) and Hemiptera (true bugs) – that still use

FIG 37. Earwigs, too, go through hemimetaboly. Here a female *Forficula auricularia* stands guard over her batch of eggs and hatchling nymphs, which are wingless, but easily recognisable as miniature versions of her.

the arguably slightly old-fashioned multiple nymphal growth process (Fig. 37). Though these insects may not be quite so diverse in terms of species numbers, they have nevertheless been highly successful on Earth and their way of life has survived right through to the present day thank you very much. The Hemiptera, with about 100,000 known species (about 50,000 of which are Heteroptera), is the most diverse among this assemblage, supporting the claim that they are still an extremely successful group of organisms. But what it does mean is that, apart from the absence of wings, the nymphs and adults of shieldbugs are structurally very similar to each other; they feed in the same way, often behave in the same way, and occur in the same places. This is an important, but often overlooked, feature of hemimetaboly. It is comical, or at least laughable, that apparently sensible and well-educated people often ask if, for example, a small beetle will one day grow up into a bigger beetle. This, after all, is the growth-increment scheme they are most familiar with in their own children, in their pets, in their livestock, and in the fluffy and feathery wildlife they see on nature documentaries. They have to be reminded of that important larva/adult disparity in structure and behaviour. However, this is exactly what *does* happen to shieldbug nymphs – they start off from the egg very small but nearly perfectly formed, then they grow bigger by gradual discrete stages into adulthood.

Now, it may seem that I've gone round the houses a bit here, simply to point out that little shieldbugs do indeed grow into bigger shieldbugs, but it is important to remember that though many people may find this a vaguely familiar and intuitive form of life cycle, it is absolutely *not* the norm in insects; it is the exception. This, then, must be the baseline understanding on which all knowledge of shieldbugs is built.

COURTSHIP

The first thing to note about shieldbug courtship is that it is so little studied, or even noticed, that most monographs on the group just do not bother to cover it at all. Apart from the oft-reported fact that coupled male and female shieldbugs stay attached for a long time (though exactly how long is seldom recorded), there are precious few reports of accurate observation (Fig. 38). I was recently sent a short video of a mating couple of *Acanthosoma haemorrhoidale* (Gary Williamson via Twitter); they had been locked together on the same leaf for over three hours when another male tried to interject. After the newcomer had wandered over the pair for a few minutes the mating female started to rock her body vigorously from side to side until he left, possibly a defence against potential damage that might have been caused to her had he tried to escalate his attentions. It's interesting that it was the female that seemed to be calling the shots here, rather than the male defending his interest; the coupled male seemed to be struggling with his footholds during the incident.

Although mating pairs are frequently seen during the day, some species may actually prefer nocturnal trysts. Ramsay (2016) noted numerous pairs of *Pentatoma rufipes* after dark on street maple trees in Oxford in July 2010 and again in July 2011. None were present the following days. This seems to be the first time mating aggregations of this species have been observed – maybe because few entomologists are out patrolling road traffic islands with a torch in the dark. In a North American rhopalid, *Jadera haematoloma*, one captive mated pair remained coupled for 128 hours (Carroll 1987). This is far longer than is required

FIG 38. Shieldbugs are well known to remain as mating pairs for many hours. I am often asked what are the 12-legged creatures sitting on woundwort leaves in springtime. They are coupled *Eysarcoris venustissimus*.

for sperm transfer, and this regular over-long mating has been interpreted as mate-guarding. This behaviour, well known in dragonflies (Odonata) and flies (Diptera), is usually associated with one male preventing other males mating with the female to which he is attached, thereby removing or diluting his sperm input into the offspring. In this attractive New World rhopalid species males regularly outnumber females, particularly in more northern populations, where more males develop to adulthood than females and males overwinter more successfully than females. With large numbers of males vying for a lower number of females, multiple matings are easily facilitated, and although this may benefit the female by ensuring sperm diversity and avoiding inbreeding, no male wants to share mixed paternity of a relatively limited number of offspring nymphs. From personal experience (limited and anecdotal, I admit), male rhopalids seem to outnumber females in Britain too. Sex-ratio and mate-guarding studies ought now to be possible in some British species of this group, especially given that *Stictopleurus punctatonervosus* and *S. abutilon* are becoming so common and widespread, often occurring in very large numbers on brownfield sites in southern England.

Although not obvious in our limited British coreid fauna, the swollen hind femora of certain exotic coreid genera (*Acanthocephala, Narnia* and *Euthochtha,* mentioned by Emberts *et al.* 2016) are used in male-on-male jousts, presumably over females as well as territory. Whether they are also used in courtship or mate-guarding is apparently not reported. *Narnia femorata* is a well-studied American

FIG 39. The Sri Lankan *Probergrothius varicornis* (Pyrrhocoridae). Left: When I was on holiday in Sri Lanka in September 1994 these bugs were running about everywhere; they were always paired, and I never saw a singleton. Right: In March 2022 I was sent almost the same image from a friend holidaying on the island. No, they do not remain coupled for 28 years, but likewise he never saw a lone bug and had not quite realised that this was a mating pair.

coreid bug, feeding on *Opuntia* prickly pear cactuses. Where mating has been observed there is limited courtship, but females control whether or not mating takes place by opening up their genital plates to allow male penetration. Cirino & Miller (2017) report that females prefer the odour of males which develop on the *Opuntia* fruits, especially red ripe ones, so those foodplant territories are certainly worth fighting over.

Studies on shieldbug mating sequences are rather few and far between, but a few foreign studies give hope that similar observations ought to be possible amongst our own fauna. Drickamer & McPherson (1992) videoed six North American pentatomids at close quarters, then analysed the videos in slow motion, noting: male antennal tappings on the female body or head, head-butting, jostling, wrestling and just plain grabbing hold, and in response the receptive female lifting her abdomen to allow intromission (or not, if she was unreceptive). They concluded that the courtship behaviours were probably species-specific, highly variable between species, and mostly tactile, with antennal touching and testing being the most important means by which male and female communicated their joint interest in each other.

Elsewhere in the Hemiptera, the noisy cicadas of warmer climates are well known for their loud communication buzzes, whines and whirrs. So it's perhaps not too surprising to discover that het bugs also interact using sound. One of the first to be noted was *Nezara viridula* (see review by Todd 1989), which creates sounds at about 80–120 Hz, by rubbing together the abdominal tergal plates combined with dorsoventral vibrations of the whole abdomen. These are inaudible to humans, and appear to be only audible to the bugs themselves through the leaves and stems on which they are sitting. This is different from the very loud and clear 'songs' of cicadas, grasshoppers and crickets, which can detect the messages through the air using conventional sound-receiving organs analogous to ears. By degrees the sound signals of shieldbugs have been analysed, and by watching the movements and behaviour of the communicating bugs some meanings can be ascribed to the various vibrations. Ota & Cokl (1991) noted that males and females made different calls when moving near each other on ivy stems and leaves. As the males walk they deliver intermittent calling signals, but once they detect a female making a rhythmical call, a pulse of about 0.75 seconds about every 2 seconds, they turn and walk towards her, employing another type of call. If a male comes to a branch in the stalk it pauses, resting its front legs across the junction, and waits for the female to call again before choosing which branch the sounds came down, and moving off towards her. Eventually male and female meet and mate. It is always the males that move towards the females, never the other way round.

Throughout the world different *Nezara* populations have different sonic structures, varying the tone and frequencies of the calls, as if they have developed different dialects across the globe. As more shieldbug species are studied they too are revealed to have these calling signals, mostly around 100 Hz, and these are all thought to be sent and detected through the plants on which the bugs are walking (Virant-Doberlet & Cokl 2004). Zorovic (2011) implied that the sound receptor nerve cells in the shieldbug's body were already attuned to the specific frequencies involved, and that these nerves were physically different from similar neurons in insects like crickets which perceive 'conventional' airborne sound signals.

It turns out that 'singing' through plants is quite widespread in shieldbugs. In the predatory *Picromerus bidens* there are at least four different male signals, three seemingly in response to a uniform female song, and one a rivalry signal to other males (Cokl *et al.* 2011), each differing in the arrangement of the pulses in the sound bursts and the interval between the bursts. Again, it is thought that the signals (at frequencies around 100–200 Hz) are transmitted from one individual bug to the other over several metres through the plant substrate, rather than through the air.

Picromerus has barely developed scent glands (abdominal or thoracic), suggesting that sound, rather than airborne pheromones, now plays the major role in mate location. There is also a suggestion that sensitivity to vibrations in the plant might have helped predatory bugs find their prey by following the vibrational cues given off by caterpillars gnawing at the leaves, but sounds are also known from purely herbivorous species too. It's a shame that the sounds are undetectable to humans unless the insect is placed onto the membrane of a loudspeaker attached to a microphone and amplifier.

LIFE CYCLES: THE EGGS COME FIRST

Shieldbug eggs are mostly laid in batches of a few dozen. Some are rimmed with a narrow circular lip or a series of small projecting points like lacy castellations. Some seem marked with emoji smiley faces (Fig. 40). Along with the bright yellow egg clusters of ladybirds, and the delicately sculptured eggs of some butterflies, shieldbug eggs are a common source of wonderment to the general naturalist, and photos of them are frequently uploaded to online insect identification sites because they are so eye-catching and distinctive (Fig. 41). The eggs of heteropteran bugs (particularly shieldbugs) have been well studied, especially since the seminal review by Southwood (1956), not just for their interesting structure, but also

FIG 40. Emoji-faced eggs of *Pentatoma rufipes* ready to hatch. There are often 14 in pentatomid batches.

FIG 41. Coreid eggs. British Coreidae seem to be very secretive when it comes to oviposition, and I could not find any suitable pictures; however, this excellent shot by Tristan Bantock shows eggs of the Mediterranean *Plinachtus* (formerly *Gonocerus*) *imitator* on the Mastic tree (*Pistacia lentiscus*), Mallorca.

for their contribution to bug classification – they are often cited as one of the characteristics that confirm the Pentatomomorpha as an overarching shieldbug grouping and a valid division of land bugs. Another interesting and useful review of shieldbug eggs is given by Javahery (1994).

The eggs vary from nearly spherical in the Pyrrhocoridae through flat 'cauldron-shaped' in some Alydidae and kidney-shaped in Rhopalidae to tall egg-shaped in Acanthosomatidae, and squat cylindrical barrel-shaped in most Pentatomidae and Cydnidae. They have a ring of tiny spines or narrow knobs around the top, a bit like a skinny balustrade or banister rods without the handrail. These are called micropylar processes, and vary from two relatively large S-shaped structures in some Rhopalidae to about 65 tiny hairs in *Eurydema*.

They are particularly pronounced in the Asopinae (Fig. 42). When the eggs are still inside the ovaries the micropylar processes have a narrow channel in them, the micropyle, through which the sperm passes to fertilise the egg, but when the eggs are laid the channels close and the micropyles take on the task of gas exchange between the developing embryo and the outside world.

In the Alydidae and Thyreocoridae eggs are mostly laid singly, but bugs in other families lay clusters of about 7–28 in some Pentatomidae and Scutelleridae, 35–55 in Acanthosomatidae, and 45–120 in Cydnidae. Maximum fecundity of about 300 eggs is reported in *Picromerus bidens* (Larivière & Larochelle 1989), though a captive North American rhopalid *Jadera haematoloma* laid 510 eggs over several

FIG 42. Fourteen eggs of *Troilus luridus*, showing the long curved micropylar processes typical of the Asopinae.

FIG 43. Rhopalid eggs, laid by captive female *Liorhyssus hyalinus*.

days under laboratory conditions (Carroll 1987). There is variation within and between different species in each family, but egg batches are often in multiples of seven (Figs. 40 and 42). This reflects the fact that most shieldbugs have two ovaries, each with seven egg-producing tubules (ovarioles) in which eggs mature more or less synchronously. Famously, many *Eurygaster* species (some of which are considered cereal pests in parts of the world) lay 14 eggs in a neat double row down a grass blade. Larger batches are achieved by the bug holding the eggs in, until each ovariole has a backlog of mature eggs lined up inside it. Regular shieldbug egg clusters of 28, 35, 56 and 84 are achieved like this, although the occasional extra egg from a particularly active ovariole, or an egg being knocked off the leaf, will interfere with this numerical precision. In the well-studied international pest species *Halymorpha halys*, eggs are usually laid in batches of 28, with a total of up to 244 in the female's lifetime, suggesting that towards the end of her life she is not firing on all cylinders (Nielsen & Hamilton 2009).

The eggs are laid in a loose ball into the soil in some Cydnidae, but in a semi-regular array onto open leaf surfaces in most other groups. Here they are glued into position by a cement-like secretion from various accessory glands just inside the genitalia, and sometimes by secretions from the gut too. The ovipositor of most shieldbugs is a very short slightly flexible telescopic tube, but an apparently unique long external blade-like ovipositor is present in the obscure Cameroon shieldbug *Birketsmithia anomala* (Leston 1954); this species is thought to use its tool to insert its eggs (of typical shieldbug form) deep into the flowers of the tree-like Asteraceae (Compositae) that grow in the area.

Shieldbug eggs are usually very pale when first laid, but within a day or two they mature to dirty green, cream, yellow or beige, sometimes marked with brown. The eggs of *Eurydema oleracea* and *Piezodorus lituratus* are particularly striking since they are ringed with dark bands (Fig. 44). The eggs of *Zicrona caerulea* turn black, but this seems to be some tanning effect of air on the cement adhesive that covered the eggs when they were laid. Normally any deep blackening of the egg is a result of egg-parasitism by minute wasps in the subfamily Scelioninae and a sign that the shieldbug embryo has been destroyed by the developing wasp larva within.

All being well, that emoji face appears on the top of the shieldbug egg after a few days. The two red dots are the eyes of the developing embryo and the grey, brown or black 'smile' is a reinforced sharp ridge of tough embryonic skin that forms an egg-burster. This structure is generally T-shaped in Pentatomidae and Scutelleridae, but Y-shaped in Acanthosomatidae, Cydnidae, Thyreocoridae and other groups. The eggs hatch in about 6–15 days, but this varies between species and also depends on ambient temperatures. In his study

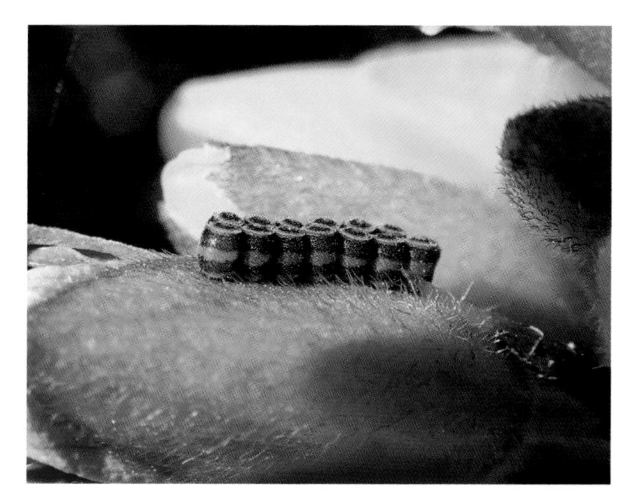

FIG 44. Distinctive barred eggs of *Piezodorus lituratus* on gorse. There are 13, so it looks as if one of the standard 14 of the clutch has been dislodged.

of 35 northern-hemisphere shieldbugs, Javahery (1994) found experimentally that temperate genera like *Palomena, Picromerus* and *Piezodorus* developed at constant temperatures of about 23–25 °C, whilst more subtropical genera like *Aelia, Carpocoris, Dolycoris, Eurydema* and *Eurygaster* required 27–28 °C for the embryo to mature. In nature, the eggs of *Picromerus* take nine months to develop because this species overwinters in the egg stage. Though the eggs are laid in summer, embryonic development stops in August and will not proceed again until the egg has registered a cold period usually approaching or reaching freezing point. This chilling is needed to break its diapause, an enforced hibernation torpor that maybe helps coordinate the emergence of the nymph the following year with the spring appearance of its caterpillar prey.

HATCHING AND FIRST FEED

When it is ready to hatch, the shieldbug embryo (sometimes called the prolarva at this stage) starts to swallow air and uses its inflating gut as a tough taut internal bladder against which it can increase the pressure of its haemolymph. Using rhythmic muscular movements it starts to expand its body, pushing the hard egg-burster up against the top of the eggshell. The shell bursts in a ring around the inside of the ring of micropyles and the lid flips open. Empty eggshells usually have the hinged lids still attached and look like a stack of empty tin cans. This flip-top lid is sometimes called an operculum, but it is subtly different from the true operculum found in other bug groups. In the Cimicomorpha, the sister group to Pentatomomorpha, and which includes other terrestrial het bug families, the top of the egg is thickened, but is ringed with a circular bar that covers a line of weakness caused by a narrow thinning of the eggshell hereabouts. As the cimicomorph prolarva expands its body in the egg, the edge of the operculum ruptures in a neat circle all along the ring seal and the lid lifts off. But there is no pre-formed weak ring around the top of a pentatomid egg, and although the lid still flips up it requires the extra pressure supplied by the sharp egg-burster to get the tear going. The difference is perhaps analogous to the ring-pull opening on a tin of baked beans, versus the perforations between postage stamps – the split/tear outcome is the same but is achieved in slightly different ways. Consequently shieldbug eggs are sometimes technically described as having a pseudo-operculum. There is no pseudo-operculum in acanthosomatid eggs, and the egg-burster simply causes a rupture that splits the rather tall narrow egg longitudinally down the side.

As the prolarva starts to shimmy free, it also moults, leaving behind its embryonic skin, complete with the egg-burster, inside the empty flip-topped eggshell. The angular shape of the egg-burster, lodged against the rim of the empty shell, emphasises what Javahery (1994) describes as a kettle-shape for some shieldbug eggs – a softly rounded and slightly over-inflated flat-topped and flat-bottomed cylinder, often with different upper and lower diameters like a late-twentieth-century electric kettle. The orientation of the pointed egg-bursters indicates the orientation of the eggs (and the orientation of the embryos within) when they were laid, often in double rows, each of seven eggs facing away from another row of seven. This is another reflection of the two ovaries, each with the standard seven ovarioles, producing eggs in strict regimented order.

The hatchling is now a nymph, and as expected of a hemimetabolous insect it is already recognisable as a tiny shieldbug (Fig. 45). Although it lacks any semblance of wing buds yet, it already has the right number of legs and antennal segments, compound eyes, and fully functional sucking mouthparts like its parents. Its first meal involves sucking some of the fluid from inside and outside the egg. Here it imbibes bacteria from the faecal smear delivered by its mother, along with the leaf-sticking cement from when the egg was originally laid. By this means the young nymphs pick up important bacterial symbionts that will live inside the gut and help them digest much of the plant material that they will eat during their lifetimes. Other reports suggest that these new hatchlings also take a bit of moisture by drinking dew. Oddly, first-instar nymphs of the normally predatory *Picromerus bidens* are also reported as sucking a bit of plant sap for this initial drink, so the 'fact' that first-instar nymphs do not feed is likely to be less than hard and fast.

There has been much research on the microbial symbionts (from symbiosis = 'living together') living inside herbivorous shieldbug guts. Most shieldbug species have blind-ending tubules on the posterior mid-gut, called gastric caeca. These are lined with small folded pockets (called crypts) packed with the bacterial symbionts. In plant-feeding pentatomids these symbionts are members of the class Gammaproteobacteria, and there is evidence of cospeciation, with each shieldbug lineage having its own identifiable bacterial lineage, showing parallel evolution between host and gut flora. Oddly, the Coreoidea (and Lygaeoidea) seem to have a relationship with another class, the Betaproteobacteria, which is not vertically transmitted between mother and eggs. Instead the gut flora appears to be acquired from the soil by early nymphs. Meanwhile the Pyrrhocoridae house a completely different group, the Actinobacteria. Echoing the age-old call of the confused scientist – there is clearly much work to be done here.

FIG 45. Huddle of 19 *Palomena prasina* nymphs; having hatched from the 28 eggs, some nymphs have either wandered off, or been taken by predators, bad luck or missed footing.

FIG 46. Huddle of five-day-old Hawthorn Shieldbug (*Acanthosoma haemorrhoidale*) nymphs.

FIG 47. Egg of *Dicranocephalus medius*, the hatching nymph as it emerges, and later in the day after it has darkened up. Captive rearing by Yvonne Couch.

INSTARS

The word instar is a very useful and important bit of curiously precise entomological jargon. It means one of the often several discrete stages in the growing part of an insect life cycle (Fig. 48). As it grows, the shieldbug nymph does not simply grow gradually bigger. It is constrained by its tight-fitting chitin-based

OPPOSITE: **FIG 48.** Nymphs of various common shieldbug species. (a) Early instar of *Palomena prasina*. (b) Fourth (left) and fifth (right) instars of *Palomena prasina*, showing the clear development of the wing buds. (c) Fifth instar of *Nezara viridula*. (d) Fifth instar of *Eurygaster testudinaria*. (e) Fifth instar of *Pentatoma rufipes*. (f) Fifth instar of *Eysarcoris venustissimus*. (g) Fifth instar of *Coreus marginatus*. (h) Fifth instar of *Gonocerus acuteangulatus*. (i) Fifth instar of *Acanthosoma haemorrhoidale*.

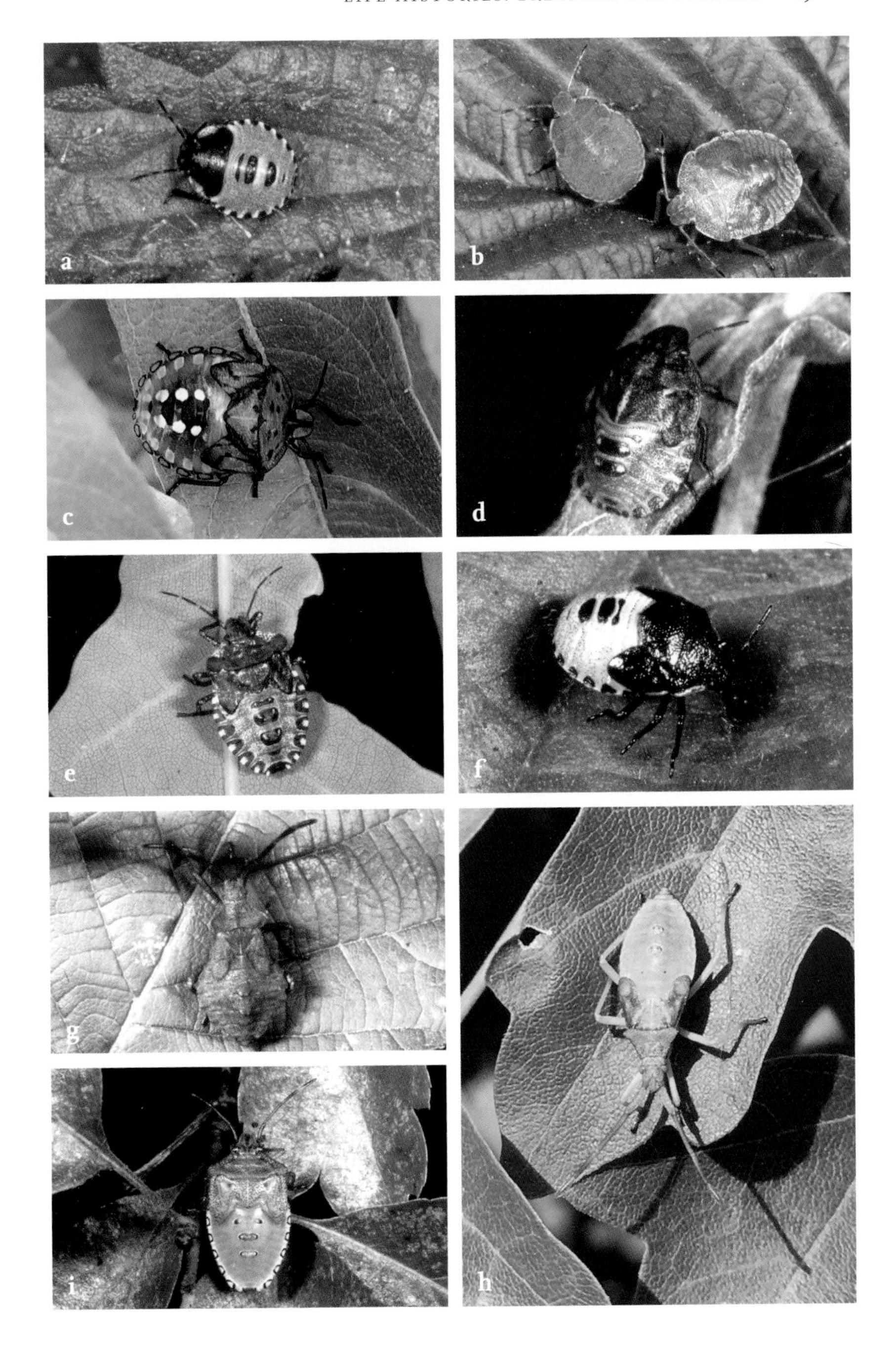
a
b
c
d
e
f
g
h
i

skin and after a bit of feeding it is quite literally bursting at the seams, unable to expand any more. At this point it splits open its old skin and discards it, having grown a new slightly more flexible one underneath it. The new skin is larger and baggier and will allow the nymph another bout of gluttony before it needs to moult again. In other insects, caterpillars, grubs, maggots, larvae and naiads also pass through variable numbers of instars before the final change into an adult. In some publications the word stadium (plural stadia) is used instead of instar.

Apart from its egg-juice drink (and maybe a few other tidbits – see above), the first instar normally does not feed, but within a few hours it moults, shedding its skin to produce the second instar. Thereafter each species of shieldbug goes through a series of precise moults, and these instars can be identified by size, shape (particularly the gradual development of the wing buds), colour and pattern. In addition to the in-egg embryo prolarva, most shieldbugs pass through five instars before finally emerging as a winged adult bug (Fig. 49). In the past some reports claimed six instars, but these appear to have included the remains of the prolarva moult, left inside the egg on hatching, or accidentally allocating the rather variable first-instar nymphs into two groups (Bouldrey & Grimnes 1995). In field and laboratory studies head width is taken as a good proxy for instar number; although the flexible abdomen may enlarge as feeding of the instar progresses, the hard head capsule remains the same size until it is cast off at the next moult. Wing buds normally start to appear in the third instar.

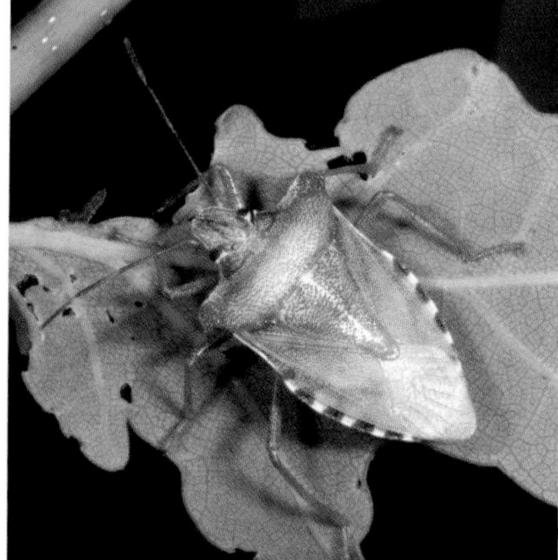

FIG 49. Part of the moulted final-instar shed skin of *Pentatoma rufipes*, and the pale adult that has just emerged from it. The white hair-like strands are the casings of the internal breathing tubes, the tracheae. It will take several hours, perhaps days, for the adult's full dark colours to develop.

The significance of that hemimetabolous, incomplete, shieldbug metamorphosis is really made apparent by our relatively good understanding of shieldbug nymphs. Because they feed in the open, often in company with the adults, they have long been known, and recognised, and for many common species they are well observed and well reported. Even some of the oldest identification guides (e.g. Donovan 1792–1817, Curtis 1823–40) described many of the 'larvae' as they were then called, commenting on their forms, colours and feeding behaviours. When Butler (1923) produced his monumental book on the life histories of the Heteroptera, he was able to include a wealth of description and commentary on the immature stages, unparalleled in many other insect orders except perhaps in the caterpillars of butterflies and some moths. This didn't happen much with the more popular groups, like beetles. It is only very recently that, say, ladybird guides started to include detailed notes on their larvae, and identifying most beetle larvae to species level is often the highly technical preserve of the expert entomologist.

Each shieldbug species has its own timeline life cycle of egg-laying, hatching, instar progression and final adulthood, and these often vary according to local weather conditions (temperature and rainfall are both important), regional climate and foodplant availability. In northern Europe (including the British Isles) most shieldbugs are thought to have just one generation per year. On the whole shieldbugs overwinter as adults; known exceptions are *Pentatoma rufipes*, *Odontoscelis* species and *Coptosoma scutellatum*, which winter as nymphs, and *Picromerus bidens* and *Myrmus miriformis*, which overwinter as eggs.

Second-instar bug nymphs start to feed immediately. Part of its heritage as a hemimetabolous creature is that it already has the same well-adapted sucking rostrum mouthparts as the adult. In the normal course of events shieldbug nymphs grow quickly – about one week each for instars 2 to 4, and about a fortnight for instar 5 (obviously much longer for overwintering *Pentatoma* and *Odontoscelis* nymphs though). The other important hemimetabolous trait is that, without the need for a chrysalis stage, the shieldbug moults directly from fifth instar to adult. In insects with complete metamorphosis like butterflies and moths, the chrysalis stage involves a complicated rearrangement of the entire insect's internal and external structure, effectively reducing the final instar's innards to a gelatinous soup in which the adult structures, including legs, antennae, eyes, wings, body segments and genitalia, have to grow from scratch – well, from small groups of embryonic cells. This takes time, often several days and frequently several weeks. This is also a time when the inert and rather defenceless chrysalis is very vulnerable to attack from diseases, adverse weather, predators and parasitoids.

FIG 50. Sequence showing *Cyphostethus tristriatus* hauling itself from its fourth to its final nymphal stage. Photomontage by Maria Justamond.

In bugs, though, the only inert stage is the egg. Apart from a brief period after each moult where the new soft and sometimes pale skin needs to harden and darken, shieldbugs can remain active their entire lives, although they slow down to a stop during hibernation. Allowing for the egg stage, a typical shieldbug can get through its life cycle in about eight weeks. Given that most of the British Isles are green and lush and warm from April to September, what is to stop a shieldbug going through two or more generations a year? Plenty of other insects don't stop and wait until next year before getting on with things. This is one reason put forward for the sudden huge population explosions in aphids, for example.

In fact, some shieldbugs can go through multiple generations in a year, but usually only in warmer regions of the world, rather than in temperate northern Europe. The change from more northern univoltine (one generation a year) behaviour to southern multivoltine (multiple generations) is well known in British butterflies, but less studied in shieldbugs. Saulich & Musolin (2012) reviewed 43 pentatomid species from around the northern temperate regions of the globe, and their paper makes interesting reading. Several species which have just a single generation in Britain and Ireland are able to complete a second generation further south – *Arma custos* has two generations in the southern Caucasus, and *Graphosoma italicum* has two or more generations around the Mediterranean, as do *Dolycoris baccarum* and *Aelia acuminata*. *Dolycoris* is particularly interesting, because it is so very widespread across the northern hemisphere and adapts its life cycle to local conditions. On hot dry Cyprus it has just a single generation and goes into moisture-conserving shelter (aestivation) during the summer, but on warm but moist oceanic Japan it sometimes manages three generations a year. Not surprisingly, one of the best-studied shieldbugs is *Nezara viridula*, now a worldwide species and a problematic agricultural pest in many areas. In southern Europe and its native North Africa *Nezara* completes two generations, in Japan it can have three, broadly overlapping generations each year, and in India and Brazil it is continually breeding, all year round. In Britain it appears to have just one generation, but with its obvious worldwide life-cycle flexibility this might be expected to change in long hot years, and a partial second generation ought not to be completely impossible.

This is a dangerous gamble for a shieldbug, though. Shieldbugs, like all insects, cannot actually predict the weather later in the year, and although a local population may have benefited from sunshine and warmth early in the season and all the nymphs may have fed through to adulthood in good time, there can be no guarantee of continued good luck. It would be a foolhardy evolutionary strategy to just go for it hammer and tongs every time the Gulf Stream or El Niño threw some random sunny spring days across northern Europe. Even though the second-generation eggs might hatch and the nymphs might start to feed, the likelihood is that most of them would not quite achieve maturity before the weather closed in during autumn. Not being adapted to overwintering, the nymphs would inevitably perish.

In fact it turns out that most British shieldbugs have instinctive fair-weather caution hard-wired into their physiology. Sure enough, warm weather initially increases their ability to feed and metabolise enough to grow faster, but the brakes are put on by increasing day length. By the middle of summer most species have achieved adulthood, but they now stop. They might keep feeding, but they do not mate and lay eggs until the following year. This pause (diapause in the technical literature) ensures they keep on track to the safe single-generation model. Diapause is strong, and can only be overruled by the very significantly warmer temperatures in the south of a species' range. It may not be that high temperatures can actually break diapause. More probably the southern populations have evolved and lost this period of obligate quiescence, so a second generation can occur, say, in southern Europe or around the Mediterranean. Despite 'Phew, What a Scorcher' tabloid headlines here, no shieldbug is going to be fooled by a bit of unseasonal warmth; it really needs a very good thermal kick in the metabolism over very many generations to persuade it to launch a risky attempt at a second generation.

The British fauna stops at the English Channel and North Sea, but elsewhere in the world a continuous north–south shieldbug community must have some overlap or transition area between univoltine and multivoltine populations. In Osaka, in southern Japan, the pentatomid *Dybowskyia reticulata* has one generation in cool years, but two generations in warmer ones, suggesting that it regularly takes the gamble, and oftentimes this pays off.

If it becomes more widespread in Britain, the Firebug (*Pyrrhocoris apterus*) will be worth studying in detail (Fig. 51). In France it is generally reckoned to be univoltine, but in Ukraine and Kazakhstan it frequently appears to have a partial second generation. In the Czech Republic there are attempts at a second generation in warm years. The picture is slightly clouded by the fact that the bug has an unusually long egg-laying period of up to six weeks. This can easily be seen by examining the large and obvious colonies that disport themselves at

FIG 51. If not multi-generational, then certainly multi-instarational, this massed huddle of *Pyrrhocoris apterus* comprises at least third-, fourth- and fifth-instar nymphs as well as adults. The bright colours, giving the effect of flames, certainly live up to the bug's common name – Firebug.

the base of large lime (linden) trees throughout mainland Europe during the summer – where all stages from first-instar hatchlings to adults are present together in the bright huddles. It seems that partial second broods are actually the norm through much of the bug's southern range. Musolin & Saulich (1996) observed this species in experimental cages in the forest/steppe zone of Belgorod in southern Russia. Some early eggs, laid near the beginning of May, hatched and fed up quickly enough to complete growth and emerge as adult females before 10 July, and these were immediately active – optimistically, they mated and laid eggs for that second brood. Adult females emerging after 22 July entered diapause and saved their mating and egg-laying energies until the following year. Those emerging between 12 and 21 July showed a mixture of behaviours. It seems that both types of development happen each year, but in cool years the univoltine cohort dominates because fewer bugs reach maturity before that later July cut-off, whilst in warm years the bivoltine cohort is favoured. In the cool summer of 1990 no females went on to lay eggs, and all diapaused until 1991.

So how do the bugs know if they have achieved adulthood early enough to try for a second brood? Temperature, it seems, is some sort of guide, but not quite good enough – the bugs also rely on day length. In Belgorod, at least, short spring days encourage the *Pyrrhocoris* nymphs to feed fast and mature quickly. If an individual bug has managed to complete its development mostly under shorter spring days then it goes on to mate and try for a second generation. If, however, it has passed most of its growth period under the longer lazier days of summer it somehow slows down, ticks off the months on its internal calendar, and knows to hold off its reproductive attempts until the following year. Experimental manipulation of day lengths under laboratory conditions also confirms this – short days encourage growth, but long day lengths inhibit it. Under natural conditions *Pyrrhocoris* nymphs can encounter short days at two times of year; in the spring short day lengths urge the nymphs to hurry up to try for that second generation, in autumn any more casually plodding nymphs are encouraged to speed up a bit to complete development so they can be ready for adult winter diapause. Many textbooks suggest that insects use their ocelli to sense and measure day length, but *Pyrrhocoris apterus* famously lacks ocelli. There is obviously still more work to be done here.

HIBERNATION

Hibernation is not just a case of keeping conveniently warm and dry whilst the weather all around blows, rains, snows and is generally frosty and unpleasant. In temperate Britain, it is a survival strategy to cope with a period of extreme danger – mostly from lack of food, rather than lack of balmy sunshine, but also from other natural perils. Many insects start to find winter shelter long before temperatures drop to danger levels. This is because just living out there anyway, in the wild, is fraught with all manner of wild danger – from predators, parasites, diseases, floods, droughts, fires, winds blowing you off course, bigger herbivores gnawing your foodplant down to nothing, or just leaving it too late so that when cold really sets in you don't have the energy left to find a safe shelter at the last minute. For a new adult the best strategy to achieve mating and egg-laying and to get the next generation off to a good start is simply to get out of the way quickly, hunker down, switch off metabolism to bare tick-over, and wait for the gentler and more forgiving times of spring.

In a review by Saulich & Musolin (2012), the majority of temperate-zone shieldbug species overwinter as adults. Unlike, say, British butterflies (most of which hibernate in non-adult stages), shieldbugs have those hard, tight-fitting hemelytra to protect their bodies and their flight wings and they are able to push deep down into the leaf litter, between stalks, hard up against buds, under loose bark, tight into the ivy thicket, under logs, in the root thatch, or tucked into the folded bark of a giant oak trunk. It is well known that the Green Shieldbug (*Palomena prasina*) changes colour when it hibernates (Fig. 52). Towards the end of November the familiar emerald-green shell of the insect starts to turn a deep

FIG 52. *Palomena prasina* post-hibernation brown, and a few weeks later fully green again.

brownish purple. When it re-emerges in March or April the following year it takes a couple of weeks to get itself back into its previous green splendour. An easy suggestion, though unproved, is that the perfect green camouflage against a leaf becomes pointless after autumn leaf-fall, so changing its base colour to brown better allows it to hide up in dead leaves, or against gnarled tree bark. Similarly, the Southern Green Shieldbug (*Nezara viridula*) and the European but not yet British *Palomena viridissima* undergo similar colour changes in autumn, turning a dull olive or ochre, perhaps for the same reasons. Most other British shieldbugs are already brown enough to make this change unnecessary, and even the greenish ones like *Acanthosoma haemorrhoidale* and *Piezodorus lituratus* are more muted and have camouflaged brownish patches or tints.

These colour changes may be outwardly visible signs, but internally there are real physiological changes going on. Again, it is the well-studied pest species *Halyomorpha halys* for which much information is known. On arriving at adulthood it takes a female some time to attain sexual maturity and be ready to mate, reported by Nielsen & Hamilton (2009) to be 148 degree-days (number of days × average daily temperature). In the subtropics, this is easily achievable, and the bug is almost continually brooded without interruption. But in temperate zones (the US mid-Atlantic states and the British Isles) the bug is predicted to be univoltine, with just a single generation each year. After its final nymphal moult in autumn, the accumulation of days at high enough temperatures slows to a snail's pace, and it is therefore advantageous for the bug to enter hibernation. It will resume its maturation in the following spring.

Amongst the British pentatomid species only *Picromerus bidens* winters in the egg stage, and *Pentatoma rufipes* as second-instar nymphs; all others hibernate as adults. Again, longer summer day length seems to halt *Pentatoma rufipes* as a second- (or rarely third-) instar nymph that will overwinter. It will only become active again after the cold of winter, when the increasing temperatures and day lengths of the following spring reach some critical threshold.

There are some unusual variations to the norm, though. In Britain, *Picromerus bidens* nymphs reach maturity in July and August, when the adults appear, mate and lay the eggs that will overwinter. Normally all the adults die off, but a few have been found hibernating on into the following year. This is not, however, a back-up plan by a few lucky long-lived individuals. These few remaining adults have been attacked by tachinid flies in the subfamily Phasiinae. These specialist shieldbug parasitoids develop as larvae inside the bugs but do not kill their victims until they themselves have nearly finished growing. The bug, however, has ovaries or testes (along with other internal organs) destroyed, so it does not mate as normal, or lay eggs. Instead it 'survives' the winter, effectively giving the parasitoid grub living within it a convenient and safe overwintering shelter. But it is doomed. Next spring the fly larva will finish its insidious development and pupate, and the adult fly will emerge to continue its parasitic life cycle by laying eggs on next year's *Picromerus* nymphs. Its host will perish. A few records of *Pentatoma* adults in winter are also thought to be of parasitised individuals.

SUCKING AND FEEDING

Shieldbugs suck, and they do it very well indeed. Generally they are phloem-feeders, inserting their proboscis stylets into the leaf veins and stems and tapping the nutrient-rich liquid sap being transported around the plant. This is how the common Forest Bug, *Pentatoma rufipes*, feeds, on small oak twigs, and it is easily able to penetrate even relatively woody stems of about 5 or 6 mm in diameter. However, there are also root-feeders (presumed in some Cydnidae), fruit-feeders (*Acanthosoma haemorrhoidale*, *Palomena prasina*, *Gonocerus acuteangulatus*), flower-feeders (*Graphosoma italicum*), seed-feeders (*Pyrrhocoris apterus*, *Leptoglossus occidentalis*, many Rhopalidae) and catkin-feeders (*Elasmostethus interstinctus*). As is usual in insects, many bug–plant associations are known because finding a foodplant is often the best way of discovering the insect, but precise feeding observations are scant, so it is sometimes not at all clear exactly how the insect is feeding, or on what part of the plant.

Many common British shieldbugs feed on the developing fruits of their foodplants (Fig. 53). Hawkins (2003) records *Palomena prasina* found on 128 different wild and cultivated plant species, although he is at pains to point out

FIG 53. Nymph of *Palomena prasina* feeding on an unripe blackberry.

that many of these will be casual resting places. Where feeding can be safely inferred, the bugs were mostly on the fruiting bodies. There is quite some latitude, however – he also notes that the main foodplant for the Brassica Bug (*Eurydema oleracea*) in Surrey is Horse-radish (*Armoracia rusticana*) escaped from cultivation. However, this often fails to flower, so the bug must also feed on the leaves and stems. In fact, it turns out that shieldbugs are quite versatile feeders, and they can turn their attentions, and their stylets, to various parts of the plant, depending on their need. They will also leave their standard foodplant later in the summer to feed on the developing fruits of other species. Famously the Dock Bug (*Coreus marginatus*) moves away from its spring situation on dock plants (*Rumex* species) and feeds on blackberries. Many's the time I have brought a specimen home in the sandwich box I repurposed as a forage container – a prelude to a blackberry-crumble baking session later in the day.

When a shieldbug feeds it inserts the thin stylet bundle into its food, but first it has to disengage part of the heavy protective rostrum sheath. This is not, however, a concertina squashing, nor a telescoping contraction, nor is it the equivalent of unbuttoning your cuffs and rolling up your sleeve to reach deeper into the drain to pull out the blockage of dead leaves. In fact the labium is not removed from the end of the stylets – it is the basal two labial segments which are removed and flexed out of the way. They hinge backwards, elbow-like, under the head, though the short labrum flips to vertical, partly supporting the stylet bases. The terminal third and fourth labial segments continue to sheathe the stylets fully, and they act as a supportive hollow gantry allowing the bug to push the pin-thin stylets down into the foodstuff (penetration depth P in Fig. 54). Going

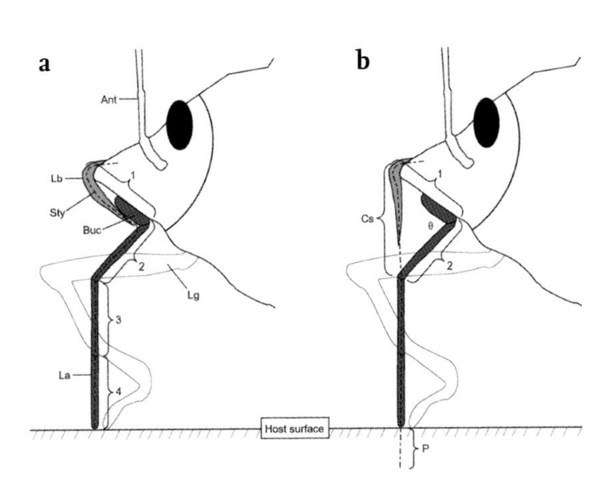

FIG 54. Diagram showing head of *Nezara viridula*, with rostrum in (a) probing and (b) feeding postures (not to scale). 1–4, segments 1–4 of the rostrum; Ant, antenna; Buc, bucculae; Cs, calculated span; La, labium; Lb, labrum; Lg, leg; P, penetration depth of the feeding stylets into the host tissue; Sty, stylet bundle, represented by a dashed line; θ, angle between segment 2 and bucculae. From Esquivel (2019).

back to my shirtsleeves analogy, this is more like undoing the epaulettes near the shoulder and removing the upper part of your sleeve (between shoulder and elbow), which now allows you to slide your cuff further up the forearm to expose your wrist. As far as I know this construction has never yet been used in any item of clothing, but if it ever is I want it to be called Jones' shieldbug labium sleeve, to distinguish it from any form of the once-popular leg-of-mutton sleeve.

One consequence of this flexing out of the way of the basal segments of the shieldbug labium is that the penetrative depth of the stylets pushing down into the foodplant is not related to total rostrum length, but is completely defined by the lengths of the first two rostrum segments (Esquivel 2019). Thus it is quite possible for species with shorter rostrums to reach deeper into a plant than those with longer overall mouthparts, simply because they nevertheless have longer segments 1 and 2. Total rostrum length can also vary within a species, and is often linked to the type of food available at any particular time. In Florida individuals of the large North American coreid *Narnia femorata* maturing in June had longer mouthparts – needed to puncture the tough green newly developing fruits of the prickly pear cactus (*Opuntia mesacantha*) on which they were feeding. Those reaching adulthood in or after August had shorter overall mouthpart lengths, seemingly because the now ripe red fruits on the cactus have thinner softer skins needing less deep penetration to get at the juicy pulp (Cirino & Miller 2017). The suggestion is that fourth- and fifth-instar nymphs feeding on the riper and more nutrient-rich red fruits achieved the requisite nutritional state to moult into adulthood earlier, so they developed into smaller adults with shorter mouthparts.

As described earlier, the shieldbug rostrum stylets form a tight-fitting bundle – about 45 μm (0.045 mm) in diameter in the well-studied *Nezara viridula*. As the thick leathery rostrum sheath of the four-segmented labium flexes out of the way, the stylets are inserted for feeding. The outer mandibular stylets have saw-toothed tips that by rapid back-and-forth movement drill into the plant. Digestive juices are then pumped down the salivary canal, one of the two tubular spaces between the maxillae, the innermost parts of the bundle. In the simplest feeding strategy the saliva released into the plant forms a gel-like plug around the end of the stylet bundle, anchoring it in place. The bug then sucks up the free-flowing plant juices, either from the phloem tubes of the stem or leaf that are busily transporting the sugary products of photosynthesis around the plant, or from the xylem tubes bringing up water and nutrients from the root system. The food canal running back up the stylet bundle is slightly wider (about 15–16 μm diameter) than the salivary canal (11–12 μm) and more oval in shape; females are slightly larger in all measurements than males. Esquivel *et al.* (2018) give details of funnel effects, tube geometry and male/female differences.

In the suborder Homoptera, phloem or xylem feeding using the salivary gel plug technique seems to be standard. This fits with the observation that large numbers of aphids cause plants to wilt as they quite literally suck the leaves dry. In contrast, shieldbugs (suborder Heteroptera) mostly engage in a more active feeding attack on a plant. Termed cell-rupture feeding (sometimes lacerate-and-flush or macerate-and-flush), this activity comprises continuous deep mechanical movement of the sharp stylet nib in and out of the plant tissue, breaking and mashing the spongy cells of the parenchyma leaf flesh. Meanwhile a more watery saliva is injected; this starts to dissolve the plant tissues and the bug sucks up the resulting soup. Depending on the shieldbug species, the foodplant, and the plant material under attack, the feeding mechanism is a variable mixture of simple sucking, physical damage (laceration) and chemical digestion (maceration).

As Hawkins (2003) suspected for *Eurydema*, shieldbugs can alter their food probings to suit whatever plant material is available throughout the year, or indeed as the need arises on the day. *Piezodorus guildinii* is a widespread pest of soya beans in North and South America, reputed to cause losses of US$600 million per year in Brazil alone (Lucini & Panizzi 2018). It uses cell-rupture feeding when attacking the hard but nutritious seed endosperm in the beans, but first takes a hydrating draught of watery food from xylem vessels in the leaves or stems. Likewise, *Nezara viridula*, another widespread agricultural pest of bean crops, feeds on the stems and leaves of the soya bean plant, but can switch to seed feeding as the beans start to mature in the pods. When I grew broad beans in the garden, *Nezara* moved in quickly and I found both nymphs and adults on all parts of the plants: stems, leaves and pods.

Much of the close analysis of shieldbug feeding has come through the technique of electropenetrography (EPG), in which a thin electrode is attached to the feeding bug, another is implanted into the soil, and a low voltage is passed between them. Measuring subtle differences in the current passing through the bug into the plant generates a wave pattern of oscillations with peaks every few seconds and subpeaks about every tenth of a second, reflecting a combination of stylet movements, salivary pump activity and changes of suction inside the bug's rostrum. Sanding down the shiny pronotum of a shieldbug and attaching a hair-thin gold electrode is quite a palaver for the average field entomologist, but since it was first suggested for aphids by McLean & Kinsey (1964), EPG has been applied to a wide variety of bugs and a large library of species-specific and plant-tissue-specific wave forms now exists. The paper by Lucini & Panizzi (2018) and references therein give much more detail of this procedure. They suggest that a variant of EPG might also be used to monitor shieldbug egg-laying.

Shieldbugs, being quite large insects with relatively large stout mouthparts, sometimes cause obvious penetration marks. If a puncture is made into a bud, the zone of laceration/maceration passes through several of the tightly furled immature leaves inside the bud, and this damage becomes visible as a series of holes and blemishes as the leaves in the bud later unwrap. If a bug has fed on a developing fruit, the rupture zone heals over, but the damage can be significant and the effect of the wound becomes amplified as the fruit matures. This is exemplified in 'cat-facing', an unsightly twisting and wrinkling seen in tomatoes, peaches and apricots frequently caused by the invasive shieldbug *Halyomorpha halys*, but also increasingly by the widespread *Pentatoma rufipes* in apple and pear orchards in northern Europe (Powell 2020). However, as suggested earlier, most British and Irish shieldbugs feed on native wild plants, and although this damage probably occurs, it has not been widely recorded – because it has not been widely noted by irate gardeners and farmers.

Strangely, none of the Hemiptera have become regularly associated with dung, a rich food source inhabited by a wide range of beetle and fly species, although the North American cydnid *Rhytidoporus indentatus* has been found feeding on the fruit and seeds in accumulated guano beneath bat roosts in caves. But Ramsay (2013) reported final-instar nymphs of both *Palomena prasina* and *Pentatoma rufipes* feeding on bird droppings. They were feeding on the white part of the piebald splatter, mostly comprising uric acid, rather than the darker faecal material. Perhaps this was a supplement to their regular diet, analogous to the behaviour of butterflies seeking minerals and other micronutrients in the moisture of mammalian dung. There are also scattered reports of various bugs from all families apparently feeding at corpses and animal droppings (Constant 2007), but again these may have been in search of moisture and mineral nutrients to supplement a regular diet. Eger *et al.* (2015) report a large number of species and specimens of South American bugs attracted to butterfly traps baited with rancid fish and shrimp carrion.

SOME SHIELDBUGS ARE PREDATORS

The pentatomid subfamily Asopinae is very unusual in that its members are all predators. Ironically the common ancestor of all the Heteroptera is thought to have been predatory. Although most modern het bug *species* are plant-feeders, this is because a small number of plant-feeding families have large numbers of species; most *families* are still made up of predators of other insects (for example Anthocoridae, Nabidae, Reduviidae and all aquatic groups; Sweet 1979). It is thought that shieldbugs (along with various other groups) evolved herbivory (possibly via fungus-feeding), and then the Asopinae re-evolved predation. The asopine rostrum is thicker than that of plant-feeding shieldbugs, and is usually described as being as thick as a front femur – incrassate is the technical term, meaning thickened, and related to the word 'increase'. In fact it turns out that many suspected predatory bugs are really omnivores. This flexibility to eat a variety of foodstuffs is advantageous, but depends on some physiological properties, in particular the ability to ramp up gut production from starch-digesting amylases to protein-digesting proteases, especially if prey is to become the major constituent of the diet. Eubanks *et al.* (2003) suggested that het bug groups that evolved omnivory had some preadaptation because originally their ancestors fed on nitrogen-rich seeds and pollen, and had wide foodplant choices rather than being fixed to a narrow range. In other words, the more adventurous eaters would eventually evolve into groups that ate a broader range of foods. Incidentally, predatory and omnivorous groups tend to have more strongly serrate stylets than pure plant-feeders.

Of the British and Irish Asopinae species *Picromerus bidens* (Fig. 55) is the best known, and because it is so widespread across the northern hemisphere (it has also been accidentally introduced into North America) it is frequently the subject of international reports. Typically it hunts for slow-moving, soft-bodied, non-hairy prey such as moth and sawfly caterpillars, leaf-beetle larvae, and fly maggots; Dolling (1991) describes it as a timid predator, gently inserting its stylets into its food rather than ferociously puncturing it. A long list of European and North American prey records is given by Larivière & Larochelle (1989), who consider it might be a useful insect for biological control of pest outbreaks,

FIG 55. A nymph of *Picromerus bidens* feasts on an adult Dock Bug (*Coreus marginatus*). The very stout rostrum is characteristic of predatory shieldbugs.

particularly *Diprion frutetorum* sawfly larvae in the pine forests of Quebec. They optimistically report that the bug is frequently found feeding on sawfly larvae on German and Swedish pines, but remain cautious about its potential impact on other species. In Europe it is often found feeding on Colorado Potato Beetle (*Leptinotarsa decemlineata*), but is nevertheless unable to satisfactorily control outbreaks. Ironically, in North America, the closely related *Perillus bioculatus* is a specialist predator of Colorado Beetle larvae, and obtains its red coloration (it slightly resembles *Eurydema*) by sequestering red carotenoid pigments from its prey. Likewise, in China, *Arma custos* has been reared for use as a biocontrol agent against the caterpillars of the Fall Armyworm moth (*Spodoptera frugiperda*), a leaf beetle (*Ambrostoma quadriimpressum*) and other agricultural pests (Fan *et al.* 2020).

FIG 56. *Picromerus* enjoying a caterpillar of the Pale Tussock (*Calliteara pudibunda*). The spiny bristles of this caterpillar are usually thought to be a defence against being eaten by birds – the hairs get stuck in the throat – but they are no deterrent against a ferocious shieldbug.

FIG 57. There's a lot going on here as two *Picromerus bidens* fight over a skewered sawfly larva, while one of them remains attached *in copula.*

The use of *Picromerus bidens* as a biocontrol agent was first mooted by the German naturalist A. C. Kühn (1775); he described them as large hard-shelled brown bugs with a pointed thorax – *coleoptratos thorace acuto.* He suggested that 6–8 specimens shut up in a room swarming with bedbugs would completely exterminate them within a couple of weeks: 'You will soon see with pleasure that these wild stink bugs seek out those filthy night vermin in all their holes and nooks and crannies and kill them until they have completely destroyed them' (thank you, Google Translate). This report is parroted down the centuries (Westwood 1839–40, Miller 1956), but it is unclear how many roomfuls of the filthy night vermin Kühn was able to clear, or whether this was just anecdotal conjecture. Certainly, under experimental conditions *Picromerus* will pick off a bedbug with ease. It works both ways, though, and as well as attacking noxious pests, adults and nymphs of *Picromerus* are known to be significant predators of threatened butterfly caterpillars, including those of Marsh Fritillary (*Euphydryas aurinia*) in Italy and the Czech Republic, and the related Scarce Fritillary (*E. maturna*) in Germany (Konvicka *et al.* 2005, Pinzari *et al.* 2019). The caterpillars of both of these butterflies gather together in huddles around a silk web to hibernate, and a bug can work its way through the cluster more or less at its leisure. Konvicka *et al.* (2005) also report the bug feeding on the adult butterflies, but only those too old and battered to fly, or one caught in a spider's web. The Marsh Fritillary is very scarce in Britain and Ireland, but whether *Picromerus* is an important predator here is not known.

Picromerus is, like many predators, an opportunistic attacker; rather than specialising in one particular prey species, it takes whatever it can find and

FIG 58. *Zicrona caerulea* taking advantage of vulnerable leaf beetles (a *Galerucella* species) that are otherwise preoccupied.

overcome. Part of its potential effectiveness as a biocontrol agent is that it will prang whatever it comes across; even if it rejects that particular prey item the stab wound, with the injection of digesting enzymes down the salivary canal, will kill the victim anyway. This is simply the predator version of the cell-rupture feeding found in herbivorous shieldbugs. If it does decide to feed, then it will stay attached, sucking, for many hours. Other British members of the Asopinae are less well observed. *Troilus luridus* is another widespread and mainly arboreal species and has been found eating similar caterpillars and beetle grubs. It has been blamed for attacking small hibernating Purple Emperor (*Apatura iris*) caterpillars during the mild February and March of 2022 (Matthew Oates, outraged personal communication). The bright metallic-blue *Zicrona caerulea* seems to specialise in metallic-blue *Altica* flea beetles and their larvae, and *Rhacognathus punctatus*, mainly a denizen of sandy heathlands, is recorded attacking the larvae of the Heather Leaf Beetle (*Lochmaea suturalis*).

TASTES CHANGE

Many shieldbugs are associated with particular plant species, and investigating a foodplant is often the easiest way to find an uncommon bug. Seeking out juniper bushes on limestone hills was traditionally the way to find both of our native juniper shieldbugs, *Cyphostethus tristriatus* and *Chlorochroa juniperina*. Likewise the Cow-wheat Shieldbug (*Adomerus biguttatus*) seems limited to a single uncommon foodplant of old woodlands. Meanwhile the 'Gorse' Shieldbug (*Piezodorus lituratus*) also feeds on broom, Dyer's Greenweed (*Genista tinctoria*) and garden laburnum, and the common Green Shieldbug (*Palomena prasina*) feeds on a very wide range of plants from nettles and thistles to garden green beans and hollyhocks. Likewise the potential new arrival *Halyomorpha halys* has been found feeding on a vast selection of foodplants on several continents – over 300 host plant species in its native range and a wide range of commercially important plants elsewhere, including ornamental shrubs, hardwood trees, soya bean, apple, pear, cherry and peach (Nielsen & Hamilton 2009), so if and when it arrives in the British Isles it could start feeding almost anywhere, and on almost anything.

Received wisdom has it that focusing on a narrow foodplant range (often a particular plant genus or sometimes just a single species) offers a degree of certainty to a herbivore. The animal 'knows' instinctively, through genetic hard-wiring (or imprinting?), which plant to seek out and lay its eggs on, and through aeons of evolutionary pressure it will have had a chance to circumvent any defence mechanisms such as deterrent or toxic chemicals in the plant. By this means it will suffer less from competition from other more casually visiting herbivores, which are less able to utilise the host plant tissues or juices. Conversely, herbivores feeding on a wide range of different plant species are able to make use of a much more widely available series of foodstuffs, but are less efficient because they have to compete with everything else feeding there too.

There is a geographic effect too, though. It is well known that near the centre of a species' worldwide range it will occur in a broader range of habitats, often on a broader variety of foodplants. The Eurasian *Odontoscelis fuliginosa* occurs well inland in southern France, with scattered records throughout the very centre of the country, but on the Atlantic coast it becomes increasingly attached to dry well-drained sandy and chalky soils, and in Britain it is severely limited to a few coastal sand dunes. In continental Europe the Cow-wheat Shieldbug (*Adomerus biguttatus*) is also recorded from Betony (*Stachys officinalis*), another plant of woodland rides (Ramsay 2019), but in a completely different plant family (Lamiaceae), and also yellow rattles (*Rhinanthus* species), whereas in Britain and Scandinavia it is only known on its namesake foodplant *Melampyrum pratense* (family Orobanchaceae) and possibly Small Cow-wheat (*M. sylvaticum*). There may be a climate effect altering the metabolism of organisms as they spread northwards to cooler zones, but there is also a likely genetic effect too. As migrants have spread out from any original heartland, during thousands, tens of thousands, or hundreds of thousands of years, outlier colonies have been established by limited numbers of colonising individuals, sometimes just a single egg-bearing female, so these populations have a shallower gene pool, and with it they have less variation with which to accommodate any climatological or habitat extremes, and are less adventurous when it comes to foodstuffs. They stick to what they know, both in terms of the immediate geographic vicinity and in terms of what they and their offspring will eat. This is also partly why organisms become rarer on the edges of their ranges – lack of genetic diversity means that they do not have much physiological latitude and they struggle against an increasingly hostile environment, where they are less able to adapt to any changes in the local conditions. Hence they become isolated in small precisely suitable habitat pockets, maintaining the genetic uniformity that continues to prevent them from breaking free to exploit other nearby niches.

Very narrow foodplant choice constitutes a severe restriction, though, and can lead the herbivore into an ecological cul-de-sac, especially if the plant is also rare, and on the edge of its range. So it was with those juniper-feeding shieldbugs. Towards the middle of the twentieth century the former grazing regimes which had previously kept the limestone hillsides suitable for the small slow-growing juniper bushes were no longer profitable for cattle and sheep farmers. With the animals gone, scrub invasion (mostly hawthorn and dogwood on the South Downs where I grew up) started to alter the character of the hillsides, as did the advent of more vigorous ploughing of the less steep slopes, and squat stubby juniper bushes became great rarities. When I found my first specimen of *Cyphostethus tristriatus* on 18 April 1976, on the precipitous scarp of Newtimber Hill, Poynings, north of Brighton, it would be close to the bug's nadir in Britain, and the only time I have ever seen it on its official native foodplant. But now a strange thing happened. *Cyphostethus* started to be found on Lawson's Cypress trees (*Chamaecyparis lawsoniana*), in parks and gardens in southern England. It had switched foodplants, albeit to a very closely related plant, in the same plant family as juniper – the Cupressaceae. It also feeds on several other trees in this family, including the Northern White Cedar (*Thuja occidentalis*; Fig. 59). Quite how or where this switch occurred is difficult to make out. Cypress-feeding by *Cyphostethus* was observed in other northern European countries during the 1980s and 1990s, and it is quite possible that this feeding adaptation arose in (or near) the Netherlands and that the cypress-feeding strain of the bug was introduced elsewhere (including into the British Isles) by the huge horticultural trade in these garden trees. Hawkins (2003) points out that the bug still occurs on juniper bushes in a narrow belt of managed chalk downland in Surrey. It may take detailed DNA analysis to determine if the juniper-feeding individuals are genetically different (implying a different geographic source) from the cypress-feeding specimens. There's a PhD in the making for someone. However, wherever the switch occurred,

FIG 59. Nymphs of the 'juniper' shieldbug *Cyphostethus tristriatus* feeding on the fruits of *Thuja occidentalis*, commonly called the Northern White Cedar though it is a cypress rather than a true cedar.

it was enough to free *Cyphostethus* from its downward-spiralling attachment to the ever-decreasing juniper bushes, and the bug took off. It is now more widespread in the British Isles than it has ever been in recorded history, with records right across England and Wales, and up into Scotland, Isle of Man and Ireland.

A sorrier fate befell the other juniper shieldbug, *Chlorochroa juniperina*. It too reached its lowest point during the twentieth century, but has not been seen since it was last recorded in Lancashire in 1925, and is now thought to be extinct here. Likewise it remains scarce throughout Europe, although it is known to feed on Crowberry (*Empetrum nigrum*) in northern Germany. This low plant is widespread in Wales, northern England and Scotland, but since the bug has mainly been recorded on the chalk hills of southern England, it perhaps never had the chance to experiment.

An even greater foodplant switch occurred in the Box Bug (*Gonocerus acuteangulatus*). For almost all of its recorded history in Britain (over 150 years), this attractive and distinctive insect was known only from a few Box trees (*Buxus sempervirens*) perched on the steep chalk slopes above the aptly named Box Hill in Surrey, where it was reputed to feed on the tough green fruits. Mythically rare, it came as a great surprise when a specimen was found at Bookham Common, some 6 km north-west of Box Hill, in January 1990, beaten from a Holly tree (*Ilex aquifolium*). By the end of that year several more specimens had been found there, including nymphs, mostly on hawthorn bushes. The London Natural History Society had been organising a long-running and very detailed ecological survey at Bookham for over 50 years. It had had huge input from local naturalists as well as visitors over the many years. My father was a regular there during the 1950s when he lived in south London, and he and I made return visits during the late 1970s. On the second Saturday of most months members and guests would gather at the society's hut (then a glorified garden shed) at 11 o'clock in the morning, then return for a lunch-time confab, exchanging news between bites of sandwiches and gulps of vacuum-flask tea. There were never fewer than six naturalists, as I recall, and sometimes over a dozen, often national experts on one group or another. More senior members got a chair to perch on, but as the youngster in the group I was expected to sit on the grass. Detailed reports of the plants, animals, fungi, lichens, in fact anything to do with Bookham, fill the pages of the society's journal. This was one of the most ecologically scrutinised plots of land in the country. It was inconceivable that this large and striking insect could have been previously overlooked there. Eric Groves, compiler of the extremely detailed 'Hemiptera–Heteroptera of the London area' (1964–89), was a frequent visitor for 40 years and would surely have found it.

Within a few years *Gonocerus* was found to be well established in this area of central Surrey, from Ashtead Common to Hackhurst Downs. It was usually feeding on hawthorn. This is a major change from the Buxaceae to the Rosaceae, although it has often been recorded from wild roses in continental Europe, and was perhaps aided by the bug's habit of feeding on the fruiting bodies, rather than the leaves or stems. The bug's spread has continued unabated, and it is now found across the whole of southern and central England, all the way to Torquay, Bristol and Yorkshire. Its British foodplant tastes also continue to widen, and it is now regularly seen on wild roses, buckthorn, apple, honeysuckle, cypress, yew and occasionally still Box bushes.

Meanwhile a few British shieldbugs, still attached to uncommon foodplants, continue to remain uncommon too. Good examples are the two scarce *Dicranocephalus* species on spurges (*Euphorbia*), *Sehirus luctuosus* on forget-me-nots (*Myosotis*), and *Canthophorus impressus* on Bastard Toadflax (*Thesium humifusum*). But there is always the chance that some subtle genetic change will enable them to switch to a less narrow niche, to expand and spread and become more common.

Some idea of how host-plant specificity can be usurped comes from captive breeding studies of various potential 'pest' shieldbug species. Hemipterists have learned from lepidopterists, who have long reared caterpillars on alternative foods if the 'traditional' foodplant cannot be easily obtained. The North American Soapberry Bug (*Jadera haematoloma*, Rhopalidae) can be reared in the laboratory using commercially available sunflower seeds (Carroll 1987), and the Western Conifer Seed Bug (*Leptoglossus occidentalis*, Coreidae) will feed on the fruits of pistachio under caged conditions (McPherson *et al.* 1990).

PEST POTENTIAL

If shieldbugs can easily change their foodplant choices there is a very real potential that some presently innocuous species, quietly feeding in out-of-the-way places on wild plants, might suddenly take a liking to a valuable crop plant. Currently shieldbugs are generally regarded with curious indifference by most British and Irish farmers and gardeners. Our standard agricultural and horticultural textbooks are thankfully free of severe shieldbug warnings. There is the occasional mention of the Green Shieldbug (*Palomena prasina*) on broad beans, but even the Brassica Bug (*Eurydema oleracea*), which is so obviously named for its association with the cabbage family, is only rarely mentioned in passing. On the whole, shieldbugs are regarded as interesting wildlife rather than noxious pests.

This might be about to change, though (Fig. 60). The recent (2003) arrival of the Southern Green Shieldbug (*Nezara viridula*) is still being monitored. Across much of Eurasia and North America this has become a widespread nuisance because, like *Palomena*, it will feed on a wide range of foodplants, including many that humans regard as their own personal private property, and it is quietly gaining a reputation as a nuisance pest. The standard stock phrase of

FIG 60. Pest proportions – *Nezara viridula* nymphs feeding on tomato pepper.

the economic entomologist – 'It's only a pest if it reaches pest proportions' – is all very well, but in today's heavily commercialised and intensively industrial farming, landowners are apt to spray first and ask questions later.

Elsewhere in the world shieldbugs are already starting to register with the local growers of all manner of crops. Apart from the quasi-notorious species already mentioned, common but 'harmless' British species like *Pentatoma rufipes* are increasingly being reviled as European orchard pests (Powell 2020).

At present the two British species of *Dicranocephalus* are very scarce and have little or no horticultural or agricultural significance here, but this genus is known to be a vector of *Phytomonas*, a protozoan disease of plants, notably in the spurge family, Euphorbiaceae, on which *Dicranocephalus* preferentially feed. These protozoa are related (albeit distantly) to *Trypanosoma*, the parasites that cause Chagas disease in humans and which are spread by 'kissing bugs', *Rhodnius* and other blood-sucking genera in the family Reduviidae. At the moment, spurges are popular and relatively trouble-free garden plants – while both *Dicranocephalus* species remain scarce in the British Isles.

Things do not have to reach biblical plague conditions for warning lights to flick on, and farmers are often quick with the pre-emptive strike to remove any possibility of trouble ahead. Likewise in the private garden, some gardeners regard even a single insect, gently probing with its proboscis, as something to be swatted immediately. At the moment shieldbugs' low public profile probably keeps them safe, but with increased publicity on *Nezara* and the Marmorated Shieldbug (*Halyomorpha halys*) in the tabloids, a sudden alertness to that distinctive shieldbug shape might sway the balance away from calm tolerance to all-out war.

CHAPTER 4

Dangers and Defences

Once they've arrived in the world all insects suffer from a perennial problem – being small, you always look good enough to eat. Insects have evolved various stratagems to avoid being eaten, and many of these are manifested in shieldbugs. Running away and hiding is an obvious one, and although they are not particularly fast on their feet, the muted leaf-greens and mottled earth browns that help them hide are everywhere to be found in the shieldbug palette. Making yourself taste foul is another defence, and this is the origin of that delicate rancid marzipan smell I've already mentioned, but which is really a bitter appetite-suppressant. Apart from some sharp corners to the pronotum, shieldbugs aren't really armed with weapons to fight off a potential attacker, but they do have some behavioural tactics available.

FIG 61. Roadkill – shieldbugs (here *Pentatoma rufipes*) are relatively small and fragile, and no amount of defensive behaviour will protect one against being crushed underfoot on the pavement.

Camouflage – blending in with the background – is perhaps the most straightforward defence available to any insect. Shieldbugs are relatively good at this and come in a variety of leaf- or bark-like colours, from green, through beige and grey, to brown, often mottled or streaked to help with the deception. The common

Green Shieldbug (*Palomena prasina*) is an excellent example of this, its bright emerald carapace often melting into the equally bright leaf on which it disports itself. But what do you do when the leaves fall in autumn? As mentioned earlier, *Palomena* now has a neat trick – it turns itself brown, possibly to blend in with the dead leaves amongst which it will overwinter, or the brown bark in the crevices of which it will seek shelter. When it emerges from hibernation the following April or May it starts to revert to its original rich green colour. Other species, including the Southern Green Shieldbug (*Nezara viridula*), also become more muted in autumn (Fig. 62) but revive their strong colours the following year.

FIG 62. *Nezara viridula*, the Southern Green Shieldbug, turns brown in winter, specially to blend in with my fence.

Another simple defence is to let go of the plant you're climbing up. This has the immediate effect of letting you drop into the rough herbage, leaf litter and root thatch down below, usually out of sight and often out of scrabbling distance of any predator. This is the infuriatingly frustrating behaviour adopted by *Aelia* and *Eurygaster* species, feeding or egg-laying on grass, which just disappear into the sward if you loom over them too closely. Flying away can be useful, but the moment of hesitation it takes to flip forward the

FIG 63. Although brightly coloured, the orange and black pattern of *Corizus hyoscyami* does help it blend in if the background is also speckled with the blacks and yellow-browns of autumnal leaf litter.

hemelytra and unfurl the membranous hind wings can be a fatal one, and since birds are prime predators flying off just brings you to the attention of a much better aeronaut. If you are going to fly you had better have a disappearing trick up your sleeve. *Coreus marginatus* and *Alydus calcaratus* fly readily to escape, and though both have dull dark brown exteriors, they also both have a bright red upper abdomen which is revealed in a bright flash when the hemelytra and wings are opened. The trick is to land, close your wings suddenly, and shuffle sideways, immediately blending into the brown stems and dead leaves of the undergrowth, leaving the red-imprinted aggressor staring at the spot where the scarlet flash vanished and wondering which portal to another universe you slipped through.

Not sitting about in a stupidly obvious spot is another good tactic. And although shieldbugs can sometimes be found in full view basking in the sunshine, they usually live on the underside of a leaf rather than on top of it. Their eyesight is so good that if a possible predator (such as a human) stumps up to them without a care they will immediately shimmy away around and under the leaf to hide. And if you can't hide under a leaf, perhaps you can hide under your mother.

MATERNAL CARE

In Britain *Elasmucha grisea* shows some of the most remarkable behaviour of any insect. After laying her batch of about 28–56 eggs, the mother shieldbug does not depart to continue her independent life; she remains with the clutch on guard duty. It may take three weeks for the first-instar nymphs to emerge; nevertheless, the mother is there waiting for them. And she continues her watch over them as they feed up and moult into second and sometimes even third instar, all the time quite literally standing atop the cluster. This goes completely against the majority insect policy of just laying some eggs in a likely spot, walking away and hoping that some of them make it through to adulthood. Certainly a few 'social' insects such as bees, ants, wasps and termites have evolved complex parental care involving nests, large colonies and division of labour, but the vast majority of insect species are loners; they mate and lay eggs, and though they may spend some time searching for just the right spot, even chewing a hole in which to hide them, they then get on with the basic prime directive of looking after number one – eating, hiding and staying alive. The eggs are abandoned to the vagaries of a dangerous bug-eat-bug world.

The Hemiptera somehow buck this trend, and complex parental behaviours have evolved several times. Males of the giant water bugs in the family Belostomatidae (tropical insects, none in Britain) carry the eggs on their backs, where they are carefully glued onto the hemelytra by the female. Likewise males (mostly) of the tiny (5 mm) southern European coreid *Phyllomorpha laciniata* carry the eggs laid on their backs until hatching time (Fig. 64). These lacy bugs have dainty leaf-like edges to their bodies, up-turned to form a natural basket-shaped depression on their backs, and they early acquired the name Golden Egg Bugs for this behaviour. This paternal behaviour is fairly passive, though, and once the nymphs emerge they are released into the wild with no further input from the father. It also seems that *P. laciniata* lays its eggs on both males and females, whoever just happens to be nearby. The evolutionary advantage seems to be that having a mobile nest for the eggs offers some itinerant protection against egg predators, which would soon polish off a stationary clutch, since the adults are edged with an up-flexed leafy flange and covered with a lacy array of spines, and can hide by blending into the hairy vegetation on which they climb. Oddly, the

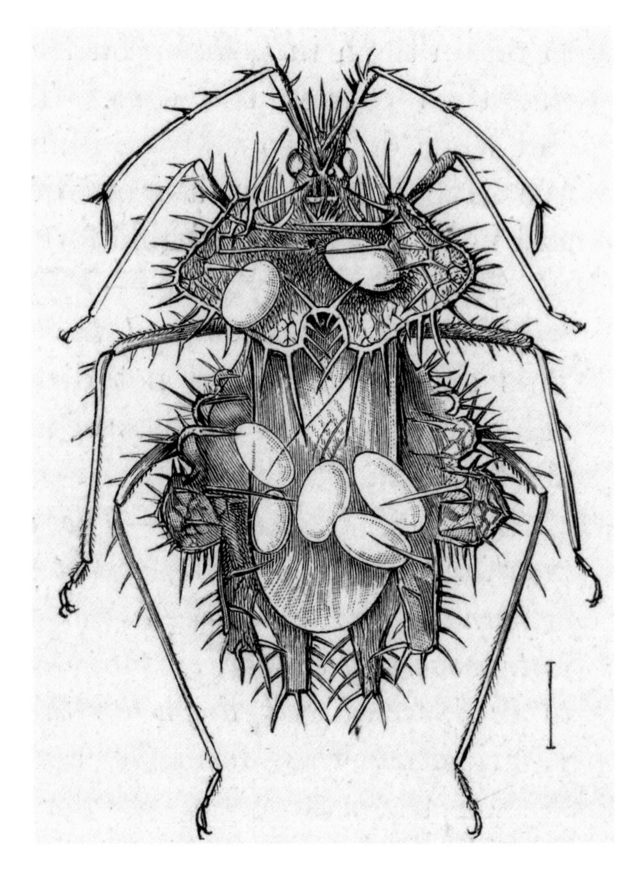

FIG 64. Engraving of *Phyllomorpha laciniata* carrying several eggs on its spiny basket-like carapace. From Sharp (1909).

very similar eastern Mediterranean *Phyllomorpha lacerata* (the only other member of the genus) does *not* appear to carry its eggs in this fashion, even though it has all the same lace-and-bristle decorations; or it may simply be that hemipterists have yet to observe the behaviour in this species. Sometimes males of *P. laciniata* are unwilling to be lumbered with any eggs anyway – actively trying to scrape off those that have been laid on them. Despite the potential advantage in helping ensure their offspring survive the vulnerable egg stage, there is an implied personal inconvenience cost to egg-carrying, leading to the suggestion that this is not simply a case of willing paternal care, and that there are conflicting features of intraspecific parasitism also involved. These cases of paternal care are well known, but are not common. On the other hand, maternal care is particularly prevalent in the shieldbugs. It is widely known in Acanthosomatidae, Cydnidae, Pentatomidae and Tessaratomidae.

We are lucky to have *Elasmucha grisea* common in the British Isles, because it shows perhaps the pinnacle of maternal care – or 'maternal solicitude' as Kirkaldy (1904a) so poetically put it. In May or June the female lays her batch of eggs on the underside of a birch leaf and guards them until they hatch in three or four weeks. She sits low over the eggs, shifting her stance if disturbed, but is unwilling to flee even in the face of curious entomologists prodding at her with fingers, pencils, twigs or grass stems. There is good evidence that she is highly successful in defending her brood, and there is a substantial body of research literature showing that survival of the eggs is reduced to zero if the female is deliberately removed, or that mortality is increased if clutch sizes are experimentally adjusted. Costa (2006) gives a good review of this and many of the other parental and social behaviours I mention in this chapter.

After mating, the slightly smaller male *Elasmucha* soon dies, but the female is on a mission. After laying, she tends the eggs (Fig. 65), and then the hatchling nymphs. Her first task seems to be to defend the eggs against a small leaf bug, *Kleidocerys resedae* (family Lygaeidae), which is extremely common on birch trees. Though ostensibly a herbivore feeding on the birch catkins, this active little bug is not above sucking out the *E. grisea* egg contents through its straw-like mouthparts, much as one would drain a lunchbox carton of orange juice, probably with associated slurping noises. Other serious predators include spiders, ants and other het bugs, particularly nymphs of the leafbug family Miridae. On the enemy's approach the mother *Elasmucha* faces the intruder, lowering her body over the eggs. She will rear up slightly, rock her body from side to side, or flick her wings. Effectively she uses her body as a shield, valiantly placing it between her babies and anything trying to threaten them. Success rates of 100% rearing are quoted, showing that this behaviour certainly pays off.

FIG 65. A female *Elasmucha grisea* stands guard over her clutch of about 50 eggs.

Occasionally two adult females are found guarding a large group of nymphs. It seems that after two females have coincidentally laid egg batches near each other on the same leaf, the hatchling first-instar nymphs instinctively huddle together to form one group; one female usually takes up position in the centre, while the other waits to the side. Arguably this might be the same instinctive urge that leads to mass cross-generational huddles in species like the Firebug (*Pyrrhocoris apterus*), where safety in numbers is thought to come from increased red and black warning coloration from the larger group. The nymphs of *Elasmucha* are strongly patterned in black and green and the family makes an eye-catching sight on the leaf.

Elasmucha grisea is not the only member of the genus to show maternal behaviour. In a useful and detailed review from the Czech Republic, Hanelová & Vilímová (2013) summarise observations of *E. grisea*, *E. fieberi* and *E. ferrugata*, all of which show familiar egg-guarding, and these observers take the opportunity to confirm that other members of the Acanthosomatidae (*Cyphostethus tristriatus*, *Acanthosoma haemorrhoidale* and *Elasmostethus interstinctus*) do not, just in case there was any suggestion that their maternal instincts had been overlooked. Throughout the world studies of other species have provided particular details that may yet be applicable to *E. grisea* as observation and experimentation continues. Presenting a crushed nymph of the Japanese *E. putoni* to the mother elicits a strong defensive response – facing off towards an attacker, body-rocking, wing-flicking – suggesting that an alarm substance is released if the nymph cuticle is punctured. In this species the mother is primarily defending her offspring against the predatory ant *Myrmica ruginodis*, which captures prey by biting it. In another Japanese species, *E. dorsalis*, the mother stays with the nymphs throughout all their immature stages. As they move about the host plant, Goat's-beard (*Aruncus dioicus*), feeding on the inflorescences and developing fruits, the mother stations herself on the stem below them, effectively acting as gatekeeper and preventing any predators from passing.

As the nymphs near adulthood the mother *Elasmucha* eventually leaves them, but her offspring will usually remain as a cluster until they have all matured and become adult themselves (Fig. 66).

FIG 66. *Elasmucha grisea* nymphs huddle together for protection. First-instar nymphs on a birch leaf (top left); fourth- and fifth-instar nymphs under a birch leaf (top right); newly emerged adults, some looking pale, and one still pulling itself out of its final moult (bottom).

FIG 67. *Trissolcus* parasitoid wasps (family Scelionidae) emerging from shieldbug eggs (probably *Graphosoma italicum*).

Egg-guarding seems to be highly protective against walk-by predators, but is less effective against egg-parasitoids. Tiny wasps in the families Braconidae and Scelionidae specialise in attacking insect eggs, and as they are not much bigger than the nutritional parcels they attack, there is enough nourishment in a single egg for a parasitoid maggot to reach maturity and metamorphose into an adult (Fig. 67). Against these insidious and extremely agile enemies the mother-as-physical-barrier method is less than completely effective. If the egg cluster is attacked by one of these wasps the mother bug lowers her body to cover as many as possible, but she does not turn to face her foe. It's as if she knows that any rotation on her part will just expose some other part of the egg mass, and she has to be resigned to losing some of her offspring. The wasps tend to target the outer eggs; they can only reach the inner ones if the adult is not present. In *E. ferrugata* (very rare in Britain, but widespread in Europe), the bug appears to factor in these potential losses to parasitoids by concentrating effort in laying larger eggs towards the centre of the batch and smaller eggs around the edges. It is these smaller peripheral eggs (her least favourite children) which will be sacrificed if a parasitoid wasp attacks, meaning that the larger central eggs have a better chance of survival.

Maternal behaviour in *Elasmucha grisea* was first reported by the Swedish naturalist Adolph Modéer in 1764, and his account was then repeated and elaborated upon by his compatriot Carl De Geer (1773). Modéer seemed to think that the main purpose of maternal protection was to defend eggs and nymphs against the voracious attentions of the father. Douglas & Scott (1865) repeated this assertion, and they added their own observations and other details sent by various correspondents. There was, however, still some doubt about the veracity of this reported behaviour, and renowned French naturalist Jean-Henri Fabre, often mooted as being the father of observational entomology, rather pooh-poohed the notion in an article he wrote on 'Les pentatomes' (1901). In a strange departure from his previous very detailed and accurate writings, he went out into his garden to look for various shieldbugs, found three different species (but not *E. grisea*) and confirmed that none of them showed any inclination whatsoever to watch over their eggs or young. They laid eggs and abandoned them, as indeed many other shieldbugs do, leaving their young to fend for themselves in an uncaring world. It took Kirkaldy (1904a) to adjudicate on the conflicting reports, and come down firmly against Fabre in the light of multiple careful and accurate reports from numerous naturalists across several different countries. Unfortunately Fabre's tone is one of ridicule, suggesting that De Geer's report of the defensive mother and potentially cannibal father should be 'relegated to the same limbo as the childish tales which encumber history'. It all got a bit testy, as Kirkaldy parried the rebuke and knocked the slightly sarcastic ball back into Fabre's garden. Sorry to mix metaphors.

The well-documented parental behaviour has given *Elasmucha grisea* its English name – Parent Bug – and credit for this perhaps belongs to Hellins (1870). His short paper is full of precise information about finding the insect and eggs and seeing the mother guard them until they hatched a few days later. He tells of her fussing about the nymphs as he prodded with his finger; she even fluttered her wings rapidly at him. Although he was simply reporting what the adult was doing, his text uses what would eventually become the accepted name, perhaps inadvertently, for the first time: 'The parent bug showed no fear, and barely moved when I touched her, only shifting her legs and sloping her shoulders and back so as to protect the side on which danger threatened.'

Maternal behaviour is also known in other members of the Acanthosomatidae throughout the world, and in some other families. Similar egg- and nymph-guarding, against similar foes, is seen in the South American pentatomid *Antiteuchus tripterus* (Eberhard 1975) and the South-East Asian coreid *Physomerus grossipes* (Nakamura 1990); meanwhile in the Brazilian pentatomid *Phloea corticata* (and other species), the first-instar nymphs are carried about on the underside of the broad paper-thin adult female (Guilbert 2003); the nymphs have disproportionately large tarsi which may help them cling to the underside of the mother.

No fully developed social behaviour has ever been recorded in British Cydnidae, perhaps because these bugs are mostly secretive soil- or root-dwelling species, infrequently observed, even though entomologists do spend a lot of time on their hands and knees grubbing about in the undergrowth. However, *Tritomegas bicolor* has been observed laying its eggs, about 100 of them, in a cavity in the ground, then standing guard over them.

Several exotic species are known to care for their young by actively foraging for food and bringing stocks of it back to the delicate nymphs huddled together in a rudimentary nest. In a well-studied East Asian cydnid, *Parastrachia japonensis*, about 100 eggs are laid in a hollow in the soil and tended by the female. In the face of a serious predator (like a spider or ground beetle) undeterred by her body waggling and flexing, she will pick up the eggs, which are stuck together 'like fresh caramelised popcorn' (Costa 2006), by putting her rostrum through the bundle and lifting them up; she then dashes off with them. Having said this, if the nymphs are attacked after they have hatched, the mother flees. There's only so much she can do. If she is left in peace, however, she places the eggs in a small depression in the ground and sets off in search of the fallen fruits (drupes) of a small hemiparasitic subtropical shrub called *Schoepfia jasminodora*, which at about 15 mm across are not much smaller than the shieldbug herself and actually weigh about three times as much as she does. A female can travel 30 metres from the home site, testing several drupes until one meets her approval; many are rejected outright. Experiments using 'trick' wooden dummy drupes smeared with juices from mature fruits show that the mother relies on olfactory cues, although weight and colour also play a role. When satisfied that the fruit meets her needs, she rolls it back to the 'nest', where the young nymphs feed on it. The nymphs take the nutrient-rich endosperm inside the developing seed, rather than the thin outer rind. Unlike the human notion of savouring fresh fruit from the tree, the bug only takes fruits dropped onto the woodland floor; she never climbs up to feed on those still growing on the plants. Depending on how quickly the nymphs feed and how quickly the mother finds suitable food, the drupe store can contain from 3–4 fruits to an average of about 30–40, and occasionally up to 100. The mother defends this cache as vigorously as she defends her young. During poor fruit years the bugs are forced to 'nest' closer under the bushes where there is less leaf litter coverage; normally they prefer to set up base towards the sunnier leafier edges of the shrub. Any supposed advantage of living underneath the foodplant is offset by the constant threat from other *Parastrachia* mothers pilfering from the sometimes limited fruit stocks. The nymphs are provisioned by the mother through several instars, sometimes to adulthood, before they make their own way off across the forest floor (Costa 2006).

On hatching, shieldbug nymphs of many different groups often stay together in a huddle, even in the predatory subfamily Asopinae which includes the ferocious *Picromerus bidens* and *Troilus luridus*. Not enough is made of cannibalism in biology textbooks, but danger from this normal and natural instinct must be balanced by safety in numbers – until such time as it isn't. Most nymphs soon disperse, and although they can frequently be found near each other on a foodplant, they have already started to live independent, hopefully less cannibalistic, solitary lives.

Remaining longer together in a nymphal huddle is a frequent behaviour in strongly patterned nymphs, which are thought to advertise their distastefulness by their bright warning (aposematic) colours. Such clusters occur in the Firebug (*Pyrrhocoris apterus*) and are well known in many other species including, for example, the North American Milkweed Bug (*Oncopeltus fasciatus*, Lygaeidae), the Boxelder Bug (*Leptocoris trivittatus*, Rhopalidae), and the Polka-dot Bug (*Pachycoris klugii*, Scutelleridae). The initial huddle relates to maternal protection, but the nymphs remain together in a dense group even if the original parent is lost or moves away (Fig. 68).

The evolution of maternal egg- and nymph-guarding is thought to be something to do with shieldbugs' egg-laying habits. Most of them exhibit semelparity, the female laying all her eggs at one time. Arguably, *Phyllomorpha laciniata* lays all its eggs in one basket. Anyway, an insect exhibiting the alternative strategy, iteroparity, laying eggs in several batches over its lifetime, and having ovaries that can produce a constant supply of eggs, can balance the loss of any one batch against the future potential of others. In this case it can afford to be a bit relaxed about depositing some eggs and just leaving them to their fate. But if a creature has only one chance at producing offspring during its lifetime it might be advantageous to invest time and energy in protecting the eggs, and young ones, for as long as possible. This, at least, gives some rationale for many different hemipteran bugs showing parental care, even if this behaviour is not universal in the group. Semelparity is also thought to be part of the reason for similar protective parental care evolving in some earwigs, cockroaches, dung beetles, burying beetles and wolf spiders.

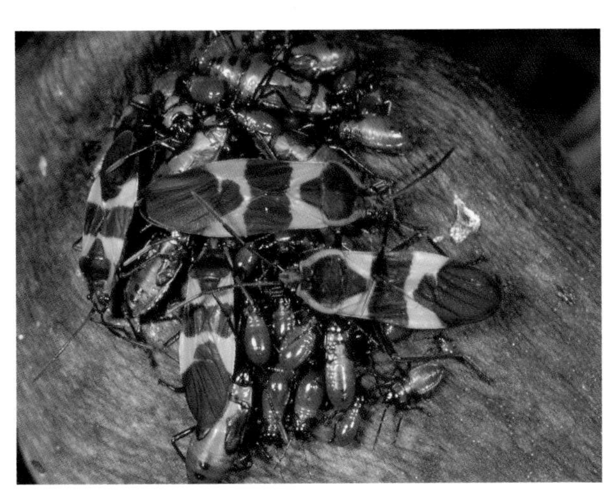

FIG 68. North American Boxelder Bugs (*Leptocoris trivittatus*, above), and Milkweed Bugs (*Oncopeltus fasciatus*, below) remain in familial huddles until almost fully grown, even as here when not feeding. It is thought that this emphasises their bright warning colours.

SHIELDBUG ALCHEMY

Apart from dousing the thumb of a zealous entomologist to stain it brown, a shieldbug's chemical armoury is really a defence against being attacked or eaten. Birds and mammals are reckoned to be the usual vertebrate enemies of insects, but it's also a bug-eat-bug world out there with an invertebrate adversary around every corner. Even without an audible bombardier beetle explosion of toxic chemical spray, or a sharp stabbing wasp sting, the effect of foul-smelling liquid oozing out of your food is a lesson quickly learned. Many distasteful insects reflex bleed if they are threatened – ladybirds, oil beetles, bloody-nosed beetles. They do not wait to be chomped by beak or tooth to get their lesson across; they exude a taster of what is to come if the aggressor thoughtlessly continues to bite or grind. And this is usually enough. The would-be predator immediately tastes the foul liquor, spits out the insect and spends some time trying to clean the unpleasant tang from its tongue, quietly regretting its action. The intended prey, on the other hand, slightly dented maybe, and covered in saliva, is at least able to get up and hobble away to live another day. And it doesn't take the firm pressure of upper and lower mandible crushing down on pronotum and hemelytra to elicit the response, it just takes a sudden movement or a jarring nudge. Tapping cypress boughs over a beating tray whilst looking for *Cyphostethus tristriatus* is often enough to fill the breeze with their scent, which they eject in mid-air as they are tumbling down onto the sheet. In the interests of scientific advancement I have taken it upon myself to taste shieldbug exudation, and I can confirm that it is disgusting. No need to pop a shieldbug in your mouth, just pick one up roughly and then lick your fingers. It's slightly less appetising than toilet cleaner.

The chemistry of shieldbug scents is remarkably complex, and every study seems to throw up a new cocktail of compounds. The key thing to remember is that the shieldbug is shunting out defensive smells for the *immediate sensation* in the mouth or beak of its adversary on that first tentative chomp, rather than a pondered savoury deeply thoughtful chew. Thus the volatile molecules are small, giving them low evaporation points. This is borne out by chemical analysis of some of the compounds. Despite a huge diversity in chemical structures, the majority are short or medium unbranched carbon-chain substances: acids, aldehydes, ketones, alcohols and esters. For the less biochemically minded these are similar to the 'simple' organic molecules like acetic acid, ethanol, ethyl acetate and acetaldehyde studied in GSCE or 'O' level chemistry, but instead of two carbon atoms most bug scents usually have 4, 6, 8 or 10 carbons in the chain. The most frequently reported compounds are in the hexanal (6 carbon atoms), octenal (8 carbons) and decanal (10 carbons) families. These are thought to be synthesised from 2-carbon base units such as occur in the simple (and ubiquitous) acetic acid (vinegar) and ethanol (alcohol). With the exception of tridecane (13 carbons), odd-numbered constituents are rarely detected. For the more biochemically minded it is interesting that many of these simple compounds have a single double bond somewhere in the chain, but the thermodynamically more stable *trans* isomer predominates over the less stable *cis* form. Isomers are molecules which share the same chemical formula, and are often manufactured by the same metabolic processes, but because of the three-dimensional way in which the atoms are joined together they have slightly different shapes – often mirror images of each other, or analogous to left and right gloves. This becomes highly relevant when it comes to species-recognition scents (see below). All of these small molecules have low evaporation points and easily start to vaporise at normal outdoor temperatures. I recently (mid-December 2021) found a specimen of *Nezara viridula* getting ready to hibernate. Air temperature was less than 5 °C but I still got a whiff of annoyance as I coaxed it onto my hand for a photo session. Good overviews of the glands and some of the chemicals involved can be found in the reviews by Staddon (1979) and Aldrich (1988), and the technical review by Kment & Vilímová (2010) reproduces a large number of high-power scanning electron microscope pictures of glands and associated surfaces and orifices.

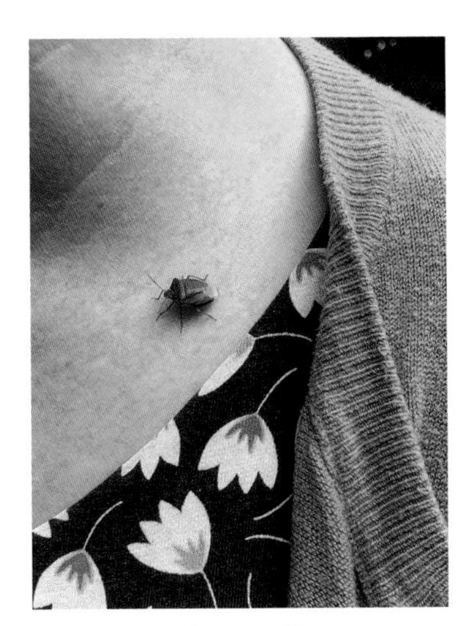

FIG 69. Apart from a tickle as one walks across the skin, shieldbugs are mostly considered harmless and non-threatening, and will not exude obnoxious chemicals unless annoyed.

Aldrich's review, still a standard on the topic even after 35 years, lists 81 frequently identified aromatic molecules isolated from various species of hemipteran; they range from hydrogen peroxide (H_2O_2) to double cyclic dihydroxychromones and mixed sesquiterpenes like caryophyllene. Most of these were from the usual suspect 'stink' bugs in the Pentatomoidea, but even the five studied Lygaeidae species were still producing 6-, 10- and 13-carbon substances, suggesting that these secretions are important right across the Heteroptera. The blends secreted by bugs are usually genus-specific, often species-specific, and the underlying assumption is that although they now also have defensive qualities, they perhaps evolved from mating pheromones which allowed the bugs to find and recognise each other. This is becoming highly relevant when it comes to designing pheromone traps to find and monitor pest species.

It's not that each species necessarily creates its own specific chemicals, more that the highly varied cocktail mixes often contain the same or similar compounds, but in varying proportions and in different combinations. These cocktails can be concocted by different rates of enzyme action as, for example, a reservoir of standard hexyl acetate in one part of the gland is split by an esterase enzyme and the resulting alcohol oxidised by a dehydrogenase. Combinations and concentrations can change as a bug ages, and male and female outputs from the same species may be dissimilar. In a study of seven South American pest shieldbugs Moraes *et al.* (2008) isolated over 30 discrete 'defensive' chemicals. Many of these were shared by multiple shieldbug species, but by comparing concentrations of each chemical in the metathoracic glands, they were able to characterise different blends for each species. The mathematics is complex; I rather facetiously described the authors' canonical variates analysis as giving robustness and statistical significance to gut feeling, but a better analogy is an algorithm that works out the best way of arranging the contents of supermarket shelves into different meals. There is some overlap within closely related groups, but it does seem that each species has its own particular distinctive chemical meal menu.

Although the same simple 6-, 8- and 10-carbon defensive molecules are being generated right across the Heteroptera, mating pheromones are likely to be more complex. To a tiny insect the world is effectively an infinite void full of danger, and finding a potential mate out there somewhere is a task beset with potentially deadly threat. It has long been known that many moths find each other by releasing highly specific chemical scents onto the wind, attracting the other sex from scores or even hundreds or thousands of metres away. This gave rise to the technique called 'assembling', used since Victorian insect collectors placed a freshly emerged female moth in a wire cage to attract the males, which would flock in their dozens, and could then be netted and added to the collection. High species specificity is necessary to avoid potential mistakes – if one species is incorrectly attracted to another there is a concomitant waste of time for both of them, and constant potential danger as they try and work out whether they are compatible with each other, or not.

So far, mating pheromone studies in shieldbugs have been focused on (i.e. funded for) potential pest species. The first sex pheromone to be isolated was that for the internationally widespread and often troublesome *Nezara viridula*; it consists of a precise mix of two isomers: *cis*- and *trans*-bisabolene-epoxide (Aldrich *et al.* 1987). These two very slightly different molecules have subsequently been found to occur throughout the genus *Nezara* and also in closely related genera, and are highly indicative ingredients in the mix. They range from 100% *trans* in *Chinavia impicticornis* to 5% *trans* 95% *cis* in *C. hilare*. Ratios were geographically sensitive too, with a nearly 50/50 mix for Kyushu Japanese *N. viridula*, 45/55 in Wakayama Japanese specimens, 75/25 in Mississippi, and 80/20 in Australia. There is clear genetic control of these proportions, and in experimental crosses between specimens from Japan and Mississippi, the resulting *N. viridula* offspring showed intermediate ratios (Moraes *et al.* 2008). This is now a very active area of shieldbug research. Once particular pheromones are identified, they can be synthesised for use in pheromone traps, either to selectively catch and kill the bugs or to monitor their appearance, abundance or spread. At present it seems that it is the male which normally releases the sex pheromone, and the females who flock in. Moraes *et al.* (2008) list 12 relatively complex molecules as sex pheromones from the seven shieldbugs they studied, and although they give a useful review table showing which other shieldbugs around the world have had pheromone studies by other research groups, they bemoan the fact that this is still only about 1% of the world's known species.

As well as being used defensively, or helping to find a mate, the species-specific chemical cocktails might have evolved as aggregation markers or as alarm signals. Another possibility, exemplified by hydrogen peroxide and phenolics in aquatic bugs, is that some may have started out having antiseptic qualities. They may also be some defence against fungal diseases. Many of the hexa- and octo- compounds made by het bugs are also produced by ants as trail or alarm pheromones, and it may have been against ants, still major enemies of bugs today, that these substances first found defensive use. The metathoracic gland is well developed in ant-mimicking bugs, perhaps helping them blend into the ant colonies in which they often live or walk.

Despite my insistence that I think shieldbugs smell of marzipan (almonds apparently contain salicyaldehyde and benzaldehyde, the latter of which has been found in shieldbug exudates), others have suggested washing-up liquid, rotten apples with a little muskiness (Butler 1923), mown grass and the branded lubricant WD40. Although it is a commercial secret, the formula for WD40 is thought to contain mostly 9-carbon to 14-carbon alkanes, so it is perfectly in line with known het bug gland outputs. According to Aldrich (1988), individual chemicals secreted by the coreid *Leptoglossus* have odours of cherries, vanilla, cinnamon and roses, although he also admits that *Alydus* secretions are just plain 'rancid'. Radioactive labelling of metabolites confirms that the bugs mostly manufacture their own scents, instead of sequestering toxins found in their foodplants (see below). Again, the assumption is that many of these scents started out as components of pheromones allowing the bugs to find each other for mating, but that they have become enhanced and concentrated for defence.

GLAND DESIGNS

One very curious facet of shieldbug defensive scent production is that nymphs and adults prepare their chemical deterrents in different parts of the body. The small, round, domed nymphs of shieldbugs are often mistaken for ladybirds because the abdomen is frequently brightly coloured red, yellow or green and is marked with dark blobs or spots. The dark areas surround the dorsal abdominal glands, often abbreviated to DAGs (Fig. 70). These are usually located on the upper surface at the junctions of abdominal segments 3/4, 4/5 and 5/6 in shieldbugs, but other groups can have a gland at junction 6/7; across the Heteroptera gland numbers vary from four (very rare) to one or no glands (in some water bugs). Paired glands secrete the scent compounds into a median reservoir formed by an in-folding of the outer cuticle, so technically the potentially toxic substances are physiologically outside the body of the bug. The reservoir, which may be single or divided, has muscles that stretch it, flattening it to eject its contents, and other muscles open the single or paired ostioles (pores), normally held shut by the elasticity of the cuticle.

Generally the dorsal abdominal glands are only effective if they are exposed to a potential attacker. In adult bugs the abdomen is covered by the wings and the sometimes very large scutellum, and any defensive secretions under the hemelytra would be ineffectual. Dorsal abdominal glands of most shieldbugs cease to function towards the end of the final instar, shortly before the final

FIG 70. Second and third instars of *Palomena prasina*. The paired abdominal glands are easily seen, first as a dark oblong patch, then as separate slightly raised mounds.

moult into adulthood. Strangely, when it turns adult, a hemipteran develops new glands on the underside of the metathorax (the third thoracic segment), just ahead of the hind coxa. These metathoracic glands (MTGs) are easily visible in side view (except in some supposedly 'scentless' plant bugs, Rhopalidae[4]) and often present a deep channel (the peritreme) and a spout or ear-like structure, surrounded by a large area with a matt or velvety appearance. Long called the evaporatorium, the function of this region of the cuticle is still unclear. It may not necessarily aid evaporation, but it at least restricts the exuding chemicals within a limited region of the bug's body, and is perhaps better described as a scent accumulation surface. This may be a protection against any toxic or caustic properties of the chemicals which might otherwise damage the bug's delicate body. The low-molecular-weight compounds readily spread and penetrate insect cuticle. Rather than allowing potentially damaging chemicals to dribble all over the bug's delicate body, the structures constrain the liquid within a defined toughened area of the underside – small enough to concentrate it for best aromatic effect, but large enough to offer high rates of evaporation.

Under high-power scanning electron microscopy, the surface of the evaporatorium is seen to be covered with small domes like mushrooms, 5 micrometres (μm, a thousandth of a millimetre) high, 9 μm across and about 10–15 μm apart, linked by ridges and grooves into an intricate lattice (Kment & Vilímová 2010). Similar study of the hardened integument of the dark spots surrounding the nymphal abdominal glands also shows a series of minute channels and lattice patterns which are thought to have similar function. The emerging scientific field of microfluidics has sought to replicate these structural features by using laser ablation on metal and polyimide foils, and measuring fluid flow (up to 1 mm/s) over and past teardrop-shaped structures. The shape of the underlying integument structures in the duct between the scent reservoir and the evaporatorium, only 6–12 μm long, clearly influences the passive flow of liquids in a favoured direction (Hischen *et al.* 2018), something that engineering companies might seek to emulate for biomedical applications.

Of course, nothing in nature is fixed or certain, and there are examples of nymphal abdominal glands remaining active into adulthood. In the North American predatory asopine shieldbug *Podisus maculiventris*, a single huge abdominal gland (at junction 3/4) remains active into adulthood, and produces pheromone chemicals (still 6-carbon hexanals but with added 10-carbon

---

4 A curious feature of some *Rhopalus* species was described by Remold (1962, quoted by Staddon 1979) where the scent fluid is directed to a small trough high up on the pleuron where a droplet accumulates; the bug can then transfer the chemicals to an attacker using its rear tarsus.

molecules like linolool and cyclics like terpineol and benzyl alcohol) to attract females. The females have a much smaller 3/4 gland. This feeds into the intuitive, but still relatively poorly studied, idea that the chemical scents of shieldbugs (indeed all hemipterans) evolved from sex-communication signals that enable females to find males, or vice versa.

SHIELDBUGS – NOT TO EVERYONE'S TASTE

Many brightly coloured bugs feed on toxic foodplants, and are known to sequester the plant toxins for their own distasteful purpose. They also tend to gather together in groups – increased numbers emphasising their bright colours. Certain exotic leaf-footed Coreidae feed on Passifloraceae (passionfruit family); they take cyanolipids from the seeds and convert them to cyanogenic sugars that can be squeezed through weak spots in the cuticle by reflex bleeding. And firebugs (family Pyrrhocoridae) often feed on Malvales (the plant suborder containing mallows, cotton, linden and cacao) which also contain well-known toxins. The brightly coloured Asian species *Dysdercus cingulatus* (the Red Cotton-stainer) accumulates the 30-carbon phenolic toxin gossypol from the seeds of cotton, on which it can be a serious pest. *Pyrrhocoris apterus* is known to exude some of the usual 8-carbon octenol series, but whether it absorbs natural toxins from its tree mallow or linden tree hosts is not known. Its habit of clustering together in multi-generational aggregations, normally a behaviour that enhances warning coloration, suggests that it is advertising its distastefulness to would-be predators. Likewise the brightly coloured *Eurydema* brassica bugs (and the Harlequin Bug, *Murgantia histrionica*, of North America) have greatly reduced metathoracic glands, are brightly warningly coloured, and are thought to sequester foul-smelling mustard oil glycosides from their cabbage-family foodplants. When disturbed they give off allyl isothiocyanate, a well-known chemical responsible for the pungent taste and smell of mustard, horse-radish and wasabi, and a lachrymator (inducer of eye-watering) similar to tear gas. Birds soon learn to avoid brightly coloured and foul-tasting prey, and even mantids can learn to avoid the brightly coloured Milkweed Bug (the North American lygaeid *Oncopeltus fasciatus*), which sequesters the same cardenolide heart toxins that render the milkweed-feeding Monarch butterfly (*Danaus plexippus*) caterpillars and adults unpalatable to vertebrate predators (Berenbaum & Miliczky 1984).

Traditionally birds are rated as the most important predators of insects, and these would be expected to eat shieldbugs if they get the chance. It is presumably against these enemies that shieldbugs have evolved their chemical defences,

FIG 71. *Graphosoma italicum* is usually regarded as being immune to predator attack, its bright colours warning of its distasteful nature and allowing it to sit brazenly in the open.

whether newly metabolised or sequestered from toxic foodplants. But yet again there are precious few studies to back up the theories. Testing the palatability (or not) of warningly coloured insects (Fig. 71) has usually meant offering them to predators which are duped into attacking them by painting over the strong markings using a neutral-coloured pigment. Vesely *et al.* (2006) did just this with *Graphosoma lineatum*,[5] choosing the romantic-sounding burnt sienna watercolour as a suitable odourless and non-toxic disguise, which did not seem to impede the bugs, or interfere with their chemical secretions. Wild-caught Great Tits (*Parus major*) tended to attack all colourways when first presented with them, but they did not repeat their attacks on the original wild-coloured bugs as often as the brown-painted ones. They appeared to discover that the bugs are foul-tasting, and this was more accurately learned if there were bright red and black warning colours to back up the revulsion. Oddly, closely related Blue Tits (*Cyanistes caeruleus*) shunned all the bugs, whatever the colour. This may have been because the shieldbugs were slightly too big potential prey items for the smaller Blue Tits, or because Blue Tits are known to show greater neophobia (fear of new things), so are rather picky eaters. In a study of slightly more adventurous Blackbirds (*Turdus merula*), Schlee (1986) fed 583 shieldbugs to 12 birds, which devoured aposematic species far less (16.3% of experimental presentations) than those that were cryptically coloured (63.1%). *Palomena prasina* and *Coreus marginatus* had

5 Incidentally, an interesting paper by Tietz & Zrzavy (1996) models the post-embryonic development of the strong longitudinal colour pattern of *Graphosoma*, linking it mathematically to the much more muted blotched patterns of *Odontotarsus*.

the highest edibility rating, and younger birds were more gung-ho than older birds, suggesting that age brings learned experience and knowledge to the dinner table. In a survey of Heteroptera-eating birds, Exnerová *et al.* (2003) recorded 17 pentatomoid species eaten by 14 central European bird species, although no *Graphosoma* was found eaten by any bird.

An interesting oblique experiment was carried out by Johansen *et al.* (2010), partly proving that birds could definitely learn to recognise the striped coloration of *Graphosoma*. Wild-caught Great Tits were taught to find and eat dead specimens which had been hollowed out and which contained a sunflower seed glued into the abdomen, making them completely palatable. This was not to test how repelled the birds were by any defensive chemicals inside the shieldbugs, but to test the birds' ability to find the slightly paler and browner pre-hibernation colour form compared to the full bright post-hibernation colourway. In Sweden, where the experiment was carried out, new adults of *Graphosoma* are at first more muted, and only develop their strong crimson and black stripes the following spring. It was suggested that this aided their camouflage against the fading brown flowers and seed heads at the end of the summer and into autumn. Sure enough, once the birds had been trained up they found the bright red-and-black spring versions much faster than the well-hidden sombre autumn ones.

To make observations of insect predators of shieldbugs, mantids have been subjected to the delightfully named 'puff test'. Tiny amounts (2 μl) of isolated

FIG 72. Mating pair of extreme colour morphs of the Common Tortoise Bug (*Eurygaster testudinaria*). This species varies from plain brown to mottled pink and purple. One explanation may be that this helps some individuals avoid birds, which often hunt by adopting a fixed search-image in their brains – targeting one colour or pattern, but missing others.

gland chemicals are dropped onto a sliver of filter paper which is inserted into a glass pipette with a flexible silicone bulb at one end. A mantid is placed in the centre of a test arena, the tip of the pipette is carefully brought up to within 1 cm of its head and the bulb is puffed three times. The degree of aversion is evaluated by estimating how far away the mantid moves to escape the unwelcome smell. Although Noge *et al.* (2012) are quick to report that 'no compound or compound blend killed the mantids at the test dosage', the fact that the poor beasts regularly scuttled off 15–20 cm when subjected to various hexenals and octenals suggests that these are, truly, repulsive chemical scents. And it has long been known that some shieldbug chemicals can actually kill other insects. Conradi (1904) reports that three specimens of the North American pentatomid *Euschistus fissilis* placed in a glass tube containing 22 Boll Weevils (*Anthonomus grandis*) killed them all within 10 minutes. His similar experiments also killed blowflies and stable flies within 9 minutes, and 'quieted a centipede' in 15 minutes, though that later recovered.

It may be that a predator just needs to get a whiff of the stink-bug stink to put it off. Certainly *Cyphostethus tristriatus* fills the air with its aroma if it is knocked out of the cypress tree onto the beating tray. The simple act of dislodging it from its perch into freefall is enough to agitate it into defensive mode. Even the nominally harmless *Pentatoma rufipes* has caused consternation. Massee (1937) reports how farm workers refused to pick from certain cherry trees in 1935 because the 'infestation' of smelly bugs was so great. Monteith (1982) reports a remarkable defence in the tiny Australian plataspid *Coptosoma lyncea*. These small globose black bugs aggregate together in the dry season, often many dozen on a leaf, and with estimates ranging in the millions per tree. If disturbed they suddenly take to the wing, buzzing remarkably loudly for their small size, and releasing a cloud of volatile defence chemical into the air – glandemonium, perhaps. Sadly, nothing like this is reported for the European, and recently British, *Coptosoma scutellatum*.

And I'm still not sure whether one of my cats really tried to eat a shieldbug one night in April 2021. He came in skittish and anxious, his left eye was swollen and closed and he smelled very strongly of that oh-so-distinctive shieldbug defensive chemical. I can only think that he maybe found one in the garden and tried to play with it, as cats so cruelly do. Maybe as he snatched at it with his claws the chemical was exuded onto his paws, and when he came to clean and groom himself it was smeared around his eye which reacted with irritated inflammation. He finally sulked off and sat grumpily on a chair, facing away from us. A couple of hours later his eye had opened, but there was still the lingering scent of diesel and marzipan. It took 24 hours to finally dissipate. I still don't really believe it, but I can't think of another explanation.

MIMICRY

Although the resemblance of some small, spotty, domed shieldbug nymphs to ladybirds has already been mentioned, perhaps the most striking instance of mimicry in the British bugs covered in this volume is that of *Alydus calcaratus*. The adult bug is sleek and elegant, a rich dark chocolate brown all over, but this is only really evident in set museum specimens. Although I started this chapter with the claim that shieldbugs are rather pedestrian in their movements, in life *Alydus* is an agitated blur – a constant buzz of frantic energy as it jerks and flits its way through the low herbage or dashes across a patch of bare ground. As it runs it also half flies, flipping open its wings for short sharp airborne darts between erratic turns and twists. As it opens its wings, *Alydus* exposes its bright red upper abdomen. In this rapid jerking movement it very closely resembles a spider-hunting wasp in the family Pompilidae. The deception really is quite remarkable, and many times I have watched a pompilid in curious awe, before I can focus my eye and remember that it is, actually, a bug. Pompilid wasps, often of similar body length, occur in the same warm, dry, well-drained, sunny locations as *Alydus*, actively hawking, half running, half flitting, through the rough grass after their spider prey, mostly ground-dwelling crab spiders. These wasps are reputed to have relatively powerful stings, although they are quite slim compared to yellowjacket wasps or honeybees, and various hoverflies are thought to gain protective advantage by mimicking them. Elsewhere in the world many species of Alydidae and Coreidae resemble stinging Hymenoptera and are thought to gain similar protection from their close resemblance to these wasps (Fig. 73).

Alydus is doubly remarkable, because its nymphs are also superb ant mimics (Fig. 74). The large head, narrow thorax, constricted waist and slightly bulbous abdomen combine with their jerky movements and rapid antennal twitching to give the precise appearance and behaviour of an ant. I have regularly scooped one out of the sweep net into a glass tube, momentarily confused by its appearance. The unicolorous early instars are just the right size for confusion with Black Pavement Ants (*Lasius niger*), and the fifth instar, with its sometimes reddish thorax and short wing buds, could pass unnoticed in a crowd of Red Wood Ants (*Formica rufa*). There is some mystery as to the bug's life history, with suggestions that it may develop in ant nests, perhaps even eating the ants. This seems highly unlikely. Elsewhere in the world no species in the superfamily Coreoidea is known to be predatory, and other members of the Alydidae are well-known plant-feeders, although they have sometimes been found attracted to carrion, and they are said to have a strong smell of decaying meat or dog dung. This is likely to

FIG 73. A rather ichneumon-like coreid (*Holhymenia* species) from Costa Rica. Parasitoid wasps in the family Ichneumonidae are usually recognisable by their slim form, narrow clear wings, black-and-yellow bodies, and often white-tipped antennae. Many have powerful stings.

FIG 74. Remarkably ant-like final-instar nymph of *Alydus calcaratus*.

be chemical defence based on repugnance. Their antiness may give them some protection from predators – Oliveira (1985) reported that members of the South American alydid genus *Hyalymenus* obtained some protection from being eaten by the mantids that occurred on their host plants, but only in the ant-mimicking nymph stages, not as adults.

OTHER ENEMIES: INSECT PREDATORS, PARASITES, PARASITOIDS

Despite cryptic or warning coloration, chemical defences and protective maternal care, shieldbugs still succumb to many insidious enemies, and key among these are the parasitoids that attack them. Unlike a parasite (head louse, flea, or tapeworm for example), which feeds off its host but does not kill it, a parasitoid eats its host alive, usually from the inside, and although it may allow the host to continue feeding and growing for some time, it eventually proves fatal. British shieldbugs are particularly prone to attack from parasitic flies in the family Tachinidae, subfamily Phasiinae. The Tachinidae are a large group of mostly very bristly or hairy, dark grey, black or mottled flies, and the majority lay their eggs on the caterpillars of moths or butterflies, although close relatives also attack bees, wasps, sawflies, beetles or grasshoppers. The Phasiinae are far less bristly than other tachinids and many are brightly coloured. The biggest and best known is the well-named *Phasia hemiptera*, which has large broad wings, strongly coloured in the male, and resembles a shieldbug when it is sitting on a flower head (Fig. 75). It's a lovely insect, and I always get a thrill when I see one, ever since I first saw them in Plashet Wood near Uckfield, in Sussex, in the early 1970s. Unfortunately it's not at all common in the London area. It has been recorded laying its eggs on *Pentatoma rufipes* and *Palomena prasina*, but probably also attacks other less common species. Many tachinids are considered generalist parasitoids, and those attacking moth caterpillars will often choose many different host species. The similar, but much smaller, *Phasia obesa* and *P. pusilla* are also thought to be shieldbug parasitoids, and although they are very common and widespread in the British Isles, host rearing records are few and far between. In Britain *P. obesa* has been reared from *Neottiglossa*, but European records are from *Zicrona caerulea*, the non-British *Sehirus melanopterus* and several non-pentatomids like the grass bug (family Miridae) *Leptopterna dolobrata*, ground bug (Lygaeidae) *Beosus maritimus* and rhopalid *Myrmus miriformis. Phasia pusilla* has been reared from the lygaeid ground bugs *Stygnocoris fuligineus* and *S. pedestris*, suggesting that their host choices are actually much wider than just shieldbugs.

FIG 75. A shieldbug parasitoid, the fly *Phasia hemiptera* (family Tachinidae). The broad darkened wings, especially in the male shown here, make it look very shieldbug-like.

Although rearing records are few and parasitism rates are poorly studied, males of some shieldbug species are reportedly more heavily targeted by tachinids than females, suggesting that the flies are homing in on male scent pheromones. Other British species in this group of flies include *Cylindromyia interrupta*, *Gymnosoma rotundatum*, *G. nitens*, *Cistogaster globosa* and *Subclytia rotundiventris*, all thought to parasitise various shieldbugs including *Aelia acuminata* and *Sciocoris cursitans*. These are all relatively scarce flies in Britain and breeding records are scant. Host ranges are wider elsewhere in the world, suggesting that there is probably some flexibility in the shieldbug species they target. Changes may be happening here too. Apart from *Phasia hemiptera*, which was always fairly widespread, at least *Gymnosoma rotundatum*, *G. nitens* and *Cistogaster globosa* appear to have increased very dramatically in the last 30–50 years (Figs. 76, 77). Historically *Gymnosoma nitens* was only known from Box Hill and *Cistogaster globosa* from a handful of chalk downland nature reserves in southern England. They were both given Red Data Book category 1, 'endangered', status by Shirt (1987) and Falk (1992), and were on the verge of being declared extinct here. But they are now considered widespread and relatively common on brownfield sites in the Thames estuary areas of Essex, Kent and Greater London, and elsewhere in the country (*Cistogaster* as far north as Birmingham, Leeds and

York). Although I never seem to find *Phasia hemiptera*, these three are regulars on some of the scrappy bits of derelict land I like to haunt. Whether they have increased because of genuine climatic changes, or because more entomologists are visiting these unprepossessing sites is not clear. *Aelia* in particular and *Sciocoris* to some extent have also been recorded much more frequently in the last 30 years, so the flies' increases in abundance may well be tied in to genuine increases in their shieldbug hosts. Perhaps supporting this is the appearance of 'new' species in this subfamily moving into Britain over the last 50 years: *Hemyda vittata* is now widespread in England since its discovery here in 1956, *Phasia barbifrons* has been similarly widespread since 1999, and *Opesia grandis* turned up (two males) new to Britain, near Peterborough in 2006 (Perry 2006). More are surely set to arrive soon. It may be coincidental, but a hexenol secretion from the dorsal abdominal glands of the African scutellerid *Hotea gambiae* is known to be toxic to dipteran eggs and may give some protection against tachinid eggs being laid. Some shieldbugs also have a thin waxy bloom on the underside of the body, and this may make it harder for the flies to glue their eggs in place.

Disappointingly, none of the large and colourful ichneumon wasps parasitise shieldbugs. Indeed, they, and their close relatives, only attack holometabolous groups of insects; mostly Lepidoptera caterpillars, but also Coleoptera, Hymenoptera and Diptera. Unlike the tachinid flies mentioned above, ichneumons have not made the behavioural jump allowing them to target a new host range, and they appear to stick to the groups that go through an obligatory chrysalis/pupa stage. Intuitively there doesn't seem to be a physiological or metabolic reason that stops them attacking, say, shieldbugs. More likely their ancestors probably started to attack some primordial holometabolous insect way back in evolutionary time and modern species are simply the result of generations and generations of these parasitoids tracking the same host range as they diversify to create new orders, families, genera and species. Host-finding and egg-laying behaviour may now be genetically hard-wired, so a switch to Hemiptera remains unlikely.

The same cannot be said for egg-parasitoids. As discussed earlier, it is common to find a batch of distinctive shieldbug eggs where some (or all) have failed to emerge, or where some of the cluster are dull and darkened, indicating that these eggs are no longer shieldbug-viable, but have been attacked by a minute parasitoid wasp. It was against these small wasps that the Parent Bug (*Elasmucha grisea*) was trying so desperately to protect her eggs. The parasitoids usually belong to the family Scelionidae. They are tiny, only a millimetre or two long, and so small that a single shieldbug egg can provide enough nourishment for the wasp's grub to see it through to pupation and adulthood.

FIG 76. *Gymnosoma rotundatum*, a parasitoid of various shieldbug species, was considered rare in Britain until about the 1990s; it is now widespread in south-east England.

FIG 77. The shieldbug parasitoid fly *Gymnosoma nitens* has increased dramatically in the last 30–50 years, but British host rearing records are still hovering around the zero mark. European records state *Sciocoris cursitans* and the non-British *S. helferi*, but the similarly soil-dwelling *Podops inuncta* is often abundant at the same sites as the fly.

FIG 78. Parent Bug (*Elasmucha grisea*) with two tachinid eggs, probably *Subclytia rotundiventris*, glued to its pronotum.

The Scelionidae parasitise a wide range of insect (and other arthropod) eggs. They tend to attack eggs laid in clusters (it's obviously more cost-effective in terms of searching time), and make no distinction between holometabolous moths and beetles, hemimetabolous shieldbugs, grasshoppers and mantids, or even non-insect groups like spiders. Given that their effects on shieldbug egg clusters are so frequently seen, they must exert a heavy toll on potential shieldbug populations.

A medium-sized black and red solitary wasp, *Astata boops*, specialises in attacking shieldbug nymphs (Fig. 79). Smaller and narrower than a yellowjacket wasp, it stings its prey to paralyse it, then flies or drags it back to a small burrow, about 10 cm long, usually excavated in sandy soil. Several pentatomid nymphs are stashed at the end of the tunnel ('pulled down, usually by the antennae', Lomholdt 1984), an egg is laid on them, then the brood cell is closed with a partition of soil, and another store of bugs is started. In this way up to 12 cells stocked with paralysed shieldbug nymphs are provided before the wasp goes off to try and start another burrow. Recorded prey includes *Picromerus bidens*, *Dolycoris baccarum*, *Eurydema* species, *Aelia acuminata* and (in Scandinavia) *Chlorochroa juniperina*. *Astata boops* occurs locally across much of southern England but is not common. The closely related, but slightly smaller, *Astata (Dryudella) pinguis* mostly attacks ground bugs (Lygaeidae), but also takes some pentatomid nymphs.

FIG 79. *Astata boops* with nymph of *Palomena prasina*.

Nielsen & Skipper (2015) show a lovely photo of the large furry Bee Robberfly (*Laphria flava*) feeding on *Dolycoris baccarum*. *Laphria* is well known to be a fearsome predator. Unfortunately in the British Isles it is a rare species of the Scottish Highlands. Butler (1923) makes a point of recording the rare robberfly *Laphria* (now *Coerades*) *gilvus* also feeding on *Dolycoris baccarum*. The Hornet Robberfly (*Asilus crabroniformis*) is certainly big enough to take an adult shieldbug out of the air, and it is well known to attack large flying beetles, but it seems to be drastically declining in Britain. In their worldwide analysis of robberfly prey records, Dennis *et al.* (2010) list several shieldbug encounters (Pentatomidae, Coreidae, Rhopalidae), so it seems quite likely that these common and voracious insects do regularly take shieldbugs here, it's just that few people are watching for or recording this behaviour. Other potential insect predators of shieldbugs must include dragonflies, ground beetles, mantids and ants, though these interactions are seldom seen or noted.

Occasionally shieldbugs fall victim to spiders and can be retrieved from their webs. From my own observations, the Sloe Bug (*Dolycoris baccarum*) would again seem to be a regular, and I quite often find them dead in the sweep net, still wrapped in web (Fig. 80). Care needs to be taken, though. I collected what I thought was a large silk-wrapped *Dolycoris* from a railway embankment at Earl's Court on 21 July 2010; I pinned it, labelled it, and stuck it away in my reference collection. But it was only two years later, when entomologist Penny Metal

FIG 80. *Dolycoris baccarum* found wrapped in a spider's web in a garden in Moreton-in-Marsh, Gloucestershire, 21 September 2022, and then freed a few minutes later, shortly before it flew off.

FIG 81. Headless and very dead specimen of *Nezara viridula* found trussed up in a spider's web in my front garden.

pointed out a live specimen of *Rhaphigaster nebulosa* that she'd found, new to Britain, breeding in Warwick Gardens in south London, that I realised I had made an identification error and my dead *Dolycoris* was in fact this new shieldbug species. I've also found a dead *Nezara viridula* in a garden spider's web just outside my front door (Fig. 81). *Aelia acuminata* and *Eurygaster testudinaria* are sometimes caught by the large wasp spider *Argiope bruennichi*. It makes its orb webs low in the long grass and mostly specialises in catching grasshoppers and bush-crickets, but these two grassland shieldbugs are obviously acceptable too.

CHAPTER 5

Evolution of Shieldbugs and a History of British Species

At some point, more than half a billion years ago, a primordial proto-insect hauled itself out of the ocean onto the damp edges of the land and began an invasion that has resulted in the most diverse group of animals on the planet. Quite what it looked like, even how many legs it had, is open to endless speculation, but it is reasonable to suspect that if you could see one through some kind of time-travel vision it would probably be recognisable as being very like an arthropod, although it a might have echoes of tardigrade (water bear) or onychophoran (velvet worm). Such time-travel vision does exist. It is called fossilisation, but peering that far back into the seemingly endless dark void has so far mostly resulted in hazy images which are poorly understood.

One possibility that reared its bulbous-eyed head a few years ago was *Devonohexapodus bocksbergensis*, reconstructions of which resembled a bug-eyed millipede with six legs much longer than the others, and with long thin antennae. Intuitively it looked just right, but was recently reassessed as another known marine creature, *Wingertshellicus backesi*, and relegated to the status of an ancient relative of hexapods rather than a real ancestor (Kühl & Rust 2009). The search for insect fossils continues with every tap of the geological hammer. There is still so much to discover.

The earliest reputed insect fossil, *Rhyniognatha hirsti*, from the early Devonian cherts at Rhynie, in Scotland, is only a fragment of jaws, and anyway it has recently been suggested that these are centipede or millipede jaws instead (Haug & Haug 2017). It really is very confusing. At present, a springtail, *Rhyniella praecursor*, better preserved, with more than just mouthparts, stands as the oldest known hexapod fossil and the nearest thing to an insect from these ancient beds.

Whatever its precursors looked like, though, an archetypal insect-like creature with six legs, two antennae, three body parts (head, thorax and abdomen) and chewing mouthparts was already crawling about on the land over 400 million years ago. Wings probably appeared around the same time, and indeed some of the largest flying invertebrates are known only from the fossil record – giant dragonflies with wingspans of 600 mm, ants as big as modern dragonflies. Long before pterosaurs or birds or bats, insects were the first organisms to colonise the air, and once the miracle of flight had been achieved it was probably one of the main drivers of insect evolution, leading to the bewildering diversity that still exists today. Almost certainly something clearly recognisable as a relative of het bugs would have been crawling, and flying, about at the time too.

The evolutionary descent of insects has traditionally been deduced from modern extant forms, with the few fossils used to fill in some gaps. Reconstruction of the evolutionary path of insects was initially based on comparisons of the known structures of modern insects, and making assumptions about what constituted original shared ancestral characters, and what constituted more modern adaptations. Although this has now been rather superseded by DNA analysis, the generally accepted evolutionary tree has remained pretty well established – with the occasional tweak. The monumental work on the evolution of insects by Grimaldi & Engel (2005) has been my go-to source for this next bit, although revised dates are taken from Johnson *et al.* (2018).

It seems wings evolved about 450 million years ago (MYA), leaving Earth-bound springtails and silverfish to skitter about on the ground. The two most significant insect milestones were now set to define the evolutionary emergence of the Hemiptera as a major group. Wing-folding arose about 435 MYA, and this left stiff-winged dragonflies representing a postulated 'primitive' flight system while other insects (including any shieldbug ancestors) moved on to more flexible membranes of different shapes that could be pleated or angled at rest. Then the larva/maggot/caterpillar system of the Holometabola (beetles, bees, butterflies and the like) arose sometime about 400 MYA. At some point between these two great evolutionary cusps (folding wings, and the emergence of the Holometabola) the Hemiptera and their close relatives likely diverged and went their own way. The rest, as they say, is prehistory.

BUGS ARRIVE ON THE SCENE

Despite the uncertainty about how they first evolved, the Hemiptera form a clearly defined group, along with the Psocoptera (bark lice), Phthiraptera (lice) and Thysanoptera (thrips). These all share various obscure mouthpart structures. In particular the hinges by which the mandibles articulate became simplified, eventually giving rise to the backwards-and-forwards movement of long thin piercing stylets, rather than the side-to-side secateur-like action of the stout triangular chomping jaws found in most other insects. This broader group of lice/thrips/bugs, often referred to as the Paraneoptera, share the traits of reduced tarsal segments, no cerci, and asymmetrical mandibles. The genetic relatedness of the Paraneoptera is confirmed by DNA studies (Li *et al.* 2014). The Hemiptera diverged from this assortment when their ancestors lost the maxillary and mandibular palp appendages known throughout most other insect groups. These palps are short feelers, usually with three or four small segments, attached near the base of the jaws, and they are used to manipulate food during feeding. They are prominent and easily observable in a wide range of insects, from hemimetabolous groups such as earwigs, grasshoppers and cockroaches to holometabolous beetles, bees and wasps. In the Hemiptera they are completely absent.

Anyone studying the Hemiptera today will be amazed at the differences between shieldbugs, boatmen, water scorpions, skaters, bedbugs, flatbugs, aphids, scale insects, cicadas and leafhoppers. These disparate groups seem to differ hugely from each other, and it might seem that each should have its own separate order in the classification scheme – in the past, some of them have been so classified. This emphasises the profound success and diversity of the group despite their not being part of the Holometabola radiation. This diversity of the Hemiptera has often seen them represented by a gaggle of uncertain and sometimes fluid suborders on the edge of the hemimetabolous field of insect orders, beyond mantids, grasshoppers and earwigs, and nearly related to the holometabolous beetles and snakeflies. What unites them, though, is the singular structure of the tubular mouthparts, which have become developed into a sucking beak, with the two pairs of mandibular and maxillary stylets sheathed in a long, grooved, cylindrical labium. The very earliest Paraneoptera are thought to have fed on organic detritus – scavenging rotting plant matter and fungal decay, and maybe also pollen or fungal spores, as do modern bark lice and thrips. What is thought to be an intermediate paraneopteran, *Mydiognathus eviohlhoffae*, was described from Cretaceous amber (about 99 MYA); it rather resembles a bark-louse in its wing venation, but its mouthparts are long and pointed like those of Hemiptera. However, it still retains long functional chewing jaws (Yoshizawa & Lienhard 2016), and its discovery was considered to have offered insights into the evolution from biting to sucking in hemipteroid insects.

The archetypal original sucking-mouthparts hemipteran is thought to have been a small thrips- or aphid-like bug with multi-segmented antennae and a strange curled ovipositor suspended underneath the abdomen and projecting backwards, and which may have been used to inject eggs into plant tissue. Crucially, though, it also had a clearly developed beak-like rostrum. Fossils of this extinct group, the Archescytinidae, are known from the early Permian, at least 300 MYA. The other suborders of the Hemiptera (formerly lumped together as Homoptera) diverged about this time, and the earliest true het bug is reckoned to be *Paraknightia magnifica*, from late Permian Australia, dated to about 260 MYA. There are still doubts, though – the fossil is just a bit of a bug. Unfortunately the head, with the all-too-characteristic defining rostrum, is not preserved in this fossil, but the wings appear very hemelytra-like with a distinct clavus and reduced venation much like many modern bugs. The assumption is that this was already a plant-feeder, probably sucking sap from the leaves or stems of the then-dominant gymnosperm plants (conifers and cycads), since angiosperms (flowering plants) had yet to evolve the full diversity we see on Earth today. A useful and readable review of the early evolution of the Hemiptera is given by Szwedo (2017).

IMPRISONED IN ROCK

The fossil record for the Hemiptera is relatively good (about 3,000 named species according to Penney & Jepson 2014), but the usual caveats apply. The earliest are impression fossils, made in silt-formed stone, and they are usually damaged, incomplete and difficult to interpret. If a fossil looks like a het bug it probably is one, but fossils are often just bits of animals. Today the Hemiptera are characterised by things like the presence of the metathoracic glands, dorsal abdominal glands in nymphs, reduction of the tentorium (reinforcing struts inside the head), and the presence of those frena and druckknopf wing-fastening clips. For many fossils it is impossible to see these cryptic structures. Analysis of exactly how and where the Hemiptera originated is clouded in the obscurity of these poorly-preserved remains. Later fossils, often in amber, are beautifully preserved and easy to observe in three dimensions, but by that time (100–50 MYA) they are usually identifiable to the modern lineages that still walk or crawl the Earth.

At present the oldest known confirmed and obviously hemipteran fossil is *Arlecoris louisi* in the extinct family Triassocoridae (Shcherbakov 2010). This and many of the other oldest het bugs are mainly water-dwelling species like water scorpions (Nepidae), saucer bugs (Naucoridae), boatmen (Notonectidae, Corixidae) and giant water bugs (Belostomatidae, Fig. 82). This is not to suggest that water bugs were the dominant bug form, or that land bugs evolved from aquatic forebears, just that these water bugs were better at being fossilised. It is hardly surprising that water bug fossils are common, since these insects were already in the water and therefore perhaps the most likely to get caught in the sudden silt influxes that form the sediments in which these fossils are preserved. Indeed, these groups of water bugs dominate the early Heteroptera fossil record, from about 300 to 200 MYA. There were, however, obviously land bugs about at that time too, and a fossil lace bug, *Archetingis ladinica* (family Tingidae), is reckoned to be one of the oldest, from Middle Triassic Italian rocks (Montaga *et al.* 2018).

Fossil shieldbugs are here in the mix, including some with bizarre structures like the broad spiny edges of several *Eospinosus* species, from German Eocene rocks, formed about 47.5 MYA (Wedman *et al.* 2021). Whether these projections were for protection against predators or for camouflage is impossible to tell, but they are rather reminiscent of spiked South American *Ceratozygum horridum* or the flattened Brazilian *Phloea corticata* mentioned in Chapter 4 for its maternal behaviour.

FIG 82. Fossil giant water bug, family Belostomatidae, from Brazil, about 113 MYA. Water bugs are frequent in the fossil record, perhaps mainly because they were already in the water where silt deposits occurred to create these fossil beds.

What may be a heteropteran egg (family uncertain though) is described from a German compression fossil from the Lower Cretaceous, roughly 145–113 MYA (Fisher & Watson 2015). It was discovered accidentally as a conifer leaf fossil was macerated to better expose it to view, and the egg floated away. It is in a remarkable state of preservation, and scanning electron microscopy reveals the sponge-like texture inside the cuticle and the many minute bristly hairs on the outer surface. Although it is impossible to tell which bug laid it, the authors concluded that it is very similar to modern Pentatomomorpha eggs, and they named it *Merangia horricomis* – the genus name is an anagram of Germania (the Roman name for Germany), and the specific epithet is a reference to its bristly nature.

AMBER – A MIRACLE OF PRESERVATION

Insects fossilised in amber bring better resolution and deeper understanding. Amber is a marvel: the thick resinous sap of now-extinct trees exuded from physical bark damage. The trees have long since disappeared and it is impossible to be sure what species did the exuding. Possibles include ancestors of modern New Zealand kauri trees (*Agathis* species), South American/African *Hymenaea* trees or the Japanese Umbrella-pine (*Sciadopitys verticillata*). Today modern *Hymenaea* and the South American *Protium copal* produce similar thick orange resins called copal, and nodules as large as beachballs are known. Seeping sap is attractive to a wide range of insects, because of the fermenting scents of alcohols, esters and sugars in the mix. Amber was a sticky danger, though, and it engulfed all manner of small creatures crawling on the tree trunks. As the smaller aromatic solvent molecules like those alcohols, ether and light hydrocarbons evaporated, larger cyclic terpenes started to polymerise into an almost crystalline mass like tough transparent plastic. These resinous plugs blocked any holes or scabbed over damage in the bark, but they also imprisoned any unfortunate insect victims that got caught in the sticky mess. The same resin chemicals that protected the trees from fungal or bacterial invasion also prevented these microorganisms breaking down the insect bodies, which remained entombed in a state of near-perfect preservation. The curing of the resin may have taken many years, as the nodules fell onto the soil and became buried for millennia, resulting in the almost stone-like hardness of the amber so valued by jewellers and palaeontologists today. Insect inclusions in the amber add a premium to an already expensive commodity, and there is now a lively international market for these items.

Ants are common in amber – they were obviously as active on trees 100 MYA as they are today, and may well have been selectively attracted to the sweet smells of the sap. And wood-boring beetles abound too, presumably many of them responsible for the bark damage that initiated the resin flows in the first place. But shieldbugs are not very common in amber. Today, shieldbugs are still mostly insects of the leaves rather than tree trunks. Amongst shieldbug-like species the flatbugs (family Aradidae) are fairly well represented in amber (Fig. 83). In the extensive list of het bug fossils given by Schuh & Weirauch (2020), Aradidae far outnumber other pentatomomorphs. This may be because these bugs are normally found under the fungoid bark of trees, and since amber began as a resinous defence against wood-boring insects, physical damage and fungal infection, it is no wonder that they got caught up in the runny sap flows.

As suggested above, most amber-enclosed shieldbugs are extremely similar to modern extant species. Poinar & Thomas (2011) describe a pentatomid from

FIG 83. *Aradus macrosomus* (family Aradidae) in Baltic amber, probably from 55–33 MYA. Flatbugs are more frequent than most other het bugs in amber, probably because they are bark-dwelling creatures more likely to get swept up in the amber bark exudate.

Mexican amber from about 26 MYA and place it in the still-surviving genus *Edessa* because it is so similar to modern species. A pentatomid nymph was found amongst copious plant material from Dominican amber (45–15 MYA) discussed by Poinar & Chambers (2016), but given that identifying living nymphs can be tricky, they declined to take the identification beyond family level. Cydnidae are sometimes found in amber (Wang *et al.* 2019) and these look very like modern species. It seems unlikely that these normally ground-dwelling bugs would get caught up in the sap flows, but it is possible they were hit by a falling gobbet that dropped from the trunk (pebbles and sand are often found in amber) or that they were caught in similar exudations from the roots.

Amber fossils of het bugs may be scarce, but those that do appear are sometimes fascinating. Some of the most peculiar are several coreids with hugely expanded final antennal segments. One of them, *Magnusantenna wuae*, from Upper Cretaceous Myanmar amber (about 100–65 MYA), has the final two segments of the antennae flattened into huge broad triangular plates, making the rather narrow flimsy nymph look very ungainly (Du *et al.* 2020; Fig. 84). The describers suggest that the bizarre antennae might be the result of sexual

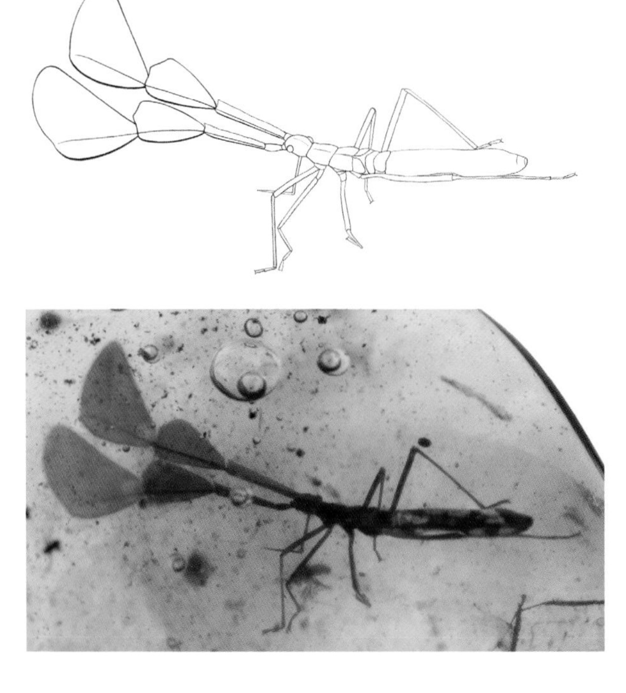

FIG 84. Amber fossil and reconstruction of *Magnusantenna wuae*, showing inexplicably large antennae.

selection or camouflage. They go on to point out that this sort of extreme morphology comes with its own evolutionary costs – there is a drain on limited physiological resources as they are channelled away from normal functional aspects of the rest of the body (a kind of evolutionary extreme-fashion design-over-content comparison), and this can be regarded as a double-edged sword (their words). This prompted Cumming & Le Tirant (2021) to run with the sword analogy, and they named a similar top-heavy coreid amber fossil (about 100 MYA) *Ferriantenna excalibur* after King Arthur's mythical weapon. Whatever the origins of these strange body forms, these species are all now extinct. Nothing quite so pronounced is known in living species, but several exotic coreids have slightly swollen antennal segments, and the early nymphal stages of *Coreus marginatus* are distinct by virtue of their over-large spiny antennae. Instead, some modern coreids have overly inflated hind legs, and look as if they are wearing flared trousers (Fig. 85). These ornaments may have evolved in a similar way to flattened antennae, and for the same reasons, but may be more manageable in terms of movement because of their lower centre of gravity.

Piecing together the more recent evolution of shieldbugs amongst the Heteroptera is usually done by trying to reconstruct phylogenetic trees of descent, aiming to demonstrate how closely related sister groups are, and offering branching diagrams to show how long ago the various modern species shared common ancestors. It's a useful and interesting study, but often fails to tell us what any of those shared ancestors might have looked like, or how they behaved. The fossil record of the Heteroptera may seem good, but 3,000 fossil species is as nothing compared to the 40,000 described living species, and reconstructed evolutionary trees still depend on a lot of assumption and guesswork.

FIG 85. Mating pair of the North American coreid *Leptoglossus phyllopus*, with both sexes demonstrating the leaf-like hind tibiae.

These phylogenetic trees of evolutionary descent are also regularly reconsidered and rearranged as new interpretations are offered. Weirauch and Schuh (2011) comment on how the seven infraorders of the Heteroptera have been variously classified and reclassified over the years. The good news is that the Pentatomomorpha are still considered a consistent and coherent unit, and so is their closest sister group the Cimicomorpha, which contains most of the other terrestrial true bug species. These two infraorders probably shared a common ancestor about 250 MYA; together they diverged from the Nepomorpha (mostly aquatic bugs like water scorpions and boatmen) 240 MYA, and they all together diverged from the Gerromorpha (semi-aquatic skaters and other water-surface groups) about 260 MYA. This descent is borne out from genetic studies of the chromosomes and DNA of extant species (Johnson *et al.* 2018, de Souza-Firmino *et al.* 2020), as is the close relatedness of the Thysanoptera (thrips), Phthiraptera (lice) and Psochodea (plant lice). This DNA-determined evolutionary background also supports the idea that the first het-bug-like creatures were scavengers, or fed on pollen and fungal spores, but later forms adapted to predatory lifestyles, before many of the more terrestrial groups (shieldbugs in particular) evolved to feed on living plants.

Within the shieldish bugs (in the broad Pentatomomorpha sense of this book), more DNA relatedness analysis carried out by Liu *et al.* (2019) confirms that the flatbugs (family Aradidae) are on the outskirts of the group, the ground bugs (family Lygaeidae, also excluded from this book) are closely related to the Coreidae and Pyrrhocoridae, and the remaining truly shield-shaped families all form a tight cluster within the superfamily Pentatomoidea. There are no surprises, but calculating how quickly (or slowly) the DNA mutates gives approximate dates when

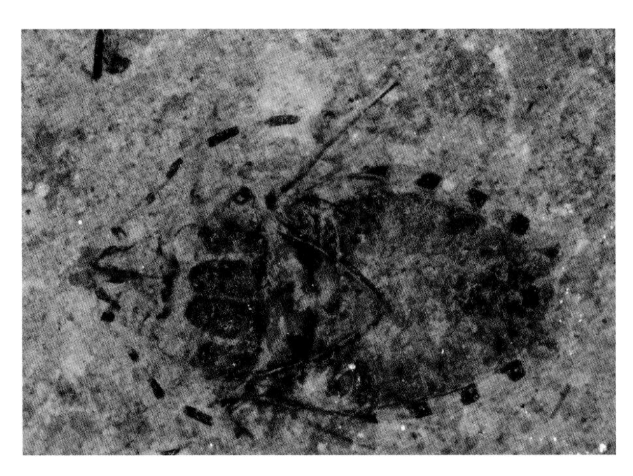

FIG 86. Fossil pentatomid (five antennal segments clearly visible) from early Eocene formation (approximately 50 MYA).

the different groups diverged, in other words when they last shared a common ancestor. Thus, the Pentatomomorpha arose sometime in the Middle Triassic, about 240 MYA. Aradoidea went their separate way about 237 MYA, Coreoidea/Lygaeoidea and Pyrrhocoroidea split from Pentatomoidea about 217 MYA. The subsequent diversification of these shieldbug groups then appears to be linked to the diversification of angiosperm flowering plants in the late Jurassic period. Just as plant-feeding beetles (leaf beetles Chrysomelidae, longhorns Cerambycidae, and weevils Curculionoidea) have become incredibly diverse as they evolved to follow the diversification of the new-fangled flowering plant species, so too did the plant bugs. Today plant-feeders dominate terrestrial Hemiptera faunas, and although they are a small group amongst the larger groups of herbivores like the plant bugs, ground bugs, leafhoppers, aphids and scales, shieldbugs have diversified across the globe too.

Evolution is an ongoing process. Famously, the appearance of melanistic forms of the Peppered Moth (*Biston betularia*) during the Industrial Revolution is taken as evidence of modern evolution in action as the increasingly common black forms remained camouflaged and undetected by predators against the soot-blackened tree trunks on which they rested during the day, whilst the typical pale mottled morphs were picked off and eaten by birds. Similar melanism, this time in Two-spot Ladybirds (*Adalia bipunctata*), has been suggested as an adaptation to cooler northern areas of its range, where their dark-bodied varieties better absorb sunlight to warm them up for daily activity. Melanism is hardly known in sheldbugs, but there is a very rare black form of the Gorse Shieldbug (*Piezodorus lituratus*) that turns up occasionally (Fig. 87). Such low-level variation at least shows that genetic variability is still there, bubbling away at the bottom of the gene pool, just in case some future climatic scenario favours darker individuals.

FIG 87. Melanistic form of the Gorse Shieldbug (*Piezodorus lituratus*). Cork, Ireland, 9 April 2015, Leon van der Noll.

THE ORIGINS OF BRITISH AND IRISH SHIELDBUGS

Shieldbugs, like other insects, hold no allegiance to national borders. The English Channel and North Sea are, however, quite significant barriers, but only in the recent scheme of things. When the last ice sheet started to retreat, about 15,000–12,000 years ago, Britain was just an outcrop peninsula of northern Europe anyway. At that time, Scotland and much of Ireland, Wales and northern England would have been covered with stifling ice, sometimes hundreds of metres thick, but even southern Ireland and England would have been cold and largely tundra – hardly conducive to insect diversity, apart from mosquitoes perhaps. But as the ice gradually retreated over the many centuries colonising plants and animals slowly spread northwards from the European heartland, and they have continued to do so ever since.

In Britain, shieldbug diversity is quite clearly skewed towards the south-east of England. Unlike in many other insect groups there are no exclusively Welsh, or Scottish, or Irish shieldbug species, though there are plenty that are exclusively English, and a fair number that are exclusively south-eastern. Indeed, Nau (1996) bemoans this pronounced south-of-the-Thames bias when commenting on the shieldbugs of Bedfordshire – hardly a remote northerly county of England. This southern and eastern focus is partly climatic, in that south-east England is generally warmer and drier, and the further west and north you go, the damper and/or cooler becomes the weather. Most insects thrive in warm conditions, and damp is often anathema to them because it encourages mould and fungal diseases, particularly during the dormant stages of overwintering shelter and hibernation. Shieldbugs are not quite so heat-loving as the obviously thermophilic bees, wasps and ants but they still do better if the weather is with them, rather than against them. This south-eastern concentration of species is perhaps an echo of the inexorable march of colonisation from further south in Europe. As generation after generation of insect colonists moved north, following the gradually retreating ice sheet, each species reached its own climatic limit. It might occasionally venture beyond this during warm years (or centuries), but here the terrain was really too harsh, and an inevitable cold period lasting tens or hundreds of years would push it back south again. This limit to an insect's range is governed by complex genetic, behavioural and physiological traits within the organism, and this is reflected in the various national distributions we see today.

The Dock Bug (*Coreus marginatus*) is a good example of a shieldbug well established in south-east England, but which fizzles out around about Manchester (Fig. 88). In the West Country and in Wales it becomes more coastal, and this

FIG 88. Though abundant in southern England, *Coreus marginatus* reaches only really to the Manchester/Hull line, north of which it is remarkably scarce.

is typical of a species struggling on the edge of its range – coastal sites tend to be hotter, drier, sparsely vegetated and with well-drained soils. Its foodplants – docks – are far more widely spread than this, so it would seem to imply that there is a genuine temperature/weather/climate effect here which restrains it. By contrast, *Pentatoma rufipes* is aptly named as the Forest Bug and is adapted to cooler, moister, more wooded conditions – so its British range extends throughout Wales and well up into the Scottish Highlands. It is easy to imagine that these two common species slowly colonised peninsular Britain from about 10,000 years ago until each met the limit of its natural comfort zone. Many other common and widespread species like *Dolycoris baccarum, Tritomegas bicolor, Eurydema oleracea, Palomena prasina* and *Eysarcoris venustissimus* have similar well-defined ranges – commoner in the south and east, but fading out at some point further north and west.

At the other extreme, many species have barely a toehold here. *Geotomus punctulatus* is only really known from one site now, Sennen Cove in Cornwall, and both *Odontoscelis* species are limited to a few mainly coastal sites around southern England. This distribution is typical of insect species with a more Mediterranean distribution elsewhere in Europe, which really start struggling on the northern and western edges of their European ranges. *Alydus, Bathysolen, Arenocoris* and *Spathocera* are intermediate – elsewhere in Europe they are more widespread and occupy diverse habitats, but here they are firmly limited to the band of sandy soils that stretch from the East Anglian Breckland, through London brownfields, then into the lowland heaths of Surrey, Hampshire, West Sussex and Dorset. Again, these sites are sparsely vegetated, well drained, hot and dry.

Even if the bug itself is not climate- and habitat-limited here, then maybe its foodplant is. Though *Dicranocephalus medius* is scarce on Wood Spurge (*Euphorbia amygdaloides*), at least that plant is fairly widely spread over England. *D. agilis* is very scarce and limited to coastal locations because its Sea Spurge (*E. paralias*) and Portland Spurge (*E. portlandica*) foodplants only occur near the sea.

Many of the shieldbugs firmly considered native to the British Isles probably first colonised more than 6,000 years ago. At about that point Britain finally became separated from Europe as the gently undulating river valley of the Fleuve Manche was flooded by rising sea levels and the English Channel formed, joining the North Sea to the Atlantic Ocean and creating the significant biological barrier it is today. Those species that got into Ireland have probably been there 10,000 years, because the Irish Sea flooded long before the English Channel was inundated.

Dangerous though they may be to small insects, the seas around Britain are not completely isolating, and there is evidence that there is still the occasional colonisation event. *Eurygaster austriaca* was reputed to have appeared on the Kent coast during the nineteenth century, but after a few sightings the insect seems to have become extinct here (Fig. 89). *Peribalus strictus* has had a fragile status in Britain for many years. Colonies appear to have been established in

FIG 89. *Eurygaster austriaca* may have nearly gained a foothold in west Kent in the nineteenth century, but it is now thought to be extinct on mainland Britain. Like its congeners *E. testudinaria* and *E. maura*, which survive well enough here, *E. austriaca* is widespread in continental Europe and may just need another chance to become established in Britain.

Kent and Sussex before the 1950s, but then these failed or subsided to the point of being undetected until a recent upsurge in records during the twenty-first century. Meanwhile, in December 2015 several specimens of the North African *Mecidea lindbergi* were found in moth traps and gardens along the Devon, Dorset and Hampshire coasts, having apparently arrived with a huge influx of exotic migratory moths wafted up on unseasonal warm southerly winds. No surviving colony seems to have been established, but this perfectly exemplifies just how possible it is for new arrivals to make landfall.

Several relative newcomers, *Stictopleurus punctatonervosus, S. abutilon, Nezara viridula, Tritomegas sexmaculatus* (Fig. 90), *Brachycarenus tigrinus* and *Liorhyssus hyalinus,* have all recently (in about the last 20 years) become firmly established in Britain. Many of these colonisation expansions are still ongoing, and the spreads can be monitored from an estimated landfall event, through local establishment, consolidation, then gradual encroachment across the country. These species all seem to have arrived in the Thames estuary and into London, which acts like a funnel for any migrants from Scandinavia or the Low Countries. These species appear to have got here under their own steam, but when the western Mediterranean ground-dweller *Sciocoris sideritidis* turned up in Purfleet, Essex, it was suggested it had arrived by human agency through the nearby container port. This is speculation, but perhaps a good guess. A colony was detected for a couple of years, but confirmation of permanent settlement will have to wait for more records in the future.

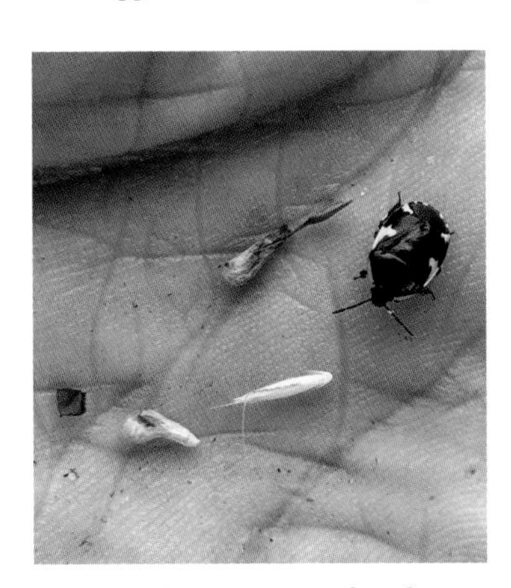

FIG 90. *Tritomegas sexmaculatus* has spread into and through London after being first found in England in east Kent in 2011. It feeds on Black Horehound, *Ballota nigra*, which grows widely in the derelict brownfield sites of the Thames estuary.

There is also the very real possibility that roads and railway lines can act as colonisation corridors. In the forthcoming Kent atlas of shieldbugs (Barnard in preparation) several species show a marked liking for the M20, particularly *Coriomeris denticulatus*, but also *Stictopleurus abutilon, S. punctatonervosus* and *Eurygaster testudinaria.* It is not clear whether this is an artefact of recording effort or a genuine ecological association, but these are all species associated with warm, well-drained and sparsely vegetated rough grassland and brownfield sites, and they are often found on scrappy road verges.

THE WORLD IS IN CONSTANT FLUX

No distribution is fixed forever, though; they all change with time. Many of these changes will be subtly imperceptible to humans as things alter across a geological time-frame, but plenty are plainly visible on a much shorter programme. It is through the modern nationwide recording and reporting of studies of bugs that these genuine trends in their statuses can be recognised. There are several stories to tell, and in the last hundred and fifty years or so shieldbugs have shown a mix of fortunes.

Despite its historically well-known localities on the North Downs and Midlands limestone outcrops, *Chlorochroa juniperina* is probably extinct in the British Isles, with no record since 1925. So too *Carpocoris mediterraneus, Jalla dumosa* and *Eurygaster austriaca* have gone, if indeed they were ever really established here in the first place. Several species remain endangered, vulnerable or nationally rare. The sometimes complex criteria for allocating these statuses are explained in the occasional national review (Kirby 1992, Bantock 2016a), but they boil down to how many genuine records there have been for any given species and whether these colonies are likely to survive or fail. Many uncommon species are listed as nationally scarce ('notable') – usually an assessment that a species has been found in fewer than 100 of the Ordnance Survey's 1 km national grid squares since a given cut-off date. These are the more uncommon species, and although they are perhaps less likely to be found, when they are found they are more likely to be reported in a short note to an entomological journal. Thus it is that every time a scarce species is found it becomes, by definition, less scarce.

As well as published records in specialist entomological journals, the concerted recording, monitoring and collating of sightings has allowed the distributions of these insects to be accurately mapped across the country by a veritable army of citizen scientists and professional naturalists. There is more on how to submit records and sightings in Chapter 9. It is impossible to guess quite how climate change will affect the British and Irish shieldbug faunas, but it will be by regular analysis of the huge databanks of records being amassed that any patterns will emerge. Judd (2011) suggested that *Liorhyssus hyalinus* already appeared to be following the standard model for British colonisation as climate patterns had changed during the period 1990–2009. There was an irregular series of a few (zero to three) records per decade from 1900 to 1989, mostly from coastal localities, suggesting that it was then an irregular migrant from Europe. But from 1990 many more sightings were made all over south-east England, with coastal records now extending around the West Country, and Wales, to Cumbria

FIG 91. *Liorhyssus hyalinus* is one of the most widespread het bugs on Earth. Since about 1990 it has become well established in Britain, but for many years it was an irregular migrant on southern English shores.

and Ireland (Fig. 91). Likewise, Shardlow & Taylor (2004) wondered whether *Nezara viridula* had finally become established in the British Isles because of climate change. During the twentieth century this originally African species had been intercepted many times in fruit and vegetable shipments into the country. Southwood & Leston (1959) had considered its establishment here as 'unlikely', but in 2003 several well-established breeding populations were found in London, and it has remained here ever since. Musolin (2011) suggested that it was gradual warming of the environment that had allowed the northward spread of *Nezara* by increasing overwintering success, shortening nymphal development times and bringing forward mating and egg-laying.

The Firebug (*Pyrrhocoris apterus*), just on the very edge of its European range in the British Isles, is likely to be affected by climate change. Throughout Europe this common bug mostly has one generation a year, but since the 1980s it has often had at least some sort of extra generation, especially in warmer years. This was mentioned in Chapter 3 when discussing generation numbers across large geographic ranges. Honek *et al.* (2020) suggest that climate change could extend the thermal window during which the second generation matures, making this life history the norm in Europe, accompanied by geographic spread northwards with improved rearing success and increasing population numbers. Until recently all British specimens of this bug were thought to be the short-winged, non-flying, brachypterous form. If there are developmental constraints on wing formation because the nymphs struggle to feed enough in our cooler season, this may explain why this species has not spread widely, even though it has

FIG 92. Long-winged (macropterous) female *Pyrrhocoris apterus*, mating with short-winged (brachypterous) male, photographed in Essex in July 2020 by Yvonne Couch.

been established here for 200 years. However, long-winged specimens were recently seen in Essex, perhaps marking a prelude to wider establishment in England (Fig. 92).

Of course, it is not just in the British Isles that these patterns will appear. Fan *et al.* (2020) modelled how climate change might affect the suitability of habitats in various Chinese provinces for the predatory *Arma custos*. This Eurasian species is sometimes used as a biocontrol agent against Fall Armyworm (*Spodoptera frugiperda*) caterpillars and other agricultural pests. An annual mean temperature of 7.5–15.0 °C and annual rainfall of 750–1,200 mm are the key limiting factors in its distribution. Using various calculated climate-change scenarios, the authors conclude that its present occurrence in east-central China will change dramatically in the next 30–50 years as the bug is edged northwards into Mongolia. Their calculations suggest it will also move north through Europe. Many of the non-British species mentioned in Chapter 8 are included because their spread across continental Europe has already been noticed. They may arrive here sooner or later.

Several native shieldbug species have also become far commoner than they were once considered to be. I think it quite likely that these are good examples of warmth-loving species which once had a relatively hard time of it here in damp cool Britain, but which have benefited from the gradually increasing summer temperatures and milder winters of the last several decades. From the 1980s onwards *Aelia acuminata* moved from being a rather local bug of chalk downland to a widespread species of rough grassland almost everywhere in southern England (Fig. 93). Likewise *Coriomeris denticulatus* and *Eurygaster testudinaria* have expanded their habitat preferences from dry chalk and damp meadows respectively and are now much more common than they were 50 years ago (Hawkins 2003). *Bathysolen nubilus* also shows a genuine range expansion in

FIG 93. *Aelia acuminata* used to be considered quite scarce, but now often occurs in its hundreds in the sweep net.

the late twentieth century, spreading from its heartland in the east Kent coastal sand hills, along the Thames estuary and also in the East Anglian Breckland. I was sent a specimen collected in a pitfall trap from a 'living roof' on the top of a tall modern building in central London in 2003.

Several bugs have undergone quite astonishing range expansions, though these are less associated with any changes in climate or weather patterns. When it switched foodplant allegiance from juniper to cypresses, *Cyphostethus tristriatus* spread wildly around the country, and the Box Bug (*Gonocerus acuteangulatus*) expanded its range from a single locality at Surrey's Box Hill to virtually the whole of England when it took on a taste for hawthorn and honeysuckle fruits rather than just those of Box bushes (Fig. 94). *Corizus hyoscyami* has also spread very widely from its coastal strongholds in south-west England, and those relative newcomers mentioned above have all spread too. They are still spreading. A review of the changing statuses of some bugs (including several shieldbugs) is given by Kirby *et al.* (2001), and these changes are illustrated in the maps published by the Shieldbug Recording Scheme (Bantock 2018).

Perhaps the most significant monitored spread has been the colonisation of almost the entire British Isles by a North American coreid, the Western Conifer Seed Bug, *Leptoglossus occidentalis,* since its arrival here in 2007. This is a large and obvious insect with a propensity to come indoors to hibernate, and many of the records have been submitted by non-specialist members of the public who found this strange creature in their houses. They sent in photos, from which it is easily identified, even if out of focus. It is now one of the most widespread shieldbug species in the country, recorded throughout England and Wales, well into Ireland

FIG 94. *Gonocerus* nymph shedding its skin. This colony was feeding on snowberry, a significantly different foodplant from its original host, in south-east Kent in August 2021.

and Scotland (the only coreid to make it north of the border) and as far out as the Outer Hebrides, Orkney and Shetland.

Leptoglossus offers a unique chance to understand how insects behave when they meet the unrestrained opportunity of a new continent, away from the predators, parasites and other pressures that have kept them in check through countless previous aeons. With *L. occidentalis* it began in the middle of the twentieth century when what was previously regarded as an insect of western North America (Mexico, California, Colorado and British Columbia) was found many hundreds of kilometres away in Iowa, and later in many of the eastern United States. It arrived in Europe, in Italy, in 1999, and within 15 years had colonised almost the entire continent. This was not initially a natural spread, but seems to have been instigated by the international conifer trade – the bugs obviously hitchhiked across the globe on Christmas trees and garden ornamentals. It had long been suggested that several separate European landfalls had occurred. The disconnected first records from Italy, Spain, France, Belgium and Britain were widely spaced, and closely linked to the port areas of Venice,

Barcelona, Le Havre, Ostend and Weymouth respectively. Using mitochondrial DNA, Lesieur *et al.* (2018) sequenced 254 individuals from 57 North American and European populations and showed that the specimens invading Europe came from the bridgehead established in eastern North America, rather than from the bug's original range west of the Rockies. The DNA also indicated that at least two separate European colonisation events had occurred – first the Italian and then the Spanish founders had definitely come their separate ways from the eastern USA. Once on this side of the Atlantic, though, the bug had no trouble flying sometimes long distances and establishing new populations under its own steam. The bug's ability to utilise native European conifer species, and the lack of any serious European predators and parasitoids, allowed it to spread rapidly. It has now made its own way across the continent from Portugal to Ukraine, Sweden, Ireland and northern Scotland.

New shieldbug species continue to appear. Whilst writing this book I was alerted to several exotics turning up, including the bright yellow South

FIG 95. Brazilian shieldbug *Grazia tincta* found dead in some supermarket grapes. This tropical species would be unlikely to survive in the British Isles, but is a perfect exemplar of how all manner of species get transported all over the world in trade.

FIG 96. A single specimen of *Pinthaeus sanguinipes* found in the Suffolk moth trap of Antony Wren by his son Aidan in September 2021.

American pentatomid *Grazia tincta* dead in some grapes from Brazil (Fig. 95), and the Mediterranean *Crocistethus waltlianus* which was seen crawling over the cauliflower display at a supermarket in Littlehampton, West Sussex. These are not natural range expansions, but perhaps the European *Pinthaeus sanguinipes* (Fig. 96) is ready to invade? It is a species at first sight rather resembling the common and widespread *Pentatoma rufipes*. I'm not sure I would have given it a second thought if I'd seen it on the edge of the beating tray. It was just a single specimen, found in a moth light trap near Lowestoft, Suffolk, in September 2021. Maybe this will be the first record of a species that will colonise the British Isles in the near future. Who knows? Nobody yet, but by actively recording, observing and studying shieldbugs, such a question might be answered within a few decades.

CHAPTER 6

History of Shieldbug Study

Shieldbugs' large size, obvious habits and bright colours now make them attractive insects to study. They must have seemed similarly eye-catching and interesting to our forebears, but confusion over what they were and how they fitted into the classification schemes of the times has sometimes made historical analysis rather tricky. Many of the well-known ancient Greek and Roman commentaries are now virtually impenetrable on this front. The likes of Aristotle (384–322 BC) and Pliny the Elder (AD 23/24–79) knew well the verminous bedbug *Cimex lectularius*, noisy cicadas, and the small homopteran scale insects that gave red dye (*Kermes vermilio*) and lacquer (*Kerria* (*Laccifer*) *lacca*) (Beavis 1988), but these were not understood to be at all related to each other; the idea that they were all part of the overarching insect order Hemiptera did not arise until modern times.

Being rather secretive and not very important agriculturally, shieldbugs were easy to overlook. They were not as numerous or diverse as the similarly hard-shelled beetles that everywhere ran on the ground, sat on flowers, hollowed out house timbers or invaded granaries, nor as intrusive as the infuriating cockroaches that infested kitchen and bakery. It is likely that all the pre-modern writers simply lumped shieldbugs in with these 'beetles', and interpreting their texts nowadays is rather challenging.

Amongst these better-known insects, they also seemingly confounded thrips, springtails, earwigs and head lice – so much so that words like *koris*,[6] *cimex*, *tiphe*,

6 Incidentally, it has been suggested that *koris* is the accidental source of the word *Aphis*. Though this familiar word for greenfly is a name so widespread and familiar as to need no explanation, it is nowhere near so ancient. It was used by Linnaeus but appears in lexicons only since the sixteenth century (Heller 1943). Even here, though, it is *adespotum*, a lovely (continued overleaf)

coccus, *tinea*, among many others, were used by ancient philosophers almost interchangeably, just as 'louse', 'gnat', 'midge' and later 'bug' would be used broadly for all manner of small mean invertebrates, and 'worm' for all manner of their larvae. The most tantalising mention of anything that might possibly be a shieldbug comes in the works of Pliny, who writes about a type of wild *Cimex* occurring on mallow, used in a medicine against earache; this might just be a shieldbug – *Pyrrhocoris* is a possibility, or maybe *Palomena*.

Unlike ants, locusts, honeybees and hornets, there are no references to heathen shieldbugs in the Bible; as mentioned in Chapter 1, the 'bug' in Psalm 91 of the Matthew Bible referred to bugbear night fears. In fact there are precious few mentions of any other insects at all in the first 1,500 or so years of the current or Christian era. This was a time when science, if it existed, was more to do with delving into the arcane imaginings of the ancients rather than examining the real world around us. Shieldbugs continued to avoid human gaze during these dark ages since they did not sing like cicadas or bite like bedbugs, they posed no threat to health or houses, and continued to be more or less irrelevant to both farming and philosophy. They were not mentioned by fabulists in their moral tales, or in psalters, prayer books or hymnals.

Shieldbugs did, however, occasionally appear as marginalia in illuminated medieval manuscripts – in particular one medieval manuscript. The remarkable Cocharelli Codex (British Library manuscript 27695), created in Genoa at some point near the beginning of the fourteenth century, is famed for its beautiful representations of plants and animals, though its text is nothing to do with wildlife – just a Christian diatribe on the vices and virtues, and some reports of historical events during the time of Frederick II (who also called himself Frederick III) of Sicily (around 1295–1337). It contains several astonishingly accurate images of shieldbugs, along with two beautifully depicted water boatmen, one with the wings slightly spread, and also many other readily identifiable invertebrates (Fig. 97). These were obviously painted from life, unlike the comical globular elephant and the bizarre zebra-striped giraffes. The artist was clearly taken with the local Heteroptera; one image on its own might be interesting from a historical point of view, but here are at least three pentatomid species in quick succession. Hutchinson (1974) discusses some of the illustrations in this and other manuscripts, notably *Pyrrhocoris apterus*, which can be identified in several other documents by its bright colours and clear pattern.

---

classical word meaning 'without master' – unowned, anonymous, rumoured, without history or explanation. There are suggestions (honoured by the *Oxford English Dictionary* no less) that it arose by a handwriting or typographical error: a rather looped kappa became alpha and omicron/rho became conjoined to produce phi, thus κορισ (koris) was mis-transcribed to αφισ – aphis.

FIG 97. Bugs from the Cocharelli Codex (c. 1330–40). (a) *Eurydema oleracea/ ornata* nymph, truly a wonderful painting (compare Fig. 98). (b) *Carpocoris* (probably *mediterraneus*); (c) and again with wings slightly spread. (d) *Dolycoris baccarum*, another plainly identifiable image. (e) *Notonecta maculata*.

FIG 98. Compare this modern photo of a *Eurydema ornata* nymph with image (a) in Fig. 97.

The clarity and precision of the insect paintings in the Cocharelli manuscript is, I think, a highly significant point in entomological art history. There is nothing like it in earlier manuscripts, and this level of subtle artistic accuracy is not repeated for nearly 200 years – until Albrecht Dürer painted his iconic stag beetle in 1505. This document represents the first clear and careful depiction of shieldbugs in all history. Elsewhere, though, this was a period of dim obscurity, and most other illustrations of insects from the illuminated manuscripts of the time are ridiculously vague; even obvious insects are stylised to the point where swarms of honeybees flying around a skep could easily be bats, birds or paper aeroplanes. The Cocharelli document really does stand alone, shining in the darkness.

SHIELDBUG RENAISSANCE

Thomas Mouffet's *Theater of Insects* was the first insect book ever to appear in English (1658). It was first published in Latin in 1634 (*Insectorum sive minimorum animalium theatrum*), and this was already 30 years after Mouffet's death in 1604. In fact the title page states that Mouffet was also the compiler/editor using texts from three other naturalists – Edward Wotton of Oxford, Swiss scientist Conrad Gesner and Mouffet's physician colleague Thomas Penny – so the book really represents a consensus entomological summary of knowledge from around the 1580s and 1590s. Mouffet's overall insect coverage is erratic: there are long entries for well-known creatures like beetles, bees, ants, butterflies and scorpions, but smaller groups are rather lost in the sometimes interminable screed. Nevertheless he lists three distinct and identifiable shieldbugs – though, of course, he calls them 'wood wall-lice of the sheath-winged kinde'. They are illustrated by woodcuts and he comments on their characteristic smell, the fact that they fly in the heat of the day 'but neither long nor far', and copulate 'tail to tail, and are almost a whole day about it'. Despite the crudity of the illustrations and the vagueness of the descriptions, these three are fairly clearly *Pentatoma rufipes, Palomena prasina* and *Graphosoma italicum,* though another less clear illustration 70 pages later, in his section on bedbugs, has a definite *Eurydema* vibe to it. And as if to confirm the confused classification of the times he also includes a woodcut of what looks like *Pyrrhocoris* (though it could be *Corizus* or *Lygaeus*), but without discussion in the text, among his 'lesser beetles' section earlier in the book (Fig. 99). Apart from some general comments about the females being bigger than the males, and the distinctiveness of their stink, he makes few

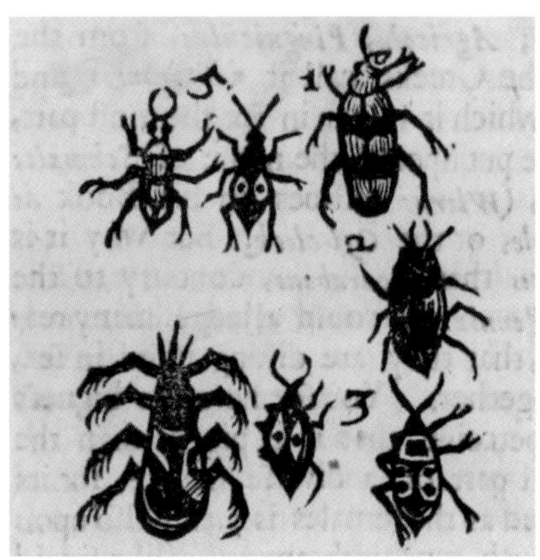

FIG 99. Wood wall-lice of the sheath-winged kinde, from Mouffet's *Theater of Insects* (1658) (above). These look like (left to right): *Pentatoma rufipes* under and upper sides, *Palomena prasina* under and upper, *Graphosoma italicum.* Also, some of his 'lesser beetles' (left), at least one of which (bottom right) appears to be *Pyrrhocoris apterus.*

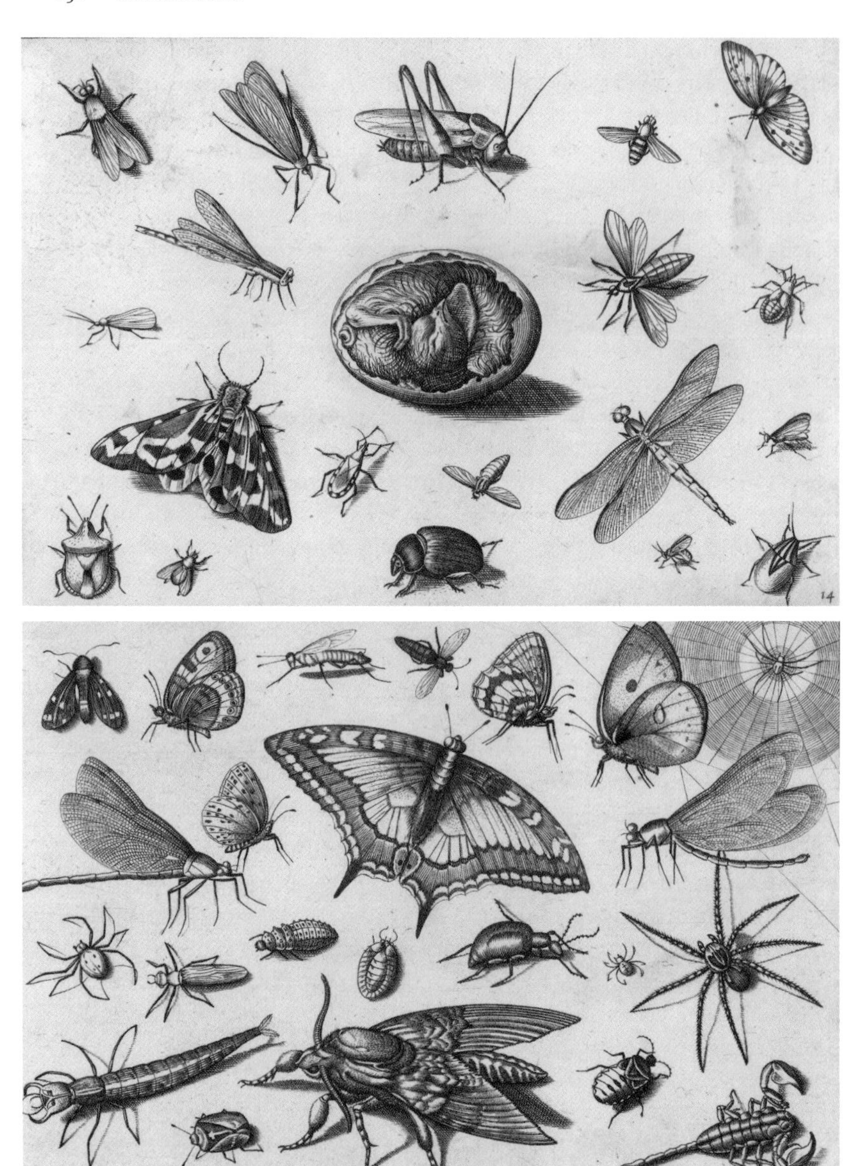

FIG 100. Plates from Jacob Hoefnagel's 1630 series of insect prints showing some shieldbugs. Above: bottom left a pentatomid, bottom right possibly *Aelia*, and middle right possibly *Pyrrhocoris*. Below: bottom left possibly *Eurydema oleracea*, near bottom right (above the scorpion) perhaps a nymph of the same.

observations other than to repeat Pliny's potion of shieldbug ash in rose oil for pains of the ears, and an equally dubious ointment of crushed shieldbug, olive oil, ox gall and ivy leaves against venomous bites of horse-leeches.

Mouffet's *Theater* marks a watershed in scientific publishing. There is some first-hand observation, and because of its brevity his page on wood wall-lice is surprisingly straightforward and accurate. However, much of the rest of the book is a rehashing of all those revered ancient writers – myths, magic and alchemical potions still abound. But the Age of Enlightenment was now emerging, and this new philosophy, based increasingly on reason and personal observation, was nowhere more apparent than in entomology.

In this country, the members of the newly formed (1660) Royal Society studied almost everything, but entomology (and natural science in general) featured heavily from the off. The *Philosophical Transactions* published by the society is the world's first and longest-running scientific journal, and in 1671 what is arguably the first het bug mention is possibly *Corizus hyoscyami*. A letter by Martyn Lister dated 30 May 1671 describes: 'a Cimex of the largest size, of a red colour spotted black, and which is to be found very frequently, and plentifully, at least in its season, upon henbain. I therefore in my private notes have formerly intitled it, *Cimex ruber maculis nigris distinctus super folia Hyoscyami frequens*.' The scientific name of Henbane is *Hyoscyamus niger*, and Lister's Latin translates as 'red and black bug found frequently on henbane'. Lister's letter continues concerning the red colour of the crushed eggs 'as lively a vermilion or *couleur de feu*, as any thing I know in nature; *cochneil* scarce excepted when afflicted with oyl of vitriol [sulphuric acid]'. Jim Flanagan (Facebook group, 16 November 2022, who drew this letter to my attention) suggests that Linnaeus took Lister's name to make it *Cimex hyoscyami*, from which the modern name is descended.

Dutch scientist Jan Swammerdam's work was all about very close personal observation, using that novel instrument the microscope. His *Book of Nature* was published in English in 1758, but was a translation from the Dutch original *Bybel der natuure*. That was published in 1737, 57 years after his death in 1680, so it represents the state of knowledge about a century after Mouffet. It is illustrated with monumental full-page engraved plates showing whole creatures or dissected body organs, and is a beacon landmark in the scientific understanding of how invertebrates worked (Cobb 2000). This was a time when he and other natural philosophers were still struggling against the widely held notion that small organisms appeared by spontaneous generation, and many of his dissections (famously of the honeybee) showed that all these creatures had ovaries, testes and internal organs to rival the intricate offal in any of the more familiar farm animals. It is a real shame that he did not give much column space to the Heteroptera, but this probably reflects their continuing insignificance compared to all the other things he covered. He mentions them in his section on 'the second order or class of natural changes, called the nymph vermicle', contrasting with the first order where the animal simply grows (scorpion, snail, louse) and the third and fourth orders which have a chrysalis stage. He does, at least, call them 'flying or land bugs' and although none are named he mentions 'twenty-six species in my museum … beautifully adorned by nature with variety of colours; and as with their lustre and gaiety they wonderfully please the eyes, though they are very disagreeable to the smell.' We have to guess at some species: 'Among these which are in our cabinet, we reckon the cruciate, the scarlet, the red marked with black lines [almost certainly *Graphosoma*], the green [probably *Palomena*], the yellow, the globular, and that which has a sharp-pointed breast [maybe *Picromerus*?].' Of the Hemiptera only the water scorpion *Nepa* and water stick insect *Ranatra* warrant small illustrations on one of his engraved plates. Nevertheless, we get a sense that Swammerdam had a basic understanding of the Hemiptera in general while he was busy revolutionising the study of invertebrates and their internal anatomy.

By the end of the seventeenth century John Ray (1627–1705) was the foremost English naturalist of his day; his *Historia insectorum* (published in 1710, five years after his death) contains none of the superstitious anecdotal waffle parroted by Mouffet, and is wholly based on clear description and observational study. Although Ray's name appears alone on the title page it is clear (and he acknowledges it throughout) that his colleague Francis Willughby (1635–72) did the bulk of the research, and the shieldbug descriptions and observations, for example, constantly reference the younger man (Fig. 101). Though mostly in Latin, the layout and structure of the book are immediately modern and familiar as he works systematically through the different groups. Ray's book contains a very few English names of familiar creatures like woodlice, crickets, the common (head) louse, butterflies and even a few obscure ones like the shepherd spider (harvestman) and oil beetle, but none for his shieldbugs, yet again confirming that they had never been considered significant enough by the general English populace to warrant English names.

Following on from the section on dragonflies, Ray's book contains Latin descriptions and comments on maybe 13 putative shieldbugs (still named as *cimices sylvestres* – wood cimexes), many of which are identifiable to modern species. The group is characterised as having broad shoulders, wings half crustate and half membranous, meeting along the back, with a long straight proboscis (not curled as in butterflies and moths) held underneath the body. Following Mouffet, he includes *Palomena prasina* and *Pentatoma rufipes*, and also seemingly

FIG 101. John Ray (left) summarised English entomological knowledge at the end of the seventeenth century, and though written mostly in Latin his book *Historia insectorum* is still relevant today. Much of the work on shieldbugs was done by his colleague Francis Willughby (right).

Picromerus bidens (sharp shoulders, lacks the chequered edge to the abdomen), *Coreus marginatus* (body slim, but shoulders prominent, antennae and legs long), *Tritomegas bicolor* (body round, resplendent black with white spots), three possible colour variants of *Eurydema oleracea* (blue-black with yellow spots, or red, or darker). Elongated red and black species may be *Corizus hyoscyami* (form larger) and *Pyrrhocoris apterus* (head, antennae, feet black ... two black spots on the red, the lower one the larger), and possibly its nymphs too. Finally a short-winged bug, foul-smelling, domed, antennae as long as the body, may be a pentatomid nymph. He later continues with 11 numbered species listed as *musca cimiciformis* (cimex-like flies), which are presumed to be smaller heteropteran species like leafbugs (family Miridae), and also several water bugs (including water boatman *Notonecta*, water scorpion *Nepa cineraria* and water stick insect *Ranatra linearis*), before moving on to grasshoppers.

SHIELDBUG STUDY IN THE SCIENTIFIC AGE

The true modern era of entomology began with the 10th edition of Carl Linnaeus's catalogue of living organisms, *Systema naturae* (1758), now taken as a baseline for the scientific nomenclature of plants and animals. The arrival of Linnaean binomial names, genus and species, means that many of Linnaeus's species are immediately familiar to modern readers – for example his specific names still stand for 28 of the shieldbugs listed in Chapter 8. Linnaeus (1707–78) relied heavily on wing structure for his broader classification of insects, and he coined the name Hemiptera (half-wing) for the order of bugs (including the usual aphids, leafhoppers, cicadas etc.), but for the true bugs he used only three genera – *Notonecta* (water boatmen), *Nepa* (water scorpions) and *Cimex* (everything else). It fell to his Danish student Johan Christian Fabricius (1745–1808) to subdivide the overly broad *Cimex* into a list of new genera (28 of them); and 15 of his newly described shieldbug species are still extant in the British fauna. Fabricius took a more holistic view of comparative insect structure, so when he published the first monograph on the Hemiptera (1803), he also created the term Rhyngota to reflect the distinctive rostrum mouthparts of the order. Fabricius mistakenly included the similarly snouted fleas alongside bugs, but this was a mere hiccup, and when renowned French entomologist Pierre André Latreille (1762–1833) published his monograph a few years later (1810) this was soon rectified. Latreille was also the first to use the terms Heteroptera and Homoptera. The first bug book divided along family lines was the *Histoire naturelle des insects, hémiptères* by Amyot & Serville (1843); although none of the species in this book before you today still bears the name they gave it, and many of their subdivision names have been superseded, the overall style and structure of their classification system was robust and is still valid today.

On the shoulders of these great European entomologists there started to appear the basis for studying, classifying and describing the modern Heteroptera fauna of Britain and Ireland. Edward Donovan's famous *Natural History of British Insects* (1792–1817) was rather a collector's book of pictures; it was issued in parts – a colour plate and a page or two of text at a time – and pretty as it is there is not much descriptive detail or biological information (Fig. 102). He does mention *Cimex* (now *Palomena*) *prasinus* ('head, corselet and shells green'), *Cimex gonymelas* (*Coreus marginatus*) ('black-kneed field bug') and *Cimex haemorrhoidalis* (*Acanthosoma haemorrhoidale*) ('greenish, breast-piece terminates in a long spine'). His brief text is often little more than a rehash description from the works of Linnaeus and Fabricius, but he occasionally slips in a real comment. He was obviously taken with the *Acanthosoma*: 'It is the most elegantly coloured creature of its tribe we have hitherto found. *Cimex* (now *Troilus*) *luridus* is more beautiful in the larva, but not in the winged state.' And I concur.

A few years later the monumental *British Entomology* by John Curtis was also issued in parts between 1823 and 1840, and this is an order of magnitude more

FIG 102. *Coreus marginatus* (above) and *Acanthosoma haemorrhoidale* (below), illustrated to best effect with wings outspread on plate 218 of Donovan's *Natural History of British Insects*. It seems to have been standard practice during the eighteenth and nineteenth centuries to name almost anything with a red tail after haemorrhoids.

FIG 103. *Chorosoma schillingi*, exquisitely portrayed on plate 297 of John Curtis's *British Entomology*.

detailed and important in both its content and its coverage. Each species is illustrated by an exquisitely engraved colour plate, accompanied by one or two pages of text. The work, though, is not fully comprehensive because only one species per genus is illustrated; thus *Pentatoma* (now *Zicrona*) *caerulea* is portrayed in full metallic-blue glory, but only a passing mention is made of all those others then considered to belong to the same genus but now split up into *Palomena*, *Troilus*, *Picromerus*, *Dolycoris*, *Eurydema*, *Pitedia* and *Piezodorus*. Nevertheless the quality of the hand-coloured illustrations and of the diagrams of body parts, and the precise scientific crispness of the text, make the book usable and informative even today (Fig. 103). Curtis bemoans the fact that the Hemiptera have not attracted much attention, that they 'have been totally disregarded' in the British Isles and that no suitable works exist except 'the rare and incomplete works of our Continental neighbours'. Curtis's inclusion of about 40 species is a very good start to British shieldbug study, and he makes many interesting and pertinent observations. He reports the first British specimen of *Atractus literatus* (now *Arenocoris fallenii*) from Braunton Burrows in Devon, creates *Acanthosoma* as a new genus, and includes *Graphosoma lineatum* on the strength of specimens found in a Norfolk nursery on plants likely to have been imported. In keeping with the terminology of the day he still talks of the shieldbug 'larva' (rather than nymph) lacking wings, and the 'pupa' (final instar) having wing rudiments. One of his most poignant remarks relates to *Aelia acuminata*, a species he regarded as uncommon, which he found 'in August and September in cornfields near Niton in the Isle of Wight'; this was a time when wheat fields were full of heartsease, poppies and cornflowers – biodiversity hotspots, not sterile chemically maintained monocultures.

By 1829 James Francis Stephens, Curtis's contemporary, had produced his *Systematic Catalogue of British Insects*, which contained about 50 species in the shieldbug groups. None of these names came with any description, however, and although many are recognisable, some are forms now reduced to subspecific or mere variety status, and others have been removed as being duplicates or misidentifications. It is a real shame that Stephens' most important book series, *Illustrations of British Entomology*, never reached the Hemiptera. After stupendous work on beetles (five volumes), moths (four volumes), a few small orders (one volume) and a start on the Hymenoptera (some sawflies and ichneumons in one volume), it ceased publication. We can only dream about the Heteroptera masterpiece that never was.

At this time the beetles were second to the butterflies and moths in terms of insect popularity, but study of the Hemiptera was in the doldrums. Serious entomologists recognised this lack of interest and tried to drum up enthusiasm. Henry Tibbats Stainton, primarily a lepidopterist, drew up a list of British bugs for the *Entomologist's Annual* that he edited (1861), and the running total of shieldbugs now reached 56 species, although this included several forms now regarded as mere colour varieties (three *Aelia*, for example) and some unlikely non-British species which might have been wishful thinking on his part. The following year John Scott (1862) urged readers to shift their gaze to the Hemiptera, 'commonly called bugs', and gave some helpful tips on where to find new and interesting species: '*Arma custos*, on what and when?' he exclaims.

The next monument of Hemiptera study in Britain is the large (640 pages) book by Douglas & Scott (1865). Though it claims to be volume 1, on the Heteroptera, no second volume on the Homoptera ever appeared. Most of the British shieldbug species we know today are included in their 54 full and clear descriptions, and they give more information than many more modern books. Though 'only' black and white, the plates illustrating whole insects have a beautiful photographic quality that stands as testament to the engraver's art and skill, and the diagrams of undersides, wings, legs, rostrums and antennae are neat and crisp (Fig. 104). This was the book that transformed my understanding of the Heteroptera when I picked up a cheap copy at an entomological fair in 1975; if memory serves, it cost me about £8, then still quite a tidy sum for a 17-year-old schoolboy. The pages were clean and bright, but the whole thing was falling apart, with loose card covers; it was broken and unusable. However, I was

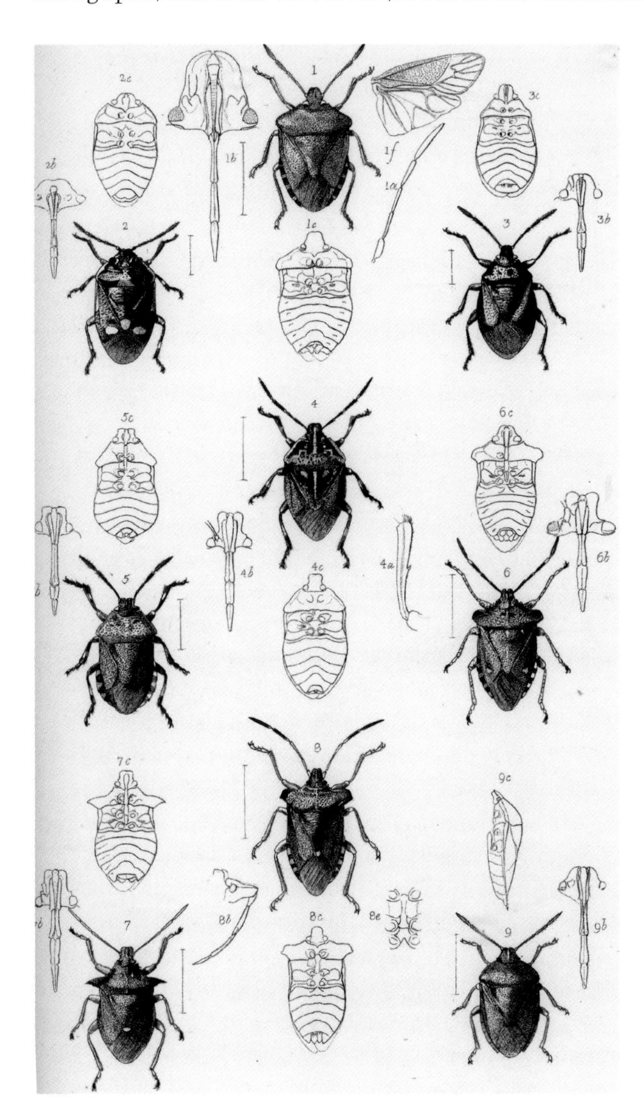

FIG 104. Plate 3 from Douglas and Scott's *British Hemiptera* (1865). The detail packed into the illustrations, including the anatomical diagrams, is superb, and this book was a major step forward in shieldbug study in the British Isles.

lucky enough to have a friend (all-round entomologist and antique collector Roger Dumbrell) repair and rebind it in gold-tooled half-leather, using his own hand-made marbled papers to cover the boards, and I immediately set to work. It was using this book that I identified my specimens of *Odontoscelis fuliginosa*, *Corimelaena* (*Thyreocoris*) *scarabaeoides*, *Podops inuncta* and *Eysarcoris melanocephalus* (now *E. venustissimus*) – smaller species not included in any of my simple picture-book guides. The bright quality of the illustrations and the superb clarity of the descriptions are reflections on the entomological emphasis of the times – collecting, identifying, naming, classifying, ordering, and arranging neatly in the museum drawer. Ecological observations are few, though, mostly second-hand reports from European authors, and the concept of precise biological recording was for another century. This is well exemplified in the entry for *Rhopalus* (then *Corizus*) *maculatus*: 'Scarce, 2 specimens, locality unknown'. It's as if they had never heard of data labels; pathetic really.

Fortunately, some British entomologists were working to rectify this. From about the middle of the nineteenth century (and continuing to this day), regular short notes of personal finds and observations make up the bread-and-butter contributions to the burgeoning scientific journals of the day. A trawl through the *Entomologist's Weekly Intelligencer*, *Transactions of the Royal Entomological Society*, *Entomologist's Monthly Magazine*, *The Entomologist* and others reveals an army of field entomologists eagerly reporting new species, new names, new behaviours, foodplant associations, local lists, additional identification pointers and odd occurrences. Several entomologists specialised in Hemiptera (and also Coleoptera, since beetles and bugs are often found together) and their names recur through the literature. John Douglas and John Scott are regulars, as are Edward Butler, George Carpenter, George Champion, James Dale, William Distant, John Power, Edward Saunders, David Sharp, William Sharp and Francis White.

Francis Buchanan White is best known for his 1883 monograph on *Halobates*, the surface-dwelling Hemiptera related to water skaters, and the only insects known to inhabit the open oceans of the world, but in 1877 he was still decrying the lack of interest in the British Hemiptera, urging lepidopterists to turn their attentions to bugs when they had exhausted all the moth species in their local area. He gives helpful tips about finding, killing and setting them in the British or Continental styles and exhorts his readers with the news that there are great discoveries to be made: 'That a great deal remains to be done in Britain is apparent from the fact that a very few workers have added to the list of British Hemiptera–Heteroptera upwards of thirty species between January 1874 and December 1876, although a great part of the country remains still unexplored.' He does, however, warn that being called a 'bug-hunter' could be rather disparaging and off-putting.

Edward Saunders (1875) was pleased to introduce *Sehirus* (now *Legnotus*) *picipes* new to Britain from specimens taken by Dr Power near Esher; he suggested they were probably intermingled with the widespread *S. albomarginatus* (now *S. luctuosus*) in collections. John Douglas, one of the editors of the journal, immediately added a footnote to the effect that, yes, he had collected some years ago, but not closely examined until then. Coleopterist William Sharp (1900) later deferred to Saunders' greater knowledge of the Hemiptera when he was sent a strange-looking pentatomid from the hills above Bangor in north Wales; it proved to be *Elasmucha ferrugata*, the first time this enigmatic shieldbug had ever been recorded in Britain. On the very next page Saunders (1900) also reports the discovery of a West Sussex specimen of the rare *Peribalus vernalis* (now *P. strictus*) he'd spotted in the collection of Bognor Regis naturalist Henry Guermonprez. Thankfully Mr Guermonprez (known affectionately as the 'Gilbert White of Bognor') was of very modern thinking and had attached a full data label to the insect: 'Slindon Woods, on hazel, Sept. 13th, 1899'. This was more like it.

Many of these notes are the contemporary equivalent of a tweet or Facebook special-interest group post – just a few words to announce a discovery, and to elicit responses from other like-minded observers. George Champion took just three and a half lines of small text to alert readers that he had seen *Sehirus dubius* and *Zicrona caerulea* (abundant) near Croydon (1870) and just five lines to announce the discovery of *Corizus* (now *Stictopleurus*) *abutilon* new to Britain (1871). Everyone and anyone could publish these notes, and they could be interactive as well as merely communicative. In the same year (1871), Welsh entomologist Edwin Roper Curzon reported numbers of *Stenocephalus agilis* on the sand hills near his Newton Bridgend home and offered his spare specimens to anyone who wanted them.

These articles make fascinating reading, not just for the scientific information they contain, but often for the social and historical context that also seeps through. I love the fact that in 1869 John Douglas, with 'a lucid interval of half-an-hour to wait for the next train to London, chains and slavery', found himself in a Dartford lane, where he 'startled a chemist by an abrupt demand for a small glass bottle without poison in it' and marched away to look for bugs to put in it. We've all found ourselves without collecting paraphernalia and had to make do with whatever alternative container comes to hand. Meanwhile John Scott set off to Caterham on a fine, clear and warm 26 March 1873 to look for *Pentatoma* (now *Chlorochroa*) *juniperina* under the many juniper bushes then growing on the chalk hillsides. At first 'my exertions were barren' but 'Suddenly, something moved on a branch of the bush at which I was at work. It was the veritable creature; and, immediately after, a second put in an appearance.' So enthused was Scott that he interrupted his wife, who had just sat down to have her packed lunch, and

FIG 105. William Lucas Distant, South London hemipterist of note.

helpfully 'showed her what to look for and gave her a bottle'. The fact that his wife had found 17 specimens by the end of the day compared to his meagre seven led Scott to suggest 'I feel convinced that the order comprising *Pentatoma* affords a fine scope of study by woman'; you can almost hear the editors (all male, of course) chuckling softly at his droll remarks.

I claim a particular personal interest in hemipterist Mr William Lucas Distant (1845–1922) because he lived for a time in Derwent Grove, close by me here in East Dulwich, south London (Fig. 105). Distant was a well-off businessman in the leather industry, but was also editor of *The Zoologist* from 1899 to 1920, and a part-time curatorial associate at London's Natural History Museum, rearranging the national collection of Hemiptera. He is primarily known for his very thorough work on the Hemiptera volumes of the grand book series *Biologia Centrali-Americana* and *Fauna of British India*. His, and other, volumes in these impressive, if rather imperious, tomes are now freely available online, but when I first saw a complete 64-volume leather-bound set of the *Biologia* on the shelves of the Natural History Museum's Zoology Library in the 1980s I was transfixed by the magnificent sight. At that time I was a subeditor in a medical publishing company in Holborn and my lunchtime strolls took me to the ramshackle second-hand book stalls then set up on nearby Farringdon Road, where one day I found a trove of *Biologia* plates (Fig. 106). There were hundreds of these elegant engraved illustrations in a large cardboard box, some already hand-coloured – obviously printers' 'overs' which had been made ready, but never bound up into the splendid (and very expensive) books, and which had no doubt been languishing in some dark corner of a publishing company during the ensuing decades. I bought the whole lot for a fiver (one of my best bargains ever) though there was no accompanying text. I managed to cobble together complete plate sets of several Coleoptera and Hemiptera volumes, and bind them up into quite acceptable books at the weekly bookbinding evening classes I attended at the London College of Printing down at Elephant and Castle. It was these engravings that I used to make my tentative identification of that fat brown *Edessa* shieldbug which stained my hand in Costa

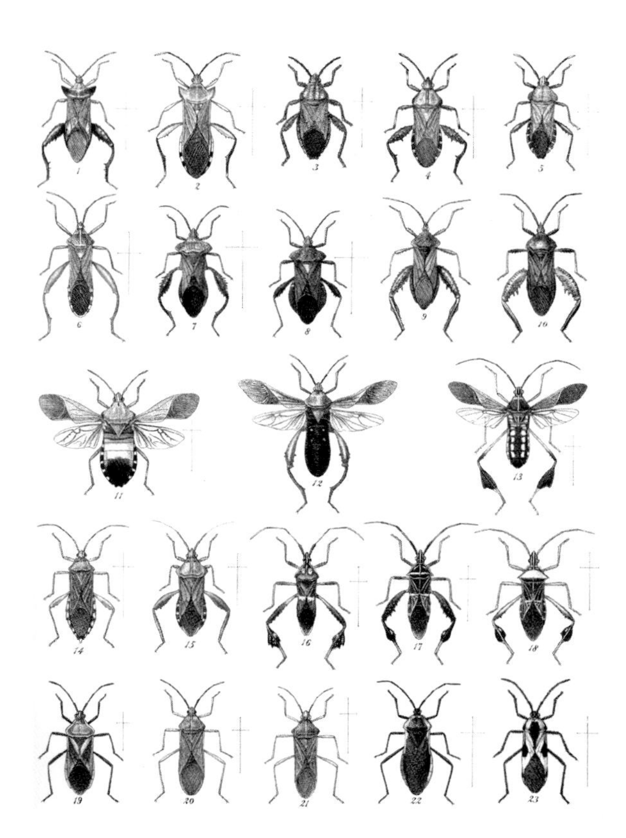

FIG 106. Plate from the magnificent *Biologia Centrali-Americana.*

Rica a few years later, and which I recently got out again to check on a fat-legged *Stiretrus* specimen from the same holiday.

Distant's commercial travels took him to South Africa, where he visited cattle farmers to check on hide quality, and researched various local tree barks for their potential in producing tannery chemicals. His *Naturalist in the Transvaal* (1892) is the usual fascinating, though pompous and (to modern ears) racist, travelogue. Between dinners at the Governor's residence and trips up-country, he collected a large number of wildlife specimens; he had the honour of having a threadsnake, *Glauconia* (now *Leptotyphlops*) *distanti*, and various insects named after him, and he described several shieldbugs new to science.

Distant is buried in Nunhead Cemetery, just a few hundred metres from where I sit typing these words, and I recently made a pilgrimage through the dense herbage to what I believe is his gravestone. Like so many of the once grand memorials here, it is leaning precariously in the dark undergrowth and the stone surface has crumbled away to the point of complete illegibility; thankfully, local historian Ron Woollacott noted the original inscription back in 1974. Also buried in the plot are Distant's father Captain Alexander Distant (died 1867, a whaler according to the *Transvaal* preface), his mother Sarah Ann (died 1889), his wife Edith Blanche de Rubien (died 1914) and two of his children, Freddie and Maggie.

During the 1870s Distant, who was later to be vice-president, secretary and council member of the Entomological Society of London, was one of the eminent entomologists who were invited to the nearby Forest Hill mansion of tea magnate (and MP for Falmouth) Frederick John Horniman, to view his expanding collection of exotic insects. Horniman had either been buying these at the many prominent natural history auctions of the day, or having them shipped directly to him by his various correspondents around the world. Eventually he would donate his entire collection of natural history specimens and ethnographic objects to the nation, and the Horniman Museum in Forest Hill (a short walk up the hill from my house) is still a gem, packed full to the gills of amazing stuff. In the build-up Horniman mingled with the scientific cognoscenti of the day, who came to drool over his manifold possessions and to name the many new species found amongst his extensive cabinets. Explorer and scientist Walter Henry Bates ('of the Amazon') named a giant chafer *Ceratorhina* (now *Cyprolais*) *hornimani* sent from a missionary station at Mongo-ma Lobah, in Old Calabar on the Cameroon/Nigeria border (Bates 1877). Lepidopterist Herbert Druce (compiler of *Biologia Centrali-Americana* moth volumes) described a large black, red and orange moth *Eusemia* (now *Heraclia*) *hornimani* from the same area in 1880. It was no surprise when Distant came across a large striking bright orange shieldbug in Horniman's cabinets, again sent from Cameroon, and called it *Tesseratoma hornimani* (1877). It was later illustrated by Waterhouse (1882–90) (Fig. 107). Distant also went on to name a

FIG 107. *Tessaratoma hornimani*, described by Distant (1877), and illustrated by Waterhouse (1882–90), who uses the terms larva (for the fourth instar, left) and pupa (for the fifth instar, right).

giant blue and black swallowtail butterfly *Papilio hornimani*, sent from highland forests near the tea plantations of southern Kenya (1879). Horniman wasn't really an entomologist (although he was elected a fellow of the Entomological Society of London in 1876), he was really a very wealthy collector/hoarder, but he maybe had aspirations to be a hemipterist. In the public natural history displays of the museum today is a drawer of shieldbugs from his original collections with several specimens of the large shining blue scutellerid shieldbug *Chrysocoris pupureus*. Pinned beside them is a prominent contemporary hand-written label claiming: 'Blue plant bug (*Callidea purpurea*) caught in the Island of Elephanta off Bombay by Mr F. John Horniman, December 5th 1894' (Fig. 108).

By the end of the nineteenth century the publishing house of Lovell Reeve dominated the high-end entomology book market with a series of highly erudite monographs illustrated by excellent hand-coloured engraved plates and

FIG 108. Drawer of exotic shieldbugs on display at the Horniman Museum, Forest Hill, south London.

with a concise formula of short tabular identification keys, precise comprehensive descriptions and details of geographical distributions based on a wide network of correspondents and informants. The *Hemiptera Heteroptera* by Edward Saunders (Fig. 109) appeared in 1892 (it was followed a few years later by the *Hemiptera Homoptera* of Edwards, 1896). Saunders' work was based on his earlier synopsis and checklist of the British Heteroptera species published in the *Transactions of the Entomological Society of London* (1875–76), and it became the default book on het bugs for the next half-century (Fig. 110). By now

ABOVE: **FIG 109.** Edward Saunders, whose book *The Hemiptera Heteroptera of the British Islands* (1892) became the standard work on British bugs for the next 67 years.

FIG 110. Hand-coloured plate from Saunders' 1892 monograph on the Heteroptera.

NAMING BUGS IS A VERY SERIOUS BUSINESS

No history of Hemiptera study at this time should pass without at least a mention of George Willis Kirkaldy (1873–1910). His interest in insects started at an early age and he became curator of the City of London School museum when he was a pupil there. After taking up work 'in the City' he worked on the Hemiptera in the evenings, until he was offered a position with the US Government Department of Agriculture at a research station in Hawaii. His interest in Hemiptera blossomed, and many of his papers on obscure details of nomenclature and classification were published in British scientific journals, although they were on worldwide species.

It seems to have only been a couple of years after his untimely death that people revisiting his work and further revising those exotic bug genera realised that Kirkaldy had been more than a little playful in his seemingly staid scientific prose. Many of his new genus names used the well-established Greek suffix –chisme (pronounced 'kiss me'), which is probably derived from similar linguistic roots to *Cimex*. What became clear, though, was that he had prefixed this quite ordinary bug-related term with the names of various women he was reputed to have been romantically involved with, thus – *Dolichisme, Elachisme, Florichisme, Marichisme, Nanichisme, Peggichisme* and *Polychisme*. These were all slipped in under the radar in a journal article putting forward various arcane nomenclatural arguments to the still jumbled world of Hemiptera names and species descriptions (Kirkaldy 1904b). In 1912 there was an attempt by British Conservative peer and entomologist Thomas de Grey, 6th Baron Walsingham, to invalidate these facetious and frivolous names, but it failed. Today Kirkaldy's slightly flippant names all seem a bit tame compared to things like beetles called *Eubetia bigaulae, Gelae baen, Gelae belae* and *Gelae donut*, braconid wasps *Heerz lukenatcha* and *Heerz tooya*, and a series of flies *Pieza rhea, Pieza pi, Pieza derisistans* and *Pieza kake*. Lord Walsingham would have had conniptions.

FIG 111. George Kirkaldy.

In the end many of Kirkaldy's names have disappeared anyway by a kind of natural attrition, as nomenclature revisions from around the world continue to regroup and rename species, but his coreid genera *Elachisme*

continued overleaf

NAMING BUGS IS A VERY SERIOUS BUSINESS *continued*

and *Marichisme* and lygaeid *Polychisme* are still valid, immortalising (for now) Ella, Mary and Polly, whoever they might have been. However, on a more serious note, Kirkaldy's major contribution to shieldbug study was his 1909 catalogue of world Pentatomidae. He was obviously a knowledgeable and expert entomologist and, for example, his illustration (1907) of the egg of the Hawaiian shieldbug *Oechalia grisea* is one of the best to be found then or now. Kirkaldy would no doubt have been the source of many more flippant bug anecdotes had he lived longer. A riding accident left him with an injured leg that never healed, and he died a few days after the last of several surgical attempts to fix it.

there were reckoned to be 55 species of shieldbugs and associates in the British Isles. The large-format coloured-plate editions were expensive, but in a clever marketing move a smaller cheaper edition without the hand-coloured plates was also available for each title. Even without these colour pictures the stand-alone text worked well and the book was perfectly usable. I still use mine. I long aspired to own a 'large paper' illustrated copy but was only able to afford one recently. *Stictopleurus crassicornis* is included, although there may be two species under this name (he mentions *S. abutilon* found by Champion), but the records of these species from Chobham, Reigate, Bournemouth and Deal would be the last reports of them in the British Isles for over 100 years. By the middle of the twentieth century they were thought to be extinct here, or probably short-lived colonies that had never become fully established.

Saunders' other legacy was to have introduced Edward Butler (1845–1925) to the Hemiptera, and this interest bore final fruit in the form of his large volume *A Biology of the British Hemiptera–Heteroptera* (1923). The preface makes it clear that this was a supplement to Saunders' volume, and not meant to supplant it. Laid out in similar style, it does not comment much on the adults, but describes in detail everything that was then known about the eggs and nymphs, with detailed descriptions of the immature stages, including some black-and-white photographs, coloured plates and thumbnail black-and-white illustrations. Full details of foodplant choices, distributions and behaviours are given. By now 61 species are reckoned as British; rather unhelpfully a numbered catalogue at the end of the book lists 63, though the numbers do not tally with those in the main text, and often different names are listed – perhaps a reflection of the statement repeated in several of the obituaries that 'his interest lay entirely in the biology and life histories, not at all on the systematic side' (Wyatt 1926). Nevertheless, this was a hugely important publication and very much marks the move away from merely collecting numbered species and arranging them in a glass-topped drawer, and towards an understanding of the living insects' biology and life histories.

MODERN TIMES

For much of the twentieth century British Hemiptera identification relied on those solid monographs, though revisions of families or genera sometimes took place in the pages of entomological journals, for example *Eurygaster* and *Corizus* etc. by China (1927 and 1941 respectively), and *Stenocephalus* (*Dicranocephalus*) by Butler (1911). The occasional new species was added – when Bedwell (1909) announced his discovery of *Odontoscelis lineola* (he called it *O. dorsalis*) near Lowestoft he was nevertheless able to remind readers that Saunders had already commented on it in his 1892 monograph and suggested its likely appearance in sandy areas of the British Isles. Of course, then up jumps George Champion (who was an editor on the magazine) to say that, yes, he had just spotted a specimen in his own collection taken at Sandown, in the Isle of Wight, back in July 1888.

The next comprehensive book on British bugs came from an unlikely source. The London publisher Frederick Warne produced (along with Beatrix Potter's stories for children) a series of small, cheap, popular natural history volumes on birds, beetles, flies, trees, blossoms, freshwater life, fishes, ferns, fungi (illustrated by Beatrix Potter) and animal life. Loosely called the *Wayside and Woodland* series, they were simple introductions to broad subject areas, written in a straightforward style and illustrated with colour and black-and-white pictures mostly reproduced from much older books. In 1959 Warne issued *Land and Water Bugs of the British Isles* by Richard Southwood and Dennis Leston (Fig. 112). Taking some of the pictures from Saunders' 1892 Lovell Reeve monograph plates, and rejigging them five or six to a page, they could have produced quite a good popular guide to some of the larger, brighter or more common bug species. Instead they complemented these selected pictures with full dichotomous identification keys and thumbnail diagrams of characters to be seen down the microscope, along with ecological notes (mostly taken from Butler 1923) on every single British het bug species. They also included some quite technical information including egg and embryo development times, instar numbers, disease vector cycles, and chromosome numbers (this was cutting-edge biological science of the time – see box on pages 170–71). Rather than just being a friendly introductory volume to the group, this book became the definitive expert identification guide to the entire British Hemiptera Heteroptera. Southwood, later Sir Richard, went on to become one

FIG 112. The default guide to British het bugs by Southwood & Leston (1959) is usually still very expensive, but can occasionally be picked up cheaply from the bargain box. This one, from the British Entomological Society's annual exhibition in November 2022, had slightly scuffed covers and some minor ink damage to the end-papers; for 15 quid I was even tempted to get a second copy.

of Britain's foremost biologists, and a top government scientific advisor, and Leston was described as 'an iconoclastic commoner of flashing intellect' by Schuh & Weirauch (2020). Southwood's Heteroptera collection is now in the Oxford Museum of Natural History (Iley 2011).

As a boy I was able to pick up most of the Warne series for a few shillings each, occasionally a few pennies at a jumble sale (and that was sometimes *d* in old money), but the bugs volume was highly sought after and highly valued. Perhaps Warne had some last-minute misgivings about the rather more technical content of the book, because in the end only about 400 copies were ever printed. I think I was eventually offered mine for £30, a year or two after I'd found Douglas and Scott's 100-year-old tome, and during the 1980s and 1990s it regularly turned up on the shelves of specialist natural history booksellers for well over £100. At the time I couldn't understand why it should cost so much, but the book's lasting usefulness and worth is demonstrated by the fact that over 60 years later it is still being offered for sale as a downloadable ebook or as a 2005 facsimile printed edition. Copies of the first edition are usually still scarce and often expensive, though: £50–150 in a quick internet search in October 2021, for instance, depending on the condition and whether the dust jacket is still attached.

Although no other book since Southwood & Leston (1959) has sought to cover all the British bugs, shieldbugs have made a leap in popularity concomitant with their large size and striking colours. Despite its local-sounding title, *Shieldbugs of Surrey* by Roger Hawkins (2003) is a book of national rather than parochial importance. Hawkins spent 25 years recording shieldbugs all over his home county, eventually collating records from dozens of other entomologists and producing an authoritative and very readable book – it includes useful distribution maps, an identification key to all British species and colour photos of those found in Surrey. It is also full of interesting comments and personal anecdotes. Hawkins' histograms of monthly shieldbug egg and nymph records and his first-hand reports of rearing them through to adulthood are widely quoted by other researchers around the world. One of the longest entries (six pages) is given over to the Box Bug (*Gonocerus acuteangulatus*), no longer the Surrey rarity of old, and now making its move across the county and out into the wider countryside. And it was finding the 'juniper' bug *Cyphostethus tristriatus* on garden cypress that kindled his particular interest in shieldbugs and led eventually to the Surrey survey and his book. I am proud to have contributed a few records from my corner of the north-east extremity of the vice-county, now subsumed by urban south London, and seldom visited by naturalists when they had the North Downs and Surrey Heaths at their feet. Hawkins' atlas is a model for twenty-first-century entomological recording and reportage. There has since followed a spate of small books and booklets on shieldbugs, including those by Nau (2004), Evans & Edmondson (2005), SNHS (2007) and Boardman (2014), with more to come; one by Barnard is in preparation.

Shieldbugs have recently become more popular groups on the continent of Europe too, with much time and recording effort resulting in several outstanding books of which the British hemipterist can now make full use. The well-known *Faune de France* series now has three volumes dedicated to the Euro-Mediterranean Pentatomoidea (Derjanschi & Péricart 2005, 2016, Péricart 2010, Ribes & Pagola-Carte 2013), and one on the Coreoidea (Moulet 1995). A popular guide to French shieldbugs is available in the form of Lupoli & Dusoulier (2015). Elsewhere, guides are available to the bugs of Finland (Rintala & Rinne 2011) and the shieldbugs of Sweden (Strid & Forshage 2012) and Denmark (Nielsen & Skipper 2015).

SHIELDBUG STUDY TODAY

Towards the end of the twentieth century there was a definite move away from finding insects for the display collection towards recording them to increase ecological understanding and for their conservation value. There is still the need

EDMUND BEECHER WILSON'S IDIOCHROMOSOMES – IT'S ALL ABOUT SEX

However shallow anyone's knowledge of genetics may be, everyone should have heard of chromosomes, and will know that they carry the genetic material in the form of DNA. And most people will have heard of the X and Y chromosomes that determine whether an organism will develop as female or male. But this was a revelation in 1905, when Edmund Beecher Wilson (1856–1939) published his findings on the chromosomes of the Hemiptera Heteroptera.

In humans each standard cell in the body has 46 chromosomes – 23 pairs. One of the pairs is unusual, though, in that sometimes the chromosome appears to have four arms in the form of an X, and sometimes only three arms in the form of a Y. It's still not completely clear, but at some point way back in deep evolutionary history a chunk of chromosome had been lopped off and lost. If, in this pair, two Xs are present the human is female, if X and Y then the human is male. Another way of representing this is:

22 pairs + XX = female

22 pairs + XY = male

The only time a human cell contains fewer chromosomes is in the production of sperm and ovum, where the pairs become separated so that only one of each is included – thus 23 single chromosomes in each sperm or ovum. This is termed the haploid condition (from Greek ἁπλόος haploos = single). When sperm and ovum meet, the 23 singletons in each are recombined to give that standard arrangement of 46 (23 pairs) again, and this is called the diploid condition (Greek διπλόος diploos = double). During sperm production in the testes a cell containing 22 pairs + XY gets divided into two, so that one will contain 22 + X, and the other 22 + Y. No matter how many sperm are produced, these are produced in equal numbers. Meanwhile in the ovary the 22 pairs + XX produce two identical haploid ova, each with 22 + X.

FIG 113. Edmund Beecher Wilson.

At the point of fertilisation a sperm fuses with an ovum. A sperm with 22 + X combining with an ovum, also with 22 + X, gives 22 pairs + XX = female. Alternatively a sperm with 22 + Y combining with a 22 + X ovum gives 22 pairs + XY = male. Since half the sperm have an X in them, and half have a Y, the chances of producing female or male offspring are 50/50. This is known as the XY sex determination system, and it occurs through almost all mammals, and in many other types of living organism.

Different organisms have different chromosome numbers, but they are always arranged in pairs, and the convention is to denote this as the diploid being twice the number (n) of the haploid. Thus the yellow fever mosquito has 3 pairs so $2n = 6$; gorillas have 24 pairs so $2n = 48$; in sugarcane $2n = 80$; and in the northern lamprey $2n = 174$. In the Heteroptera, chromosome numbers vary from $2n = 4$ in *Lethocerus* species (family Belostomatidae) to $2n = 80$ in *Lopidea marginalis* (family Miridae), although $2n = 10$ is the commonest arrangement (de Souza-Firmino *et al.* 2020).

At the beginning of the twentieth century Wilson was carefully examining het bug cells under the microscope in his laboratory at Columbia University in New York, and had seen that one of the chromosomes in the male was always much smaller than the corresponding one in the female of the same species, as if part of it had been lopped off and lost during evolutionary history. He had discovered the Y chromosome (though he termed it the small idiochromosome), and he correctly proposed that sperm containing this would result in a male bug, whilst sperm containing the larger idiochromosome (the X) would give rise to a female.

Wilson also observed that males of some species (*Pyrrhocoris* and *Alydus*, for example) had one less chromosome than the female. In this case, he surmised, the small idiochromosome had been lost altogether, leaving an odd number. Here, a sperm containing the large idiochromosome would produce a female (XX), but a sperm lacking it would become male – what nowadays would be described as XO sex determination rather than XY. Both XY and XO sex determination systems are known in bugs, indeed throughout many insect groups. Most Pentotomoidea use XY, though Pyrrhocoridae, Alydidae, Coreidae and Rhopalidae use XO, and Stenocephalidae use both (Papeschi & Bressa 2006).

Although I've invoked Wilson here, because of his work specifically on the Heteroptera, Nettie Stevens (1861–1912), working at the Bryn Mawr College in Pennsylvania, is generally jointly credited with first announcing (also in 1905) the significance of the X and Y chromosomes, with her work on the Mealworm Beetle, *Tenebrio molitor*. Entomologists – always at the cutting edge of biological science.

to firmly and correctly identify them, but a far greater emphasis is now put on reporting finds and collating records to give a national (or international) picture for each species.

Now is the internet age, and the modern equivalent of the identification guide is the online photo gallery maintained by Tristan Bantock and Joe Botting at britishbugs.org.uk, which features most British bugs and is regularly updated with new species or new information. An illustrated chart of shieldbug nymphal stages by Ashley Wood is also available there. International websites illustrating the bugs of other parts of Europe are also available, and links are provided. Tristan Bantock also runs the Shieldbugs and allies recording scheme; an atlas of the collated records was issued in 2018. The site also hosts back issues of the *Heteropterists' Newsletter*, a series of simple-format photocopied sheets originally mailed to interested hemipterists from 1983 to 1999, and its slightly glossier full-colour successor *Het News* from 2003 to 2016. A journal, *The Hemipterist*, was also set up in 2014 as a vehicle for publishing specialist articles on bugs, and all issues are freely available to download from the Google host site (sites.google.com/site/thehemipterist). Further information on how to make a contribution to future shieldbug study is given in Chapter 9.

SHIELDBUG IMPACTS ON HUMANS

One of the reasons shieldbugs escaped intimate human attention for so long is that they rarely come indoors so are not deemed disease-spreaders on the kitchen surface or potential chewers in the woollens drawer. If they do come inside houses, it is something to write home about – which is just what I did on 18 February 2011 when I put a picture of *Eurydema oleracea* on Facebook after it crawled up the power lead to my laptop onto the kitchen table. I made the usual facetious computer bug joke, and although I did not know exactly where it came from I knew it was probably just a casual visitor from the garden. Not so the report by Lo Verde & Carapezza (2018), who describe how, in the summer of the previous year, many thousands of a small black cydnid, *Macroscytus brunneus*, had invaded houses on the tiny volcanic island of Linosa, in the Mediterranean Sea between Sicily, Malta and Tunisia. This normally soil-dwelling bug arrived at the houses each evening, over a period of several weeks from July to September, flying around every porch light and street lamp and entering buildings through open windows to the lit interiors. The buzzing insects were particularly troublesome to squeamish residents and frustrated tourists – restaurant owners were forced to cover their lamps with red cloth to reduce their attractiveness to the bugs. The uneasy residents sprayed walls and floors where the bugs settled and each morning bucket-loads (literally) of dead insects were swept up and cleared away. Where there was no spraying the bugs had vanished by daybreak anyway, and there was no sign of them, but each evening they would return. There was no obvious explanation for the outbreak of this normally secretive species, other than a freak population explosion caused by as yet unknown alterations to the bug's ecology and its association with its local foodplant, a horned poppy, *Glaucium corniculatum*. It is possible that high summer temperatures and low rainfall for two years had increased winter survival and spring development rates. At least this time the shieldbugs caused just mild annoyance. In a similar event in the Amami Islands, south of Japan, in July and August 1974 the related *Aethus pseudindicus* appeared in enough numbers to contaminate food and drink with their stink fluid, and more than 70 people were treated in hospital for injuries resulting from the bugs crawling into their ears (Takai *et al.* 1975). Nothing like this has ever been reported for British shieldbug species. Yet.

Again, we are lucky here in oceanic temperate north-west Europe in that shieldbugs do not really constitute major agricultural pests. Elsewhere on the globe things are different. Numerous pentatomid shieldbugs are reviled around the world, including *Murgantia histrionica* and *Bagrada hilaris* on North American cabbages, *Nezara viridula* on a whole raft of crops, *Piezodorus hybneri* on African peas and beans, and *Antestiopsis thunbergi* on coffee. Though *Aelia* and *Eurygaster* are secretive genera in Britain, doing little or no damage to wild grasses, *A. rostrata* and *E. integriceps* are both infamous cereal pests in the Middle East (Ogur & Tuncer 2019; Fig. 114). A slender alydid, *Leptocorisa acuta*, is a pest

FIG 114. *Aelia rostrata* on an ear of barley. Around the Mediterranean this species is reckoned an agricultural nuisance, its bites causing blemishes on the growing leaves, and desiccation and wrinkling of the spike.

of rice in India and South-East Asia. The north American coreid *Anasa tristis* is a major nuisance on squash and pumpkins; often called the squash bug, it gives its common name to the entire family. The North American cydnid *Scaptocoris castaneus* is a short, squat, broad-legged shieldbug, and feeds at the roots of cotton, sugar cane, bananas, tomatoes and many other crops. Meanwhile its Central American congener *S. divergens* also feeds on banana roots. This is not a wholly negative impact, however, for it helps to control the highly destructive fungal disease fusarium wilt, which can decimate the banana plantations. A toxic gland secretion from both nymphs and adult bugs inhibits the fungus, and only high bug densities allow the trees to grow well. Sometimes the bugs are so abundant in the soil that the ground is said to stink from the scent-gland odour. These are a few instances of shieldbugs having potentially significant impact on humans. But they are still geographically far enough away to be safely ignored by British interests. At least for now.

Nevertheless, in the first week of March 2021 a new target species to monitor crept into the mainstream media consciousness, as the Brown Marmorated Shieldbug (*Halyomorpha halys*) started to appear in British tabloid headlines. Long known as an orchard crop pest of apples, pears, peaches, apricots and cherries in North America (Hoebeke & Carter 2003), this East Asian shieldbug had occasionally been intercepted at British ports in imported goods, but live specimens are now turning up in pheromone traps around the country – Essex and central London in March, then Surrey in September (Fig. 115). Worse than its fruit damage, the bug also has a strong propensity to enter houses looking for hibernation sites, and there are worldwide reports of tens of thousands of them invading buildings (Schulz 2018). These congregations are hugely disconcerting to the occupants, not

FIG 115. A Brown Marmorated Shieldbug (*Halyomorpha halys*) that crawled out of a bunch of tulips in central London in May 2022. With the huge volume of horticultural and agricultural material shipped about the world it is no wonder that these things get about.

FIG 116. Superficially resembling *Halyomorpha halys*, the Mottled Shieldbug (*Rhaphigaster nebulosa*) is also a newly colonised species here, first recorded in Peckham by Penny Metal in 2011. Here a fourth- (centre) and a fifth-instar nymph (top) huddle with two adults on a lilac leaf.

just because of the small critters scuttling over walls and ceilings, but because their lingering smell becomes overpowering. If any of the bugs get into food it becomes tainted and inedible, and if they invade food manufacturing plants (including wineries) even small numbers can have significant financial impact as whole production lines have to be stopped and cleaned and the wasted food needs to be disposed of. Such accumulations of bugs are uncommon, but any invasion of the home is an unwelcome personal violation as far as the householder is concerned. The news coverage has been tempered by helpful press releases from the Natural History Museum and a full article in a scientific journal (Powell *et al.* 2021), but lurid stink-bug headlines are probably assured. Hemipterists now wait, on tenterhooks, to see whether there is a surge of sightings of this (and other shieldbug) species, and whether the Brown Marmorated Shieldbug's infamy will rub off onto a group not normally considered insect baddies.

SHIELDBUGS IN CULTURE

It's perhaps not surprising that since shieldbugs have always been rather obscure insects, they feature less often in art, literature and culture than other better-known beasts like butterflies, beetles and bees. I sport a couple of enamel lapel badges in the form of shieldbugs (Fig. 117), but these are very niche jewellery items. I am not following an elegant tradition. This would be different if I were writing a book about sacred scarabs or cicadas – insects that have been well represented in personal ornamentation for millennia. Not shieldbugs though.

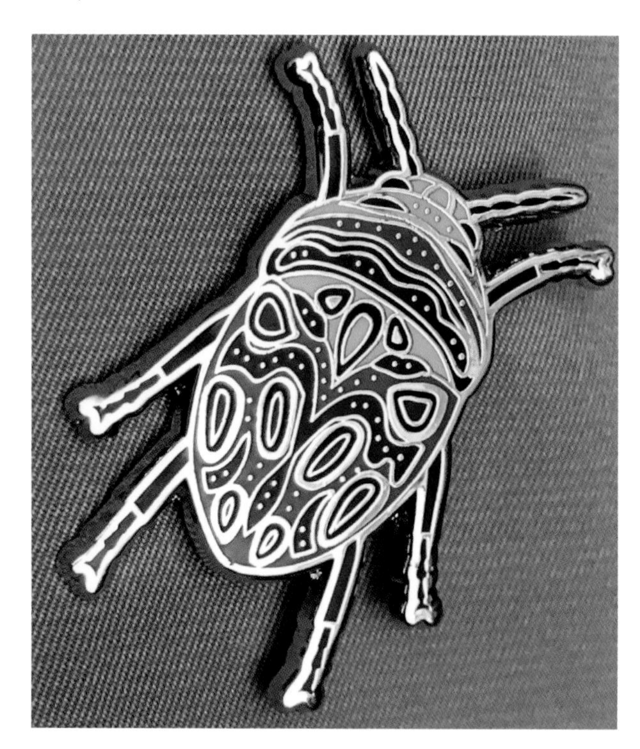

FIG 117. Shieldbug enamel badge, based on the brightly coloured African species *Sphaerocoris annulus*, as worn on the lapel of yours truly.

Likewise, shieldbugs make no appearance in the giant statuary from Egyptian, Sumerian, Greek or Babylonian archaeological digs. There are no Roman mosaic floors depicting them, nor cuneiform impressed clay tablets counting them. As mentioned earlier, shieldbugs do not feature in the Bible, Aesop's fables, or many illuminated manuscripts. In fact Neolithic cave paintings, the Parthenon sculptures, the Bayeux Tapestry, the da Vinci drawings, and the entire canons of the Pre-Raphaelites, Impressionists, Post-Impressionists, Cubists, Dadaists and Surrealists are all remarkably free of shieldbug depictions. I'm beginning to get a bit desperate trying to follow this thread. Luckily, off-beat Jan van Kessel the Elder (1626–79) stepped into the void. Kessel is noted for his careful portrayal of mixed insect and invertebrate subjects, usually painted on copper sheets used to decorate the cabinets of curiosity which were all the rage in the seventeenth century. The insects he depicts recur regularly and it seems probable that he was drawing these from life, or at least from dead specimens preserved in his own cabinet. Along with butterflies, beetles, hornets and seashells there are several shieldbugs, including a green species which may be *Palomena prasina*, something brown that might be *Arma custos*, and singles or mating pairs of *Pyrrhocoris apterus* (Fig. 118).

FIG 118. Two quirky selections of insects by Jan van Kessel, showing (top) possible *Arma custos* and (bottom) unmistakable *Pyrrhocoris apterus*.

He was more or less on his own though. There are a few sightings of shieldbugs in similar works by Jan Augustin van der Goes (Fig. 119) and Balthasar van der Ast, but works by the usual still-life painters such as Rachel Ruysch, Jan Brueghel the Elder, Jacob van Es, Ambrosius Bosschaert, and George Flegel (often awash with pretty insects decorating the vases of flowers) are disappointingly shieldbug-free. Japanese art has a long tradition of including insects, and Mori Shunkei's picture album (1820) has at least one identifiable pentatomid and one coreid amongst the usual fare of grasshoppers, beetles, spiders and butterflies (Fig. 120). A quick Google search for 'shieldbug art' throws up a mixture of modern photographs and paintings of a few brightly coloured species. However, a similar search for 'beetle art' has a surprising number of shieldbugs in there too. But you could haunt the halls of museums and art galleries the world over for centuries and not see a single genuine shieldbug representation on display.

Maria Sibylla Merian includes a lone shieldbug (possibly the Neotropical *Pachylis pharaonis*) in her revolutionary *Metamorphosis insectorum surninamensium* (1705). Like other insects in this beautiful book it is drawn in a lifelike pose, with wings outstretched in flight, but its legs look stiff, as if it is a set museum specimen, and the antennae are missing, suggesting it was drawn later from a collected individual (Fig. 121).

There are even fewer shieldbug references in literature or song (Fig. 122). Reinhardt (2018) has done a good job finding references to bedbugs in novels, plays, musicals, songs, poems and comics, but shieldbugs are too obscure to build any useful tropes upon, and given that they are constantly muddled with beetles many possible bug references probably go misnamed. I remember being slightly affronted when, as a teenager, I first read Edgar Allan Poe's *Gold-Bug* (1843) only to discover that the beast in question was obviously a scarab beetle, and not a bug. We have *Lord of the Flies*, but no *Lord of the Shieldbugs. The Love Bug* (produced by Disney Studios in 1968) was, of course, also a beetle, this time an anthropomorphic Volkswagen Beetle called Herbie. And although the studio's much later *A Bug's Life* (1998) had plenty of different insects in it, including ants, several beetles, cockroaches and grasshoppers, there are still no fully identifiable shieldbugs. Believe me, I've looked.

Shieldbugs remain very bug-like, altogether 'other', alien and strange (Fig. 123). They turn up occasionally, in strange places, often because they are brightly coloured, but not always immediately recognisable (Fig. 124). Several seemingly anatomically correct shieldbug plush toys are available from online shops – I identified *Palomena prasina* and *Acanthosoma haemorrhoidale* as well as several other slightly more imaginative fabric colourways. I was delighted when a series of dress

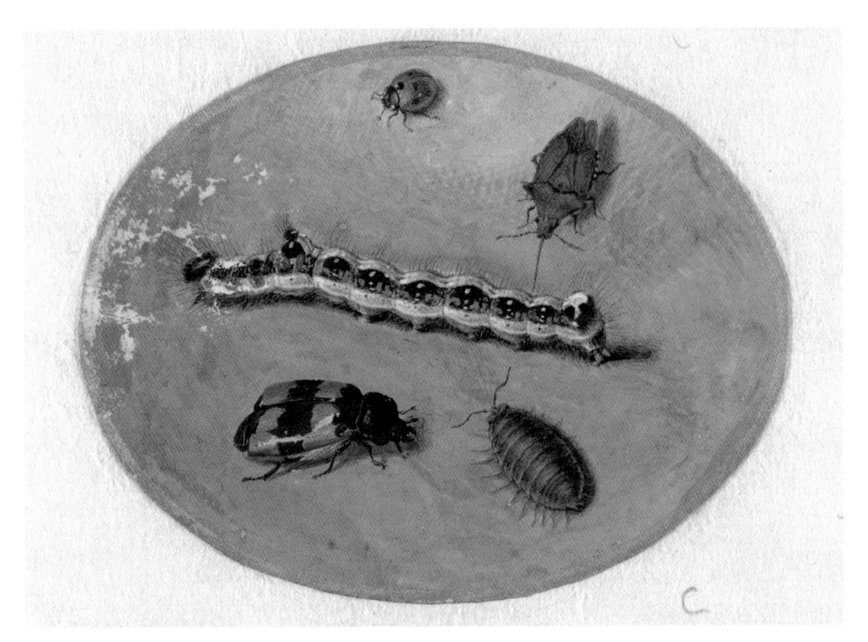

FIG 119. Lieveheersbeestje, rups, pissebed en twee torretjes (ladybird, caterpillar, woodlouse and two beetles), by Jan Augustin van der Goes, painted 1690–1700. Anatomical accuracy and scientific understanding varies. The moth caterpillar (Dark Dagger, *Acronicta tridens*), Seven-spot Ladybird (*Coccinella septempunctata*) and burying beetle are all good, but the bed-wetting woodlouse (they smell of ammonia/urea) has several more than its requisite 14 legs. The second 'beetle' is, of course, a shieldbug, and is excellent, seemingly *Arma custos*. The artist perhaps deliberately depicts its predatory behaviour, showing its rostrum extended towards its caterpillar prey.

FIG 120. A pentatomid shieldbug counterpoised by two bush-crickets by Mori Shunkei.

FIG 121. A South American shieldbug (top left), possibly *Pachylis pharaonis*, illustrated by Maria Sibylla Merian, 1705.

fabrics popped up on my Twitter feed back in the lockdown of 2020 designed by Peggy Wolven (the Vexed Muddler), based on various named exotic shieldbugs including *Poecilocoris donovani*, *Sphaerocoris annulus* (the Picasso Bug) and *Calidea dregii* (often called Rainbow Shieldbug). It's a shame she does not offer waistcoats in similar fabrics.

Many of these, and other brightly coloured pentatomids, also feature on postage stamps, miniature works of art in themselves, with the various species often chosen for their striking and design-like patterns (Fig. 125). As well as having functional purpose for mailing letters, these bright pictorial stamps also appeal to the international collectors' market. The Vietnam Post Office seems particularly enamoured of shieldbugs, with at least 10

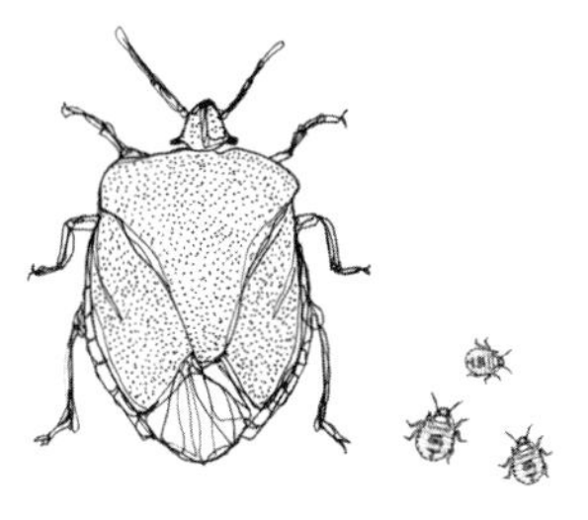

Shield bug

A shield bug as green as a leaf
Found a frond and laid eggs beneaf.
The nymphs came out waddling,
Bright colours a-modelling.
Protection 'gainst predator teef.

FIG 122. There were no shieldbug poems available, so I had to write one myself. From Jones & Ure-Jones (2021). Illustration of adult and nymphs by Calvin Ure-Jones.

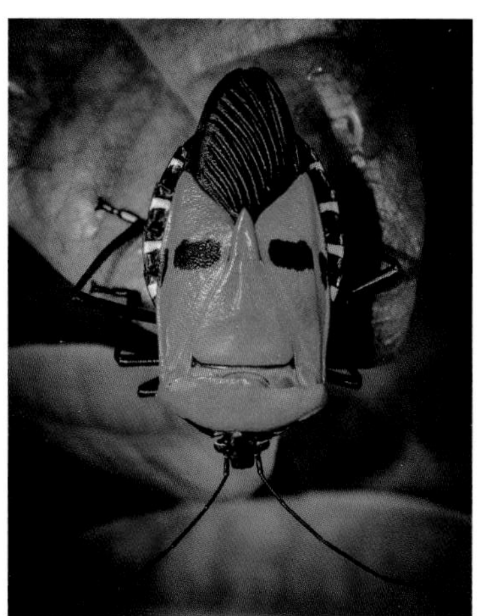

LEFT: **FIG 123.** The South-East Asian *Catacanthus incarnatus*, the well-named Man-faced Bug, with its smug smile, deep-set eyes and high teddy-boy quiff, is the very incarnation of a human face and is frequently the subject of internet memes.

RIGHT: **FIG 124.** Almost quite literally a thumbnail image of *Pyrrhocoris apterus*.

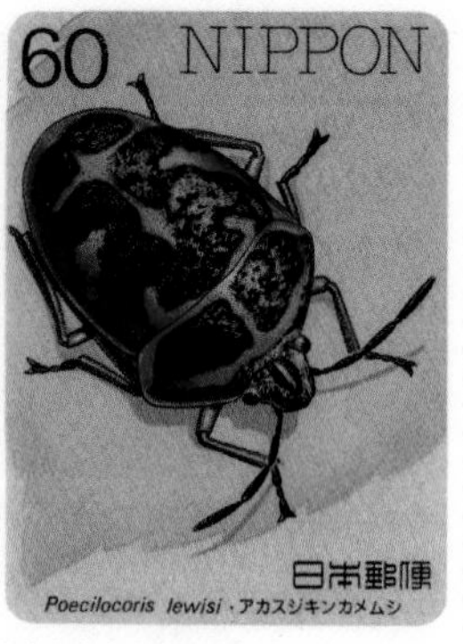

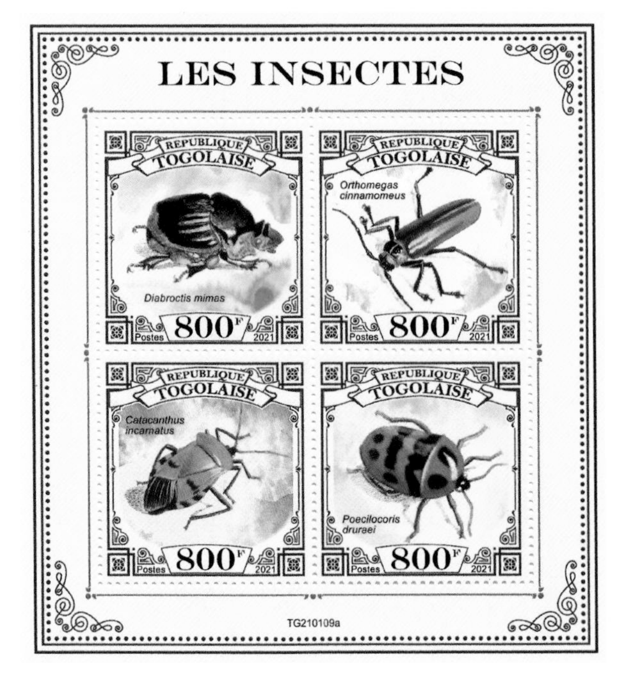

FIG 125. Brightly coloured shieldbugs on stamps from Japan (above) and Togo (below). Incidentally, the Japanese choice, *Poecilocoris lewisi*, was described new to science by south London hemipterist William Lucas Distant in 1883, and is known from much of the Far East. Oddly, none of the Togolese insects appear to be native to this African nation. The bugs are South and East Asian, the beetles South American.

species figured (along with other insects) on two sets issued during the 1960s and 1980s. Shieldbugs also appear on stamps from Argentina and Australia to Togo and Tuvalu. Perhaps my favourite is the 2017 Costa Rican stamp showing *Edessa rufomarginata* – very like the shieldbug that stained my hand that time. Sadly no shieldbugs have ever appeared on British stamps, but several familiar species like *Palomena prasina, Nezara viridula* and *Rhaphigaster nebulosa* were considered suitable for Niger in 2020 (probably because of their pest statuses), and the pretty and striking *Graphosoma italicum* has appeared on stamps from Hungary (1980) and Switzerland (2002). In a similar marketing strategy, the Bailiwick of Guernsey issued a special set of 10p coins in 2021 featuring insect wildlife of the island, and one of these was the Hawthorn Shieldbug, *Acanthosoma haemorrhoidale*. All of these items are available on eBay and other selling websites, often costing much more than their nominal issue value.

Elsewhere, shieldbugs still mostly fall below the general public's gaze. There was the possibility of a brief spell in the limelight in 2015, when the Royal Society of Biology instigated a poll to find the nation's favourite insect. Sadly the Green Shieldbug (*Palomena prasina*) failed even to make a podium place, as the Buff-tailed Bumblebee (*Bombus terrestris*) took the title. I thought I was onto something when I came across a link to a Pokemon shield bug, but this was just a reference to a computing glitch that affected the Pokemon Sword and Shield online game. Headlines about the dread plague of the Brown Marmorated Shieldbug (*Halyomorpha halys*) are occasionally offset by more upbeat news items like the rediscovery of the Cow-wheat Shieldbug (*Adomerus biguttatus*) in Scotland after a gap of over 30 years, or the naming of a new shieldbug from Patagonia – *Planois smaug* – after the dragon in Tolkein's *The Hobbit*, because it was found 'sleeping' in a museum collection for 60 years (Faundez & Carvajal 2016).

Despite my professed disdain for common names, perhaps the present trend of offering English names for shieldbugs will help propel them more into the limelight. I am spurred on by the similar use of common Danish names in the recent guide by Nielsen & Skipper (2015), where can be found such delights (if Google Translate is to be believed) as Spætat tornben (spotted thornbones, *Tritomegas bicolor*), Almindelig bispetæge (ordinary bishop's egg, *Aelia acuminata*), Slank enebærtæge (slim juniper cake, *Cyphostethus tristriatus*) and Bølleløvtæge (bully leaf beetle, *Elasmucha ferrugata*). Many brightly coloured Coleoptera are titled soldier beetles because of their bright dress uniform colours and patterns, but there are no soldier shieldbugs, though I am reliably informed that both nymphs and adults of *Pyrrhocoris apterus* are called gendarmes in France. And it came as a real heart-warming moment when I recently discovered that in Croatia *Rhaphigaster nebulosa*, a common species in southern Europe, is called Smrdljivi Martin, 'stinky Martin', for its regular appearance around 11 November – St Martin's Day. This is a name I feel I can get behind. At the same time *Dolycoris baccarum* is Smrdljivi Greta, 'stinky Greta', perhaps because its pinkish colours made it seem a female counterpart to the mottled brown, and therefore masculine, Martin.

COLLECTIVE NOUNS FOR SHIELDBUGS

Though shieldbugs mostly live solitary independent lives and are often found singly, there are several that regularly turn up in large numbers. Many animals that were hunted for food or sport acquired special terms for these accumulations – a flock of pigeons, a brace (male and female pair) of partridge, a herd of deer, a pack of wolves, a skulk of foxes. Perhaps it's time a few bugs got these group names too (Table 2). Yes, I know this is all flippant nonsense; please do not write in to complain.

TABLE 2. Suggested collective nouns for shieldbugs.

Scientific name	***English name***	***Collective noun***	***Notes***
Elasmucha grisea	Parent Bug	A family	Pretty obvious joke, even if it is a single parent. Although nymphs are frequently found in a tight huddle, the adult female is also present at the beginning of the family.
Coreus marginatus	Dock Bug	A coven	*Coreus* is a wonderfully gothic-looking creature, especially the nymphs, which resemble dead leaves and often cluster together on foodplant dock, but also bramble, in the sunshine.
Leptoglossus occidentalis	Western Conifer Seed Bug	An alarm	Slightly unnerving for householders when this large and active bug gathers in corners indoors to shelter and hibernate. It rattles ominously but harmlessly when it flies.
Tritomegas bicolor	Pied Shieldbug	A domino	Obvious for such a glossy black and white bug, though not often found in large groups. *T. sexmaculatus* is often in greater numbers on its foodplant Black Horehound.
Aelia acuminata	Bishop's Mitre	A synod	Frequently swept in very large numbers from rough grassland, though it used to be considered quite a scarce species of chalk hillsides.
Dolycoris baccarum	Hairy Shieldbug	A fluff	One of the commonest shieldbugs, and groups of the late-instar nymphs are particularly fluffy. The long hairs are easily visible with a hand lens, and even by naked eye they look soft.

FIG 126. A coven of *Coreus marginatus*, with a single *Palomena prasina* trying to join in the fun.

Scientific name	*English name*	*Collective noun*	*Notes*
Eurydema oleracea	Brassica Bug	A variety	Groups often contain several different colourways, with white, yellow, orange or red marks. They always remind me of Liquorice Allsorts, but whatever you do, do not eat.
Eysarcoris venustissimus	Woundwort Shieldbug	A confusion	Often found in mating aggregations on the foodplant in May, and conjoined pairs, with 12 legs between them, can confuse novices, who sometimes take them to be single animals.
Graphosoma italicum	Striped Shieldbug	A warning	Definitely warningly coloured, a danger signal amplified by groups of many of them huddling together on flowers brazenly in the bright sunshine. Please do not eat.
Nezara viridula	Southern Green Shieldbug	A psychedelia	A frequent 'pest' species, often found feeding on broad beans in large numbers, the bright nymphs being especially prominent with their wild pink and white markings.
Palomena prasina	Green Shieldbug	A subtlety	By comparison to *Nezara*, this species is modest and demure in its colour scheme and pattern. Its gregarious nymphs are spotted like ladybirds, so maybe they should be 'a punctuation'.

continued overleaf

TABLE 2. *continued*

Scientific name	*English name*	*Collective noun*	*Notes*
Halyomorpha halys	Brown Marmorated Shieldbug	A panic A fragrance A reek	Judging from the upset this insect is causing in North America, Europe, and now in the British Isles, large congregations are apt to be discovered and cursed with far stronger words than these.
Podops inuncta	Turtle Shieldbug	A pod	Thank you Max Barclay for this suggestion. Secretive and ground-dwelling, but often found in numbers under wooden planks or squares of roofing felt left to monitor reptiles.
Pyrrhocoris apterus	Firebug	An inferno	Often gathers in large and prominent groups of adults and nymphs at the base of lime (linden) planted as street trees. Scarce in Britain but noticeable when it occurs.
Corizus hyoscyami	Cinnamon Bug	A spice	Uncertain how it came by its English name, and though once rare in Britain this striking black and red bug is now often found in large numbers in rough grassland.
Stictopleurus punctatonervosus		An invasion	Often found in large numbers on mugwort, particularly on urban brownfield sites. A recent arrival into Britain, with its invasion starting about 1997.

CHAPTER 7

Key to British Shieldbug Species

The inclusion of this long dichotomous key to species might seem at odds with the notion that shieldbugs are relatively easy to identify. Many shieldbugs can, indeed, be identified from pictures, and in the next chapter you will see that I use the word 'unmistakable' all too often. However, many need careful examination under a lens or microscope to distinguish the different members of difficult species groups. For the novice entomologist, even skimming through pages and pages of pretty pictures can seem quite daunting, as they try and get a notion of what are the distinctive features of the various genera and species. I've put this key in the book not just for completeness, but also to give a tool to readers wishing to delve deeper into Heteroptera identification, taxonomy and classification.

Overleaf the following key is to the families and groups of British and Irish Heteroptera, in amongst which are the various species of shieldbugs. The key covers adults of all 79 British shieldbug species described in Chapter 8, plus the 11 Channel Islands species and the 16 potential colonisers. The page number after the species name is to the main entry in the catalogue of Chapter 8. This is an artificial key based on relatively easily observed characters, rather than reflecting evolutionary descent. There is, of necessity, some repetition. Lengths quoted are for most average specimens, though it is possible that some specimens at the extremes of the size ranges will not key out properly. Many of the line illustrations in this chapter have been redrawn, with permission, from various sources, including Derjanschi & Péricart (2005), Hawkins (2003) and Moulet (1995).

1 Antennae very small, hidden under the head, not visible from above .... Various families

Notonecta

Nepa

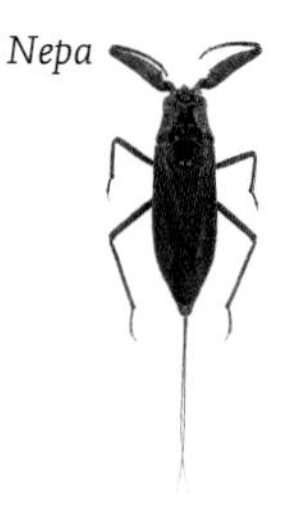

Aphelocheirus

Ilyocoris

Not shieldbugs. Several families of true water bugs, mostly fully aquatic but known to fly, including water boatman, water scorpion and water stick insect. The shield-shaped but non-shieldbug Saucer Bug (*Ilyocoris cimicoides*) keys out here.

— Antennae long and easily visible from above 2

2 Head many times longer than wide, longer than the pronotum, eyes some considerable distance from the front of the pronotum .... Hydrometridae

Not shieldbugs. Water walkers, delicate, long-legged, long-bodied, slow-moving predators on or near water.

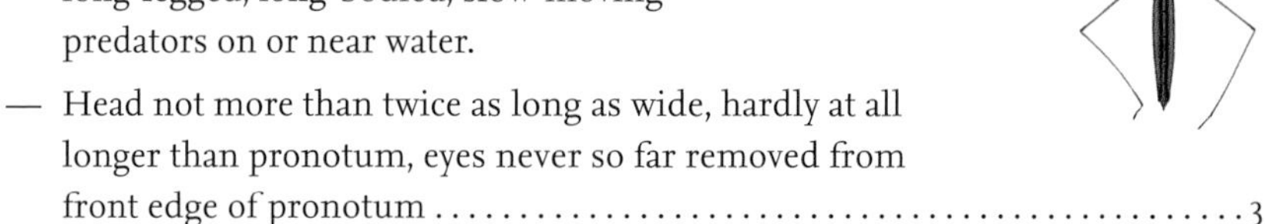

— Head not more than twice as long as wide, hardly at all longer than pronotum, eyes never so far removed from front edge of pronotum 3

3 Antennae with five good segments 4

— Antennae with only four good segments 8

Note: In some Coreidae, there is a small nodular or annular sub-segment before the final antennal segment – but treat this as a standard four-segmented antenna.

4 Tiny (2 mm), scutellum not reaching middle of abdomen, underside of body covered with dense silvery pubescence .... Family Hebridae

Not shieldbugs. Velvet water bugs – small waterside predators found near streams and bogs.

— Larger than 3 mm, scutellum large, reaching at least to the middle of the abdomen, and frequently completely covering it. Underside of body not densely silver-haired 5

5 Scutellum very large, usually oval, shaped like a fingernail, reaching nearly to the end of the abdomen, and almost entirely covering the membrane .... **Scutelleridae** (and others) 22

Podops *Eurygaster* *Graphosoma* *Thyreocoris*

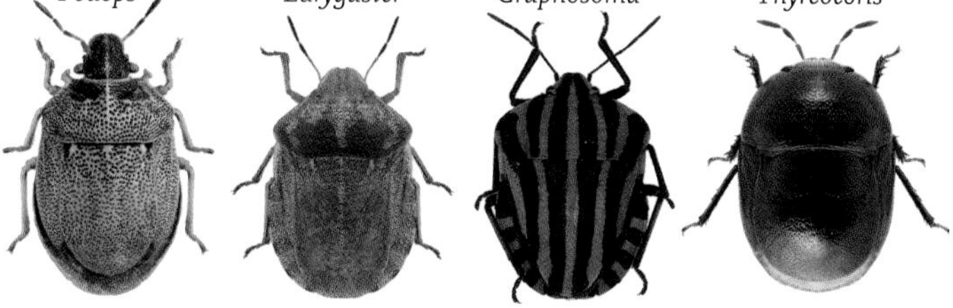

— Scutellum smaller, more or less triangular, reaching at most three-quarters the length of the abdomen, and leaving the membrane clearly visible 6

6 Tarsi two-segmented. Thorax with a longitudinal keel underneath, easily visible in side view .... **Acanthosomatidae** 31

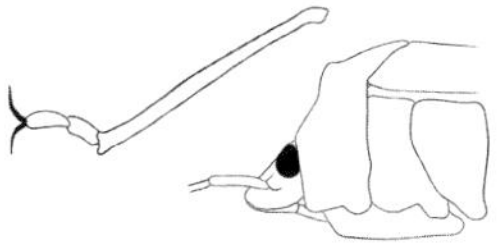

— Tarsi three-segmented. Thorax without a prominent keel underneath 7

7 Tibiae with several rows of strong spines ..**Cydnidae** 36

— Tibiae without rows of strong spines . **Pentatomidae** 48

8 Underside of abdomen densely covered with fine silver water-repellent hairs .... Various families

Not shieldbugs. Several groups of slim, long-legged insects living on the surface of fresh water, including Gerridae water skaters and Veliidae water crickets.

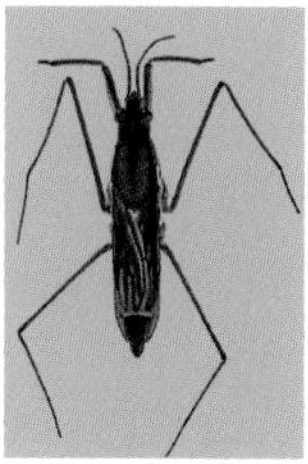

Velia

Gerris

— Underside of abdomen without a covering of fine silver hairs 9

9 Rostrum more or less curved when viewed from side, not pressed flat against the body when at rest .... Nabididae and Reduviidae

Not shieldbugs. Long-legged, predatory 'damsel' and assassin bugs.

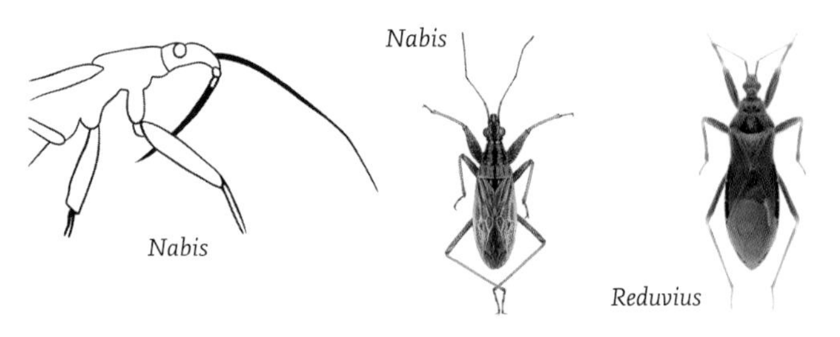

— Rostrum not curved in side view, held pressed flat against the body at rest . . 10

10 Upper body surface covered with a fine net-like pattern .... Piesmidae and Tingidae

Not shieldbugs. Lacebugs, tiny (less than 5 mm) but pretty, on plants.

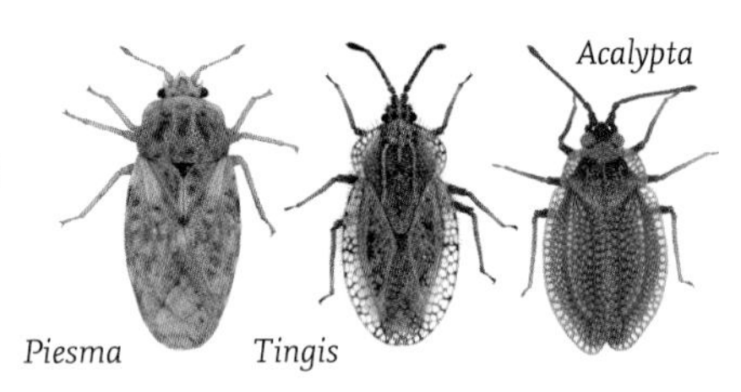

— Upper body surface not patterned with fine net-like sculpturing, but may be strongly and regularly punctured 11

11 Tarsi with two segments 12
— Tarsi with three segments 13

12 Very flat bugs, antennal segments short and robust.................... Aradidae

Not shieldbugs, but outlier family in the suborder Pentatomomorpha. Under the bark of dead and rotten tree trunks, stumps and logs.

— Not flat, antennal segments long and slim ....Microphysidae

Not shieldbugs. Minute (less than 2.5 mm) secretive bugs on lichen-covered trees.

13 Ocelli absent 14
— Ocelli present 15

14 More than 8 mm long, bright red and black, wings without a cuneus .... **Pyrrhocoridae** 91

A single distinctive species, *Pyrrhocoris apterus*.

— Less than 8 mm long, or if longer then not strongly marked black and red, wings with cuneus separated from corium by a distinct groove or fold .... Miridae

Not shieldbugs. Leaf bugs, highly varied, diverse and speciose group.

Pyrrhocoris

15 Rostrum with three segments. Small oval black bugs, sometimes patterned with white marks. Usually active, fast-running and quick to fly, mostly at edges of water .... Saldidae

Shore bugs. Not shieldbugs.

— Rostrum with four segments. Body otherwise shaped and coloured 16

Saldula

16 Less than 5 mm .... Various families

Not shieldbugs.

— At least 5.5 mm 17

17 Membrane of wings with five or fewer more or less parallel veins. If short-winged then abdomen coloured black or dark brown above.............................. Berytinidae and Lygaeidae

Not shieldbugs, but outlier families within the Pentatomomorpha. The Lygaeidae, especially, might be mistaken for some of the smaller more delicate shieldbug groups.

Rhyparochromus

Aphanus

Heterogaster

Gastrodes

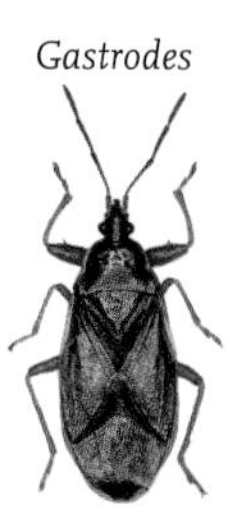

— Membrane of wings with six or more often branched veins. If short-winged then abdomen green or yellow-green with red or black markings ..18

18 Antennae, tibiae and middle and hind femora with strong dark and light bands ...... **Stenocephalidae** 92

Dicranocephalus

— Antennae, tibiae and middle and hind femora not all with strong dark and light bands 19

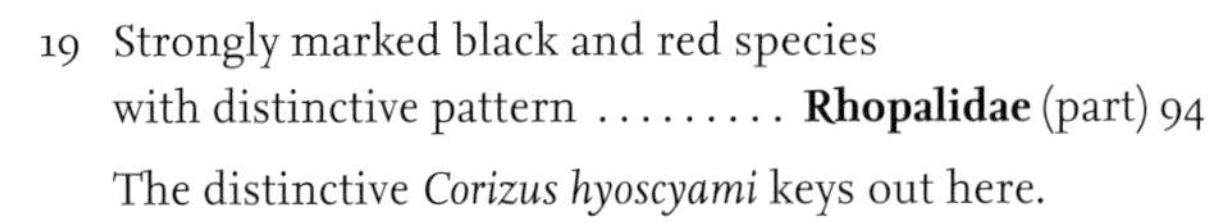

19 Strongly marked black and red species with distinctive pattern .... **Rhopalidae** (part) 94

The distinctive *Corizus hyoscyami* keys out here.

— Not strongly marked with black and red pattern ... 20

Corizus

20 Hemelytra with at least some basal areas transparent and unpunctured between the veins. Orifice of stink gland not visible in side view .... **Rhopalidae** (part) 94

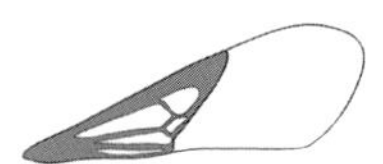

Rhopalus

— Hemelytra hardened and opaque; if basal areas between the veins appear pale and opaque then they are punctured. Orifice of stink gland visible in side view 21

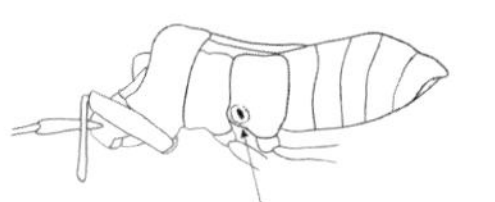

21 Head triangular, about as broad across eyes as pronotum ....**Alydidae** 107

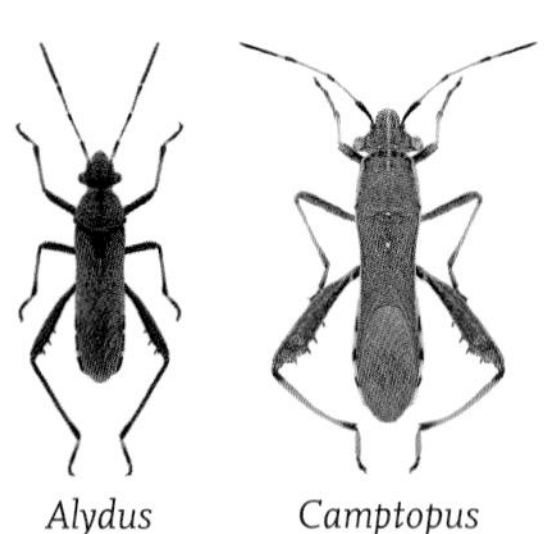

Alydus *Camptopus*

— Head more or less quadrangular, not as broad as pronotum .... **Coreidae** 109

Coreus

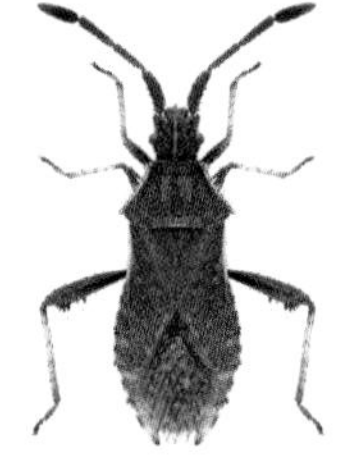

Coriomeris

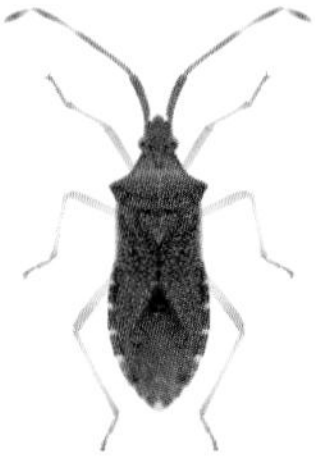

Gonocerus

Spathocera

Scutelleridae (and other families)

Several species of Pentatomidae will key out here by virtue of their unusually large scutellum, so they are included in keys to both families.

22 Mostly glossy black and shining, small, less than 4.5 mm 23
— Not glossy black. Larger than 4.5 mm 24

23 Very squat, domed, globose, broadest behind middle. Legs slim and without spines. Tarsi with two segments .... *Coptosoma scutellatum* (p. 312)

Family Plataspidae. Only recently discovered in Britain, at Hastings, Sussex, climbing on Dyer's Greenweed (*Genista tinctoria*).

Coptosoma

— Longer oval. Legs with rows of short but distinct spines on tibiae. Tarsi with 3 segments .... *Thyreocoris scarabaeoides* (p. 251)

Family Thyreocoridae. Widespread. Ground-dweller, at roots of plants, sometimes under logs and planks of wood.

Thyreocoris

24 Scutellum sinuous at apex, drawn out to an elegant but blunt point. Upper surface patterned with reddish-brown streaks against a pinkish-beige background. 8–11.5 mm .... *Odontotarsus purpureolineatus* (p. 321)

Not recorded from the British Isles, but widespread on mainland Europe.

Odontotarsus

— Scutellum broadly rounded at apex 25

25 Entire upper surface strongly marked with narrow black and red stripes ..... *Graphosoma italicum* (p. 309)

Family Pentatomidae. Recently discovered in Britain in Essex and Surrey. Usually on flowers of Wild Carrot (*Daucus carota*) and other umbellifers.

Graphosoma

— Coloured with mottled brown blotches, although dark streaks can sometimes occur 26

26 Each side of front of pronotum projecting into a small hook-, hammer-, or axe-shaped knob near the eye . . . .*Podops inuncta* (p. 311)

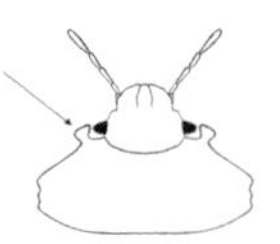

Family Pentatomidae. 5–6 mm. Brown. Slow-moving ground-dweller in grass roots, under logs, planks and boards.

— Pronotum without knobs beside eyes 27

27 Densely hairy, edges of forewings and connexivum not, or hardly, visible. Eyes vertical 28

Odontoscelis

— Not hairy, edges of forewings and connexivum not completely covered by the scutellum. Eyes not vertical 29

Eurygaster

28 Smaller, 4–5 mm. Upper surface with stripes of silvery hair contrasting with darker hairs . . . . *Odontoscelis lineola* (p. 320)

Ground-dwelling, sandy places, scarce.

Odontoscelis lineola

— Larger, 6–8 mm. Upper surface with lighter or darker brown hairs, but not appearing silvery . . . . *Odontoscelis fuliginosa* (p. 318)

Ground-dwelling, sandy places near the coast, rare.

Odontoscelis fuliginosa

29 Larger, 11–13 mm. Side lobes of the head touching each other in front of the central lobe. Scutellum with a more or less clearly defined linear ridge down the centre. Male aedeagus with 6–8 short curled or twisted spines ... *Eurygaster austriaca* (p. 317)

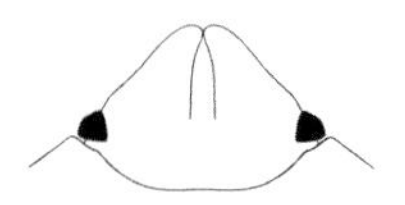

Old records only, Kent coast, probably extinct in Britain.

— Smaller, 8.5–11 mm. Side lobes of head not meeting in front of the central lobe. Scutellum may have narrow pale streak, but this is only faintly ridged. Male aedeagus with only two or four curled spines 30

30 Slightly larger, 9.0–10.5 mm. Pronotum slightly more protruding at sides. Central lobe of clypeus slightly lower than side lobes, giving the impression that it has been pushed down. Second antennal segment only slightly longer than third, at most 1.5 times as long. Female genital plates not touching the edge of the previous segment. Male aedeagus with four curved spines .... *Eurygaster testudinaria* (p. 314)

Eurygaster testudinaria

Widespread in rough grassland.

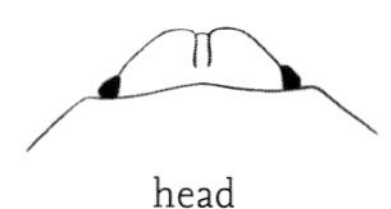

head

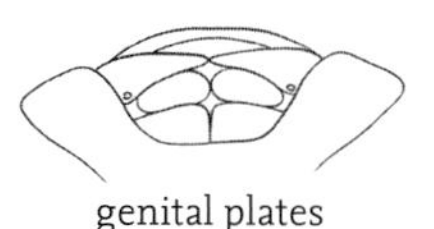

genital plates

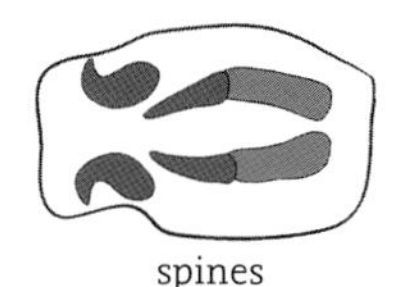

spines

— Slightly smaller, 8.5–9.7 mm. Pronotum slightly less protruding. Central lobe of clypeus and side lobes completely flat, not depressed in the middle. Second antennal segment nearly twice as long as third – at least 1.8 times as long. Female genital plates touching the edge of the previous segment. Male aedeagus with two curved spines .... *Eurygaster maura* (p. 316)

Eurygaster maura

Scarce in rough grassland.

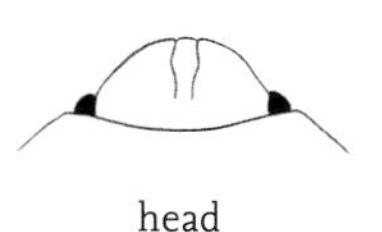

head

genital plates

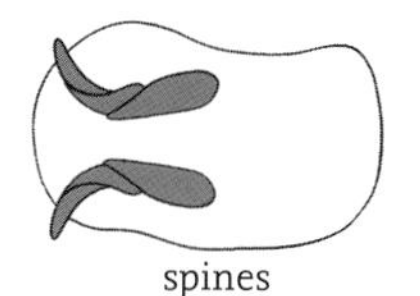

spines

Acanthosomatidae

31 First antennal segment barely reaching the end of the head, pronotum with black punctuation only near outer angles. Shining green with pinkish red boomerang-shaped marks on the forewings. 8.6–11.0 mm .... *Cyphostethus tristriatus* (p. 228)

On junipers, and cypress trees, widespread.

Cyphostethus tristriatus

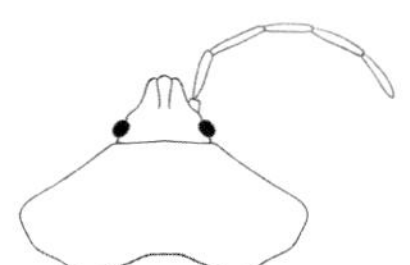

— First antennal segment reaching well beyond the end of the head, head and pronotum with black punctuation 32

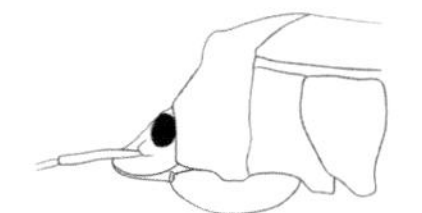

32 Larger, 12.5–16 mm. Keel on mesosternum (visible in side view) projecting forwards, to the rear edge of the head .... *Acanthosoma haemorrhoidale* (p. 226)

On hawthorn and other bushes, widespread.

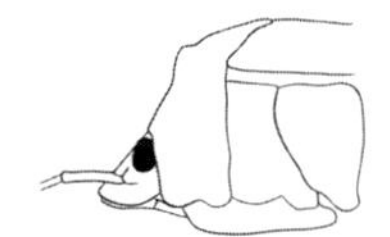

— Smaller, less than 12 mm. Keel on mesosternum not reaching so far forwards 33

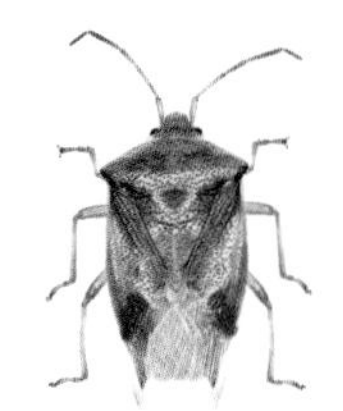

Elasmostethus interstinctus

33 Connexivum pale green. Insect green with brownish-red marks, darker during hibernation .... *Elasmostethus interstinctus* (p. 231)

On birch, widespread.

— Connexivum chequered with black marks. Insect mainly brownish 34

34 Corners of pronotum with long spines .... *Elasmucha ferrugata* (p. 233)

Very rare, uplands, probably extinct in Britain, sometimes possibly introduced vagrant.

— Corners of pronotum only slightly projecting 35

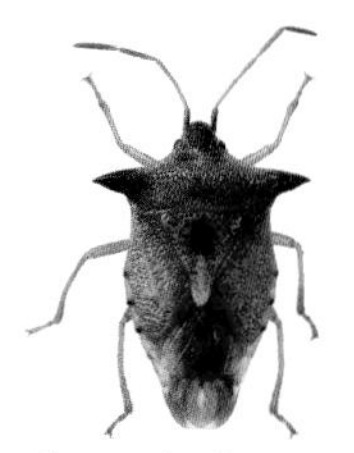

Elasmucha ferrugata

35 Underside of body not black-spotted. Front corners of pronotum with a tiny node behind the eye. Antennae variable, but usually lighter .... *Elasmucha grisea* (p. 234)

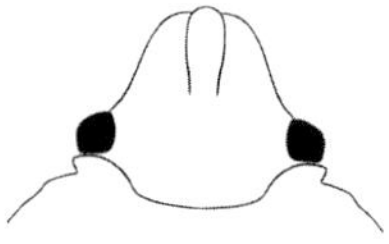

On birch, widespread, female often with batch of hatchlings.

— Underside of body black-spotted. Front corners of pronotum with small but distinct denticle just behind the eye. Antennae variable, but usually darker .... *Elasmucha fieberi* (p. 236)

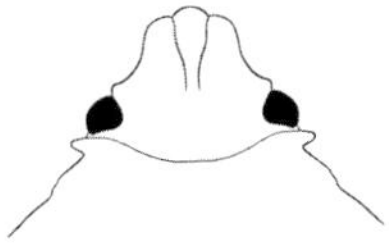

Scattered in Europe to Normandy.

Cydnidae

36 Scutellum nearly an equilateral triangle, short, leaving a large area of membrane covering about half the abdomen. Side lobes of clypeus meeting in front of central lobe. Large, 8–13 mm .... *Canthophorus aterrimus* (p. 242)

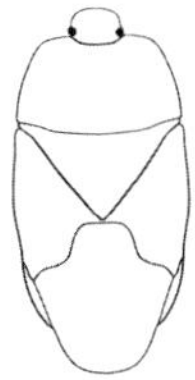

Channel Islands, but not yet mainland Britain, on spurges (Euphorbiaceae).

— Scutellum long, an isosceles triangle, leaving a shorter membrane covering about one-third of the abdomen. Side lobes of clypeus not completely enveloping the central lobe 37

37 Margins of head and pronotum with long hairs (easily abraded), and sometimes spines too 38

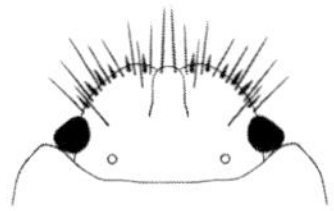

Aethus flavicornis 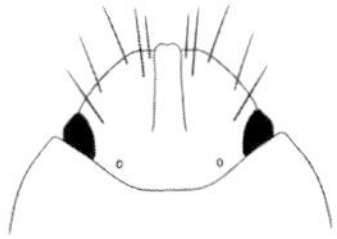 *Geotomus punctulatus*

— Margins of head and pronotum without long hairs or spines 40

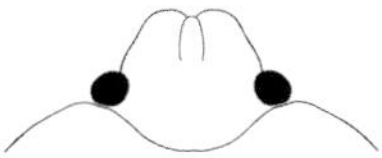

Sehirus

38 Margin of head with 12–14 minute black spines among the hairs. Body length 3.0–3.5 mm, dark brown or black, margins of insect with long brown hairs (see above) . . . *Aethus flavicornis* (p. 237)

Once recorded on Isle of Wight, vagrant or probably extinct in Britain.

— Margin of head without spines, only with hairs. Length more than 3.5 mm, black, sides of pronotum and hemelytra with a few long hairs 39

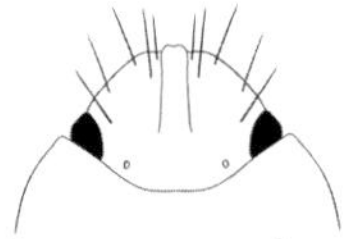

Geotomus punctulatus

39 Black with brownish tinge to pronotum and corium. Male slightly larger, more than 4.5 mm. Parameres of male genitalia pointed at tip ....*Geotomus punctulatus* (p. 238)

Rare, southern, coastal, sandy places, recently only Sennen Cove, Cornwall.

— More uniformly black. Male slightly smaller, less than 4.5 mm. Parameres of male genitalia flat at tip .... *Geotomus petiti* (p. 239)

Rare, recently discovered colonies at Dungeness, Kent.

Geotomus punctulatus *Geotomus petiti*

40 Eyes protruding, broader than long 41

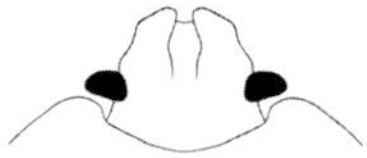

— Eyes not protruding, nearly round 42

41 Central lobe of clypeus shorter than side lobes, and clearly indented. White margin of corium reaching almost to the membrane .....*Legnotus limbosus* (p. 244)

Widespread, rough grassy places, on bedstraws and goosegrass (*Galium* species).

Legnotus limbosus

— Central lobe of clypeus about as long as side lobes and hardly indented. White margin of corium very short, not nearly reaching membrane ....*Legnotus picipes* (p. 245)

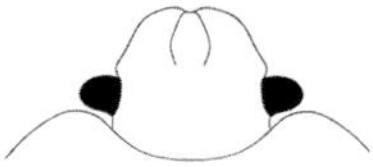

Scarce, in sandy places on bedstraws (*Galium* species).

Legnotus picipes

42 Small, less than 4.5 mm ....*Ochetostethus nanus* (p. 246)

Ground-dwelling, under plants, Channel Islands but not mainland Britain yet.

— Larger, more than 5.5 mm 43

43 Extensively marked with white, especially a large hook-shaped mark on corium and spot or streak at front angle of pronotum 44

Tritomegas bicolor

Tritomegas sexmaculatus

— Any white marks confined to a narrow rim on margins of pronotum and corium, and perhaps a small round spot on the middle of each wing 45

Adomerus biguttatus

44 White mark on front corner of pronotum broader, but not reaching much beyond halfway along the pronotal margin (see above) ....*Tritomegas bicolor* (p. 248)

Common on White Dead-nettle (*Lamium album*).

— White mark on front corner of pronotum longer and narrower, reaching three-quarters of the way to the hind angles (see above) .... *Tritomegas sexmaculatus* (p. 250)

Recently established in Kent, spreading, mostly on Black Horehound (*Ballota nigra*).

45 Upper surface almost entirely black 46

— Margins of corium with a narrow but distinct white streak. Sometimes also with a small white spot in the middle of the corium 47

Adomerus biguttatus

46 Larger, 8.5–11.5 mm. More intensely black, without any bronze reflections on the corium ....*Sehirus morio* (p. 247)

Status uncertain.

— Smaller, 5.5–8.3 mm. Corium with slight bronze reflections ....*Sehirus luctuosus* (p. 246)

Ground dwelling, usually on forget-me-nots (*Myosotis* species).

47 Black, hemelytra with a white spot in the middle of the corium .... *Adomerus biguttatus* (p. 241)

Scarce, on Cow-wheat (*Melampyrum pratense*) in open woodlands.

— Blue-black, corium without a white spot .... *Canthophorus impressus* (p. 243)

Scarce, chalk downland on Bastard Toadflax (*Thesium humifusum*).

Adomerus biguttatus

Pentatomidae

48 Scutellum very large, oval, shaped like a fingernail, reaching nearly to the end of the abdomen, and almost entirely covering the membrane (right)........49

— Scutellum smaller, more or less triangular, reaching at most three-quarters the length of the abdomen, and leaving the membrane clearly visible ..50

Graphosoma italicum

Podops inuncta

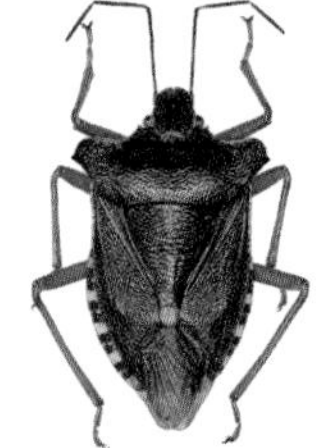

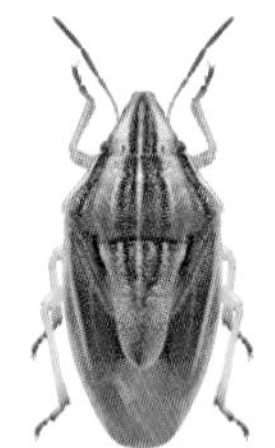

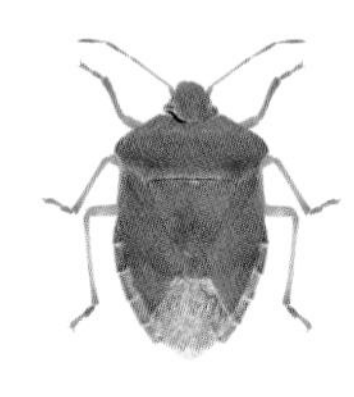

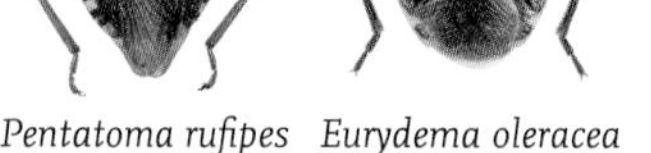

Pentatoma rufipes *Eurydema oleracea* *Aelia acuminata* *Palomena prasina* *Dolycoris baccarum*

49 Entire upper surface strongly marked with narrow black and red stripes or streaks. Front edge of pronotum without a small projection near the eye (see above).............. *Graphosoma italicum* (p. 309)

Recently discovered in Britain in Essex and Surrey. Usually on flowers of Wild Carrot (*Daucus carota*) and other umbellifers.

— Entire upper surface brown. Each side of the front of pronotum projecting into a small hook-, hammer-, or axe-shaped knob near the eye (see above).............................. *Podops inuncta* (p. 311)

5–6 mm. Brown. Slow-moving ground-dweller in grass roots, under logs, planks and boards.

50 Rostrum robust, first segment about as thick as front femur and not resting into the ridged channel underneath the head 51

— Rostrum slim, resting in a ridged channel underneath the head when at rest 57

51 Corners of pronotum not projecting, outline more or less smoothly rounded into hemelytra. Margins of pronotum smooth (right) 52

— Corners of pronotum projecting beyond base of hemelytra. Margins of pronotum with small denticles (below) 53

Zicrona caerulea *Jalla dumosa*

Picromerus bidens *Troilus luridus* *Rhacognathus punctatus* *Arma custos*

52 Metallic inky blue-black. Front femur without small tooth. Smaller, 5.5–8.0 mm (see above)............................ *Zicrona caerulea* (p. 263)

Widespread, damp grassy places, woodland edges.

— Dark brown, but with cream markings on head, pronotum and scutellum. Legs ringed with cream on tibiae. Front femur with a small tooth. Larger, 11.0–16.0 mm (see above).................. *Jalla dumosa* (p. 256)

Old records from Kent, probably extinct in Britain. Widespread in Europe.

53 Front femur with a small but distinct tooth underneath 54

— Front femur unarmed 55

54 Corners of pronotum drawn out into a sharp point. Connexivum broader and rounder, giving the rear body a more inflated shape . . . *Picromerus bidens* (p. 257)

Widespread, woods and hedgerows.

Picromerus bidens

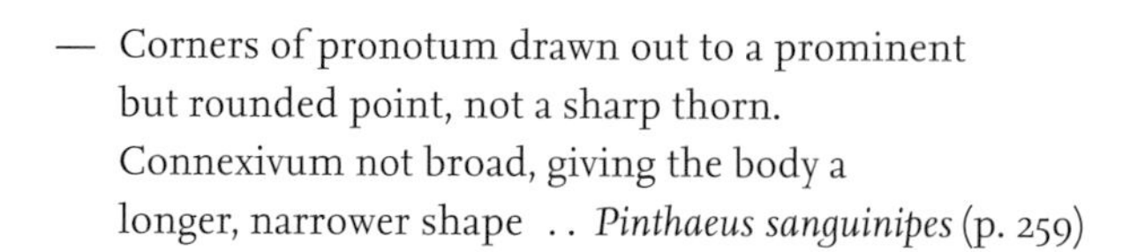

— Corners of pronotum drawn out to a prominent but rounded point, not a sharp thorn. Connexivum not broad, giving the body a longer, narrower shape . . *Pinthaeus sanguinipes* (p. 259)

A single specimen from Suffolk in 2021, widespread in Europe.

Pinthaeus sanguinipes

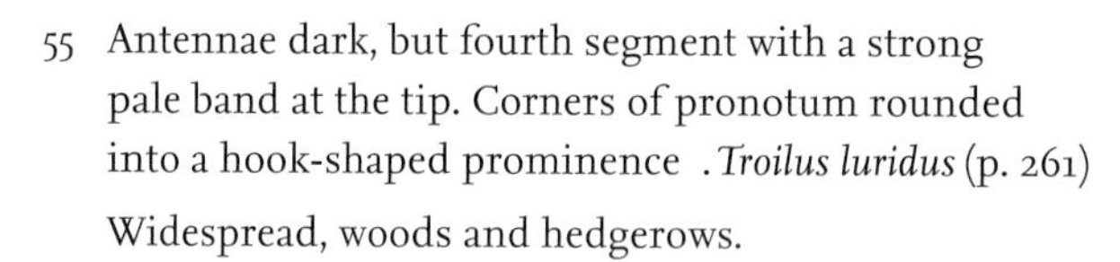

55 Antennae dark, but fourth segment with a strong pale band at the tip. Corners of pronotum rounded into a hook-shaped prominence . *Troilus luridus* (p. 261)

Widespread, woods and hedgerows.

— Antennae all dark, or mostly pale, but without contrasting pale band on segment four. Corners of pronotum not hook-shaped, nearly rectangular . .56

Troilus luridus

56 Antennae nearly entirely dark. Tibiae dark but ringed with strong pale band. Smaller, 7.5–9.5 mm . . . . *Rhacognathus punctatus* (p. 260)

Dry heaths and boggy places where heather grows.

Rhacognathus punctatus

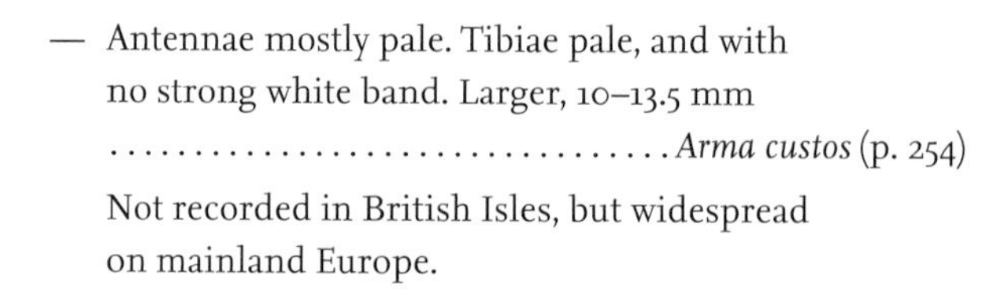

— Antennae mostly pale. Tibiae pale, and with no strong white band. Larger, 10–13.5 mm . . . . *Arma custos* (p. 254)

Not recorded in British Isles, but widespread on mainland Europe.

Arma custos

57 Sides of pronotum drawn out into a flattened flange. Connexivum broadly expanded. Head strongly rounded in front. Rather flattened insects 58

Dyroderes umbraculatus

Menaccarus arenicola

Sciocoris cursitans

— Sides of pronotum not produced into a flange. Connexivum narrower. Head less broadly rounded. Insects less flattened 62

Pentatoma rufipes *Aelia acuminata* *Piezodorus lituratus* *Eysarcoris aeneus* *Peribalus strictus*

58 Front corners of pronotum very broadly rounded, and with a large contrasting pale spot. Other contrasting pale marks on connexivum, tip of scutellum and front corner of hemelytra (see above) ....*Dyroderes umbraculatus* (p. 297)

On bedstraws (*Galium*), known from Channel Islands since 1960s, discovered in London, Hampshire and Kent since 2013.

— Corners of pronotum not broadly rounded and lacking any contrasting pale blotches 59

59 Head extremely broad and widely rounded, with short hairs along margin (see above) ....*Menaccarus arenicola* (p. 299)

On sand dunes, amongst Marram Grass (*Ammophila arenaria*) and other grasses. Known on Channel Islands and mainland Europe.

— Head not so broad, lacking hairs along its margins (see *Sciocoris* above) 60

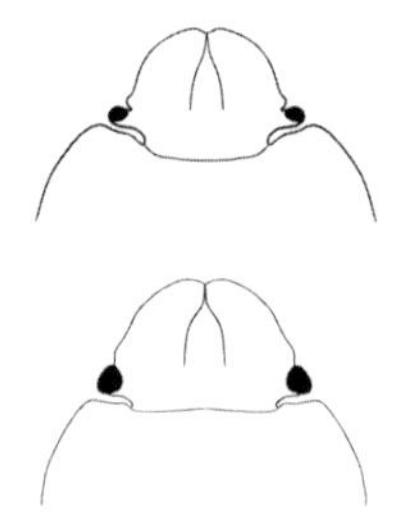

60 Eyes projecting, appearing stalked, with a small notch on the margin of the head in front of each eye. Large, 5.9–8.5 mm ....*Sciocoris homalonotus* (p. 301)

Only recently found in Britain (Kent and Surrey), since 2016.

— Eyes less prominent, head not or less notched in front of each eye. Smaller, 4.5–6 mm 61

61 Pronotum with obviously pale margins. Underside of abdomen with a central dark stripe .... *Sciocoris sideritidis* (p. 302)

Southern European species. A single colony in Essex, maybe an introduction.

— Pronotum without strongly contrasting pale margins. Underside of abdomen with central area pale, but bordered each side by a dark streak ....*Sciocoris cursitans* (p. 300)

Scarce, southern England, strongly ground-dwelling.

62 Tip of head rather drawn out to a point, but may appear blunt from above in some species because it is curved down at the tip. Check front and side view 63

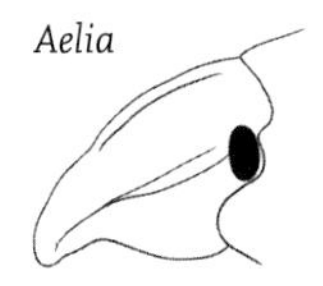
Aelia

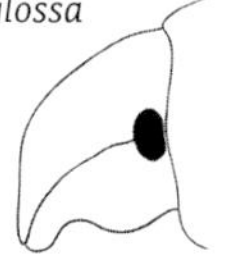
Neottiglossa

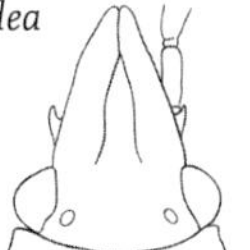
Mecidea

— Head not drawn out to a point, more rounded, blunt or square at the front, but never curved down when viewed from side 68

63 Very narrow, parallel-sided and rather flat insect. More or less uniformly straw-coloured, punctures not black. Legs and antennae rather long .... *Mecidea lindbergi* (p. 253)

Rare vagrant. A Mediterranean species; several recorded in moth traps, coast of southern England, December 2015.

Mecidea

— More convex, less narrow or parallel-sided. Brownish, mottled with clouds of black punctures and/or streaked with long dark blotches. Legs and antennae relatively shorter 64

Neottiglossa leporina

Aelia acuminata

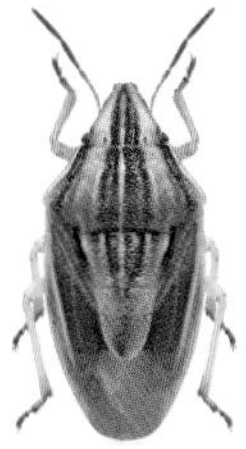

64 Viewed directly from above the head appears flattened or rounded in front because it is strongly down-turned. Tip of head about level with middle of second antennal segment (see above) 65

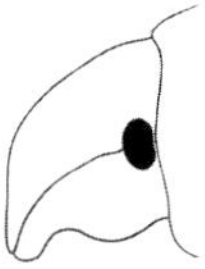

Neottiglossa

— Viewed from above the head is drawn out and conical, much less strongly down-turned, the tip about level with the middle of the third antennal segment (see above)............................ 66

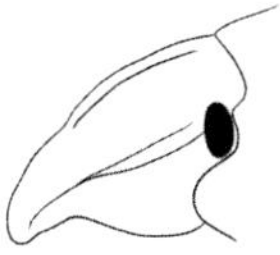

Aelia

65 Scutellum longer, clearly longer than corium. Underside of abdomen more broadly pale under the connexivum .... *Neottiglossa leporina* (p. 269)

Channel Islands, and widespread in France.

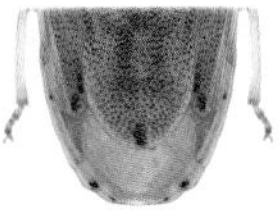

Neottiglossa leporina

— Scutellum about as long as corium. Underside of abdomen only narrowly pale under connexivum .... *Neottiglossa pusilla* (p. 268)

Widespread in grassy places.

Neottiglossa pusilla

66 Underside of middle and hind femora with paired dark spots .... *Aelia acuminata* (p. 264)

Widespread, rough grassy places.

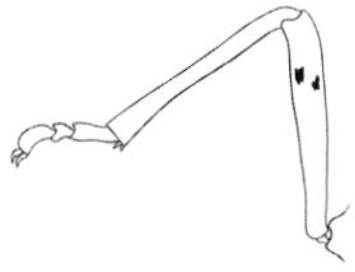

— Underside of middle and hind femora lacking paired dark spots, either with one small mark or clear 67

67 Bucculae with a sharp tooth beneath, visible in side view. Larger, 9.5–12.0 mm .... *Aelia rostrata* (p. 267)

A European mainland species.

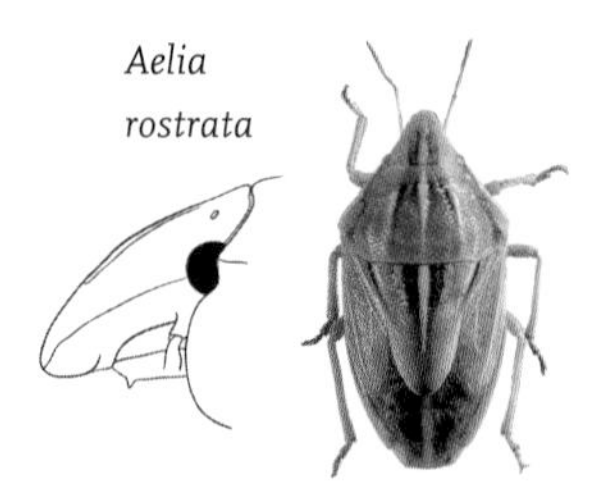

— Bucculae with at most a blunt prominence beneath, visible in side view. Smaller, 6.0–8.5 mm .... *Aelia klugii* (p. 266)

A European mainland species.

68 Strongly coloured black (sometimes with metallic blue tinges), strikingly and contrastingly patterned with red, orange, yellow or white 69

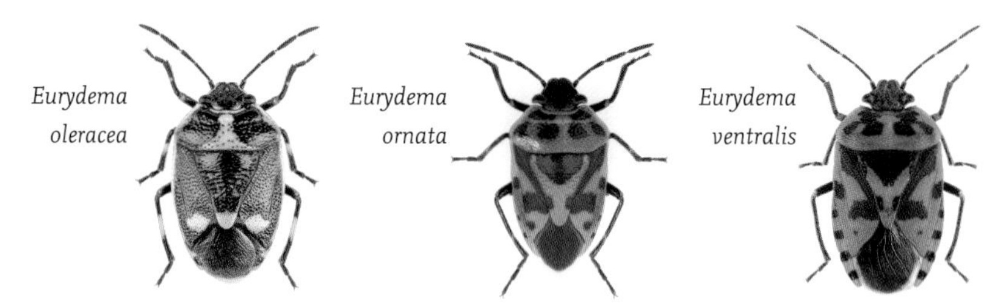

— Coloration less strikingly or contrastingly patterned, usually green, brown, grey, often mottled and without a strong pattern of bright white, red or yellow marks 73

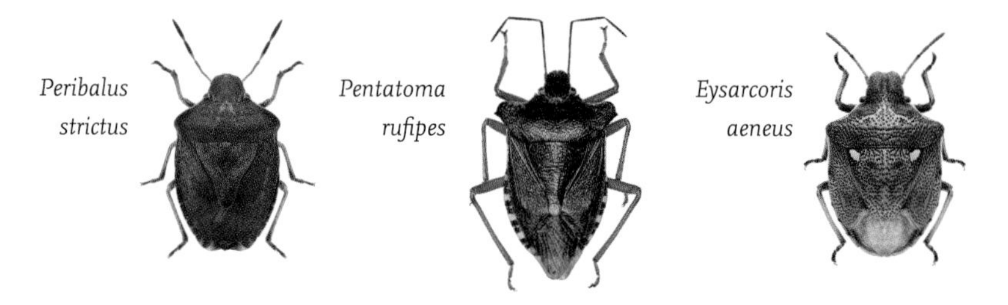

69 Apart from narrow marginal streaks along edges of pronotum and hemelytra, pattern more or less limited to central stripe down middle of pronotum and a bar of three marks made up from oblong mark near the tip of each corium and a circular spot on the tip of the scutellum. Occasionally scutellum has basal side streaks. Overall a

black or slightly metallic insect with a few pale cream, yellow or red blotches. Underside of abdominal segments pale or reddish .... *Eurydema oleracea* (p. 303)

Common and widespread on wild and cultivated brassicas, and other wild crucifers.

Eurydema oleracea

— Pattern more extensive. Overall a red, orange or yellow insect with black marks (see opposite). If upper surface is mainly black, then underside of abdominal segments is also black 70

70 Edge of corium entirely red .... *Eurydema dominulus* (p. 305)

Rare, woodland rides, now only Kent and Sussex. Widespread in Europe.

— Edge of corium with at least one black mark 71

71 Underside of abdominal segments mostly orange, red or yellow .... *Eurydema ventralis* (p. 308)

Not known from British Isles, but widespread on mainland Europe.

— Underside of abdominal segments mostly black 72

72 Corium almost entirely pale, white, yellow, orange or red at the edge, and with a black spot near the middle. Pattern on scutellum united into a single V-shaped mark.............. *Eurydema ornata* (p. 306)

Recently discovered in British Isles, mostly coastal between Devon and Sussex.

— Edge of corium mostly black, only reddish or orange near base. Scutellum black with three separated marks ....*Eurydema herbacea* (p. 308)

Known from the Channel Islands, but not yet from mainland Britain.

Eurydema ornata

73 Underside of second abdominal sternite with a long forward-projecting spine reaching between hind and middle coxae, and usually visible in side view 74

— Base of abdomen beneath without a forward-projecting spine; sometimes with a small forward-projecting denticle between rear coxae 75

Piezodorus underside

74 Predominantly green, although sometimes brown-purple on base of pronotum and hemelytra, but usually with a narrow border of yellow along margins of pronotum, hemelytra and connexivum. Antennae reddish. There is a rare all-black colour form .... *Piezodorus lituratus* (p. 296)

Common on gorse, broom and other Leguminoseae/ Fabaceae.

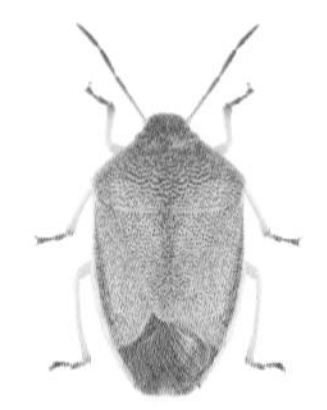

Piezodorus lituratus

— Bronze/brown mottled with cream or yellowish; connexivum chequered black and cream. Antennae dark, but strongly ringed with cream on segments 4 and 5 ..... *Rhaphigaster nebulosa* (p. 294)

Recent arrival in Britain, on various plants, mostly London area. The potential invasive colonist *Halyomorpha halys* superficially resembles this species, but lacks the abdominal spine. See couplet 88.

Rhaphigaster nebulosa

75 Small, usually less than 7 mm. Form broader and more convex 76

— Larger, usually more than 7.5 mm. Form less broad and convex 79

Eysarcoris aeneus

76 Underside of abdomen with dark metallic blotch down the centre line. Pronotum with large dark blotch behind each eye. Front edges of pronotum concave or straight (see *Eysarcoris* above and right)...77

— Underside of abdomen wholly pale brownish yellow. Pronotum without or with less contrasting dark blotch behind each eye. Front edges of pronotum swollen, convex
.... *Stagonomus bipunctatus* (p. 289)

European species. On herbaceous plants, mostly Scrophulariaceae. Small specimens of *Holcogaster fibulata* (also a European species, see couplet 89) may key out here. It can be distinguished by its more mottled, often metallic, appearance, and the fact that it occurs on conifers.

Eysarcoris venustissimus

Stagonomus bipunctatus

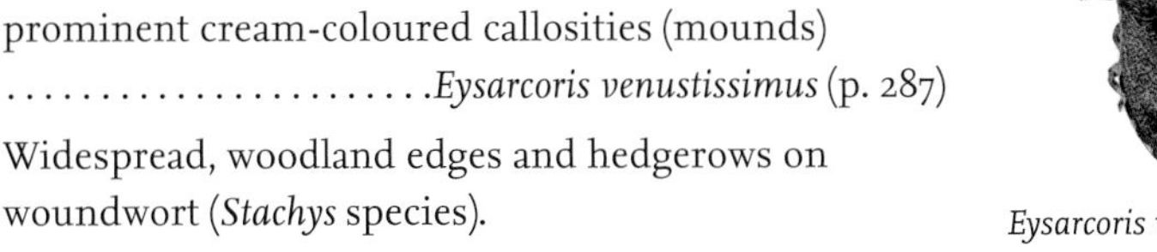

77 Scutellum pale grey with a large bronze/purple patch across the entire base, but without two prominent cream-coloured callosities (mounds) .... *Eysarcoris venustissimus* (p. 287)

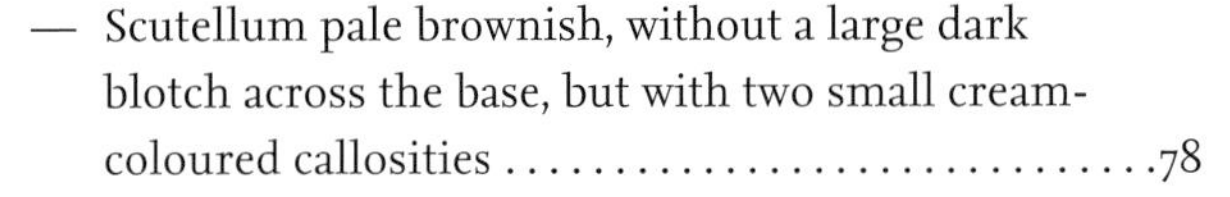

Widespread, woodland edges and hedgerows on woundwort (*Stachys* species).

Eysarcoris venustissimus

— Scutellum pale brownish, without a large dark blotch across the base, but with two small cream-coloured callosities 78

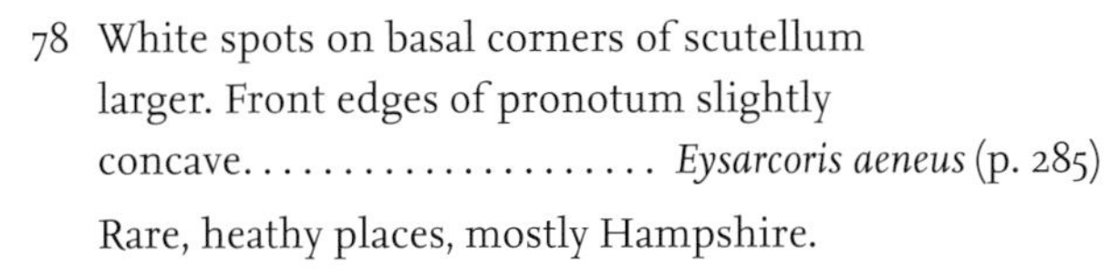

78 White spots on basal corners of scutellum larger. Front edges of pronotum slightly concave..................... *Eysarcoris aeneus* (p. 285)

Rare, heathy places, mostly Hampshire.

Eysarcoris aeneus

— White spots on basal corners of scutellum smaller. Front edges of pronotum straight, not concave .... *Eysarcoris ventralis* (p. 286)

Known from the Channel Islands and widespread on mainland Europe.

Eysarcoris ventralis

79 Colour mainly green. Connexivum not strongly chequered, although there may be slight pale or dark flecks 80

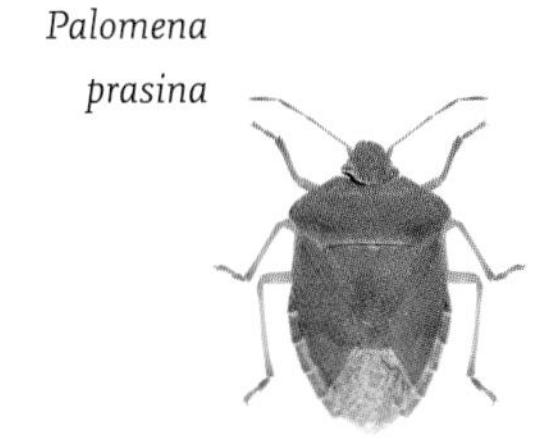

Palomena prasina

Chlorochloa juniperina

Nezara viridula

— Colour mainly brown and purple, but with some black and/or green marks. Connexivum chequered with alternating pale and dark marks 84

Dolycoris baccarum

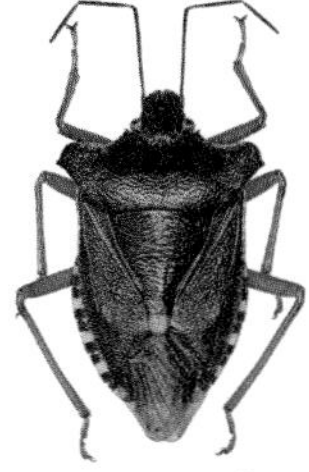

Pentatoma rufipes

Carpocoris purpureipennis

Peribalus strictus

80 Upper surface green, but with distinct yellow margins to pronotum, sometimes also base of hemelytra, sometimes these may appear orange or reddish. Rare species and forms 81

Chlorochloa juniperina

Nezara viridula form *torquata*

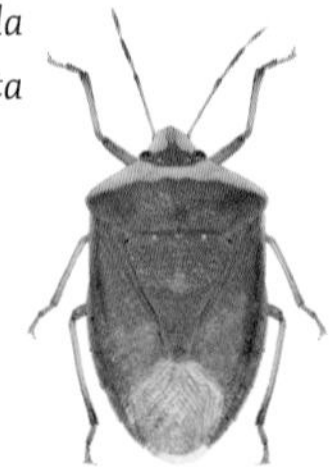

— Upper surface more or less entirely green, lacking strong yellow margins. Common 82

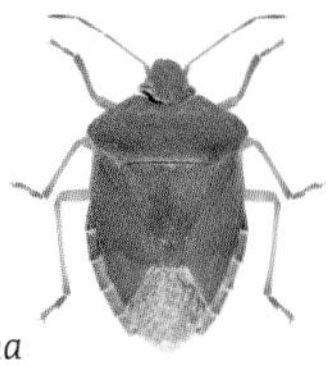

Palomena prasina

Nezara viridula

81 Generally stronger darker green. Tip of scutellum pale. Body broader and more rounded in form. Antennae dark at tip but green near base (see above) .... *Chlorochroa juniperina* (p. 275)

Very rare, on juniper, probably extinct in British Isles.

— Generally paler green. Scutellum uniformly green, but with three or five very small pale spots along margin with pronotum and a tiny black dot in each corner there. Body generally narrower and slightly more parallel-sided. Antennae green, but with reddish or dark bands around segments 4 and 5 (see above) ..... *Nezara viridula* form *torquata* (p. 290)

A rare form of this common species.

82 Base of scutellum with three or five very small pale spots where it meets the pronotum, and a tiny black spot in each corner there. Form slightly narrower and more parallel-sided. Membrane appearing pale green. Punctuation on body not black .... *Nezara viridula* (p. 290)

Nezara viridula

Widespread on various plants, first found in Britain in 2003 and spreading.

— Base of scutellum without white spots or black corner marks. Form slightly broader and more rounded. Membrane appearing black. Punctuation on body black 83

83 Leading edges of pronotum very slightly convex. Second antennal segment 1.5–1.8 times the length of segment 3 . . . . *Palomena viridissima* (p. 284)

Not yet known in the British Isles. Rare but widespread in Europe.

Palomena viridissima

— Leading edges of the pronotum very slightly concave. Antennal segments 2 and 3 about equal in length (see above) . . . . *Palomena prasina* (p. 282)

Widespread and common on various plants.

Palomena prasina

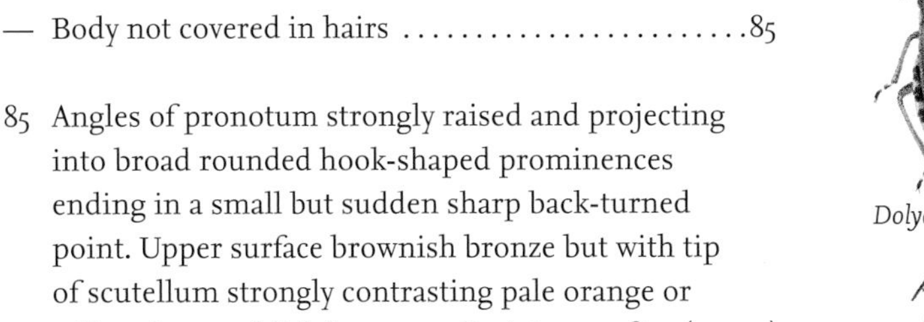

84 Body densely hairy . . . . *Dolycoris baccarum* (p. 277)

Common and widespread on various plants

— Body not covered in hairs 85

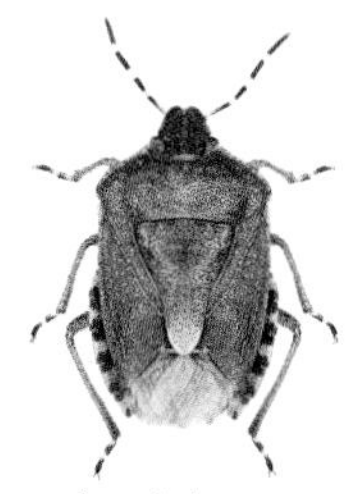

Dolycoris baccarum

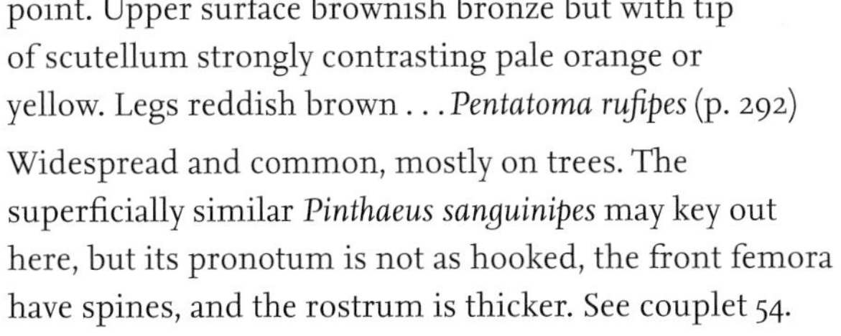

85 Angles of pronotum strongly raised and projecting into broad rounded hook-shaped prominences ending in a small but sudden sharp back-turned point. Upper surface brownish bronze but with tip of scutellum strongly contrasting pale orange or yellow. Legs reddish brown . . . *Pentatoma rufipes* (p. 292)

Widespread and common, mostly on trees. The superficially similar *Pinthaeus sanguinipes* may key out here, but its pronotum is not as hooked, the front femora have spines, and the rostrum is thicker. See couplet 54.

Pentatoma rufipes

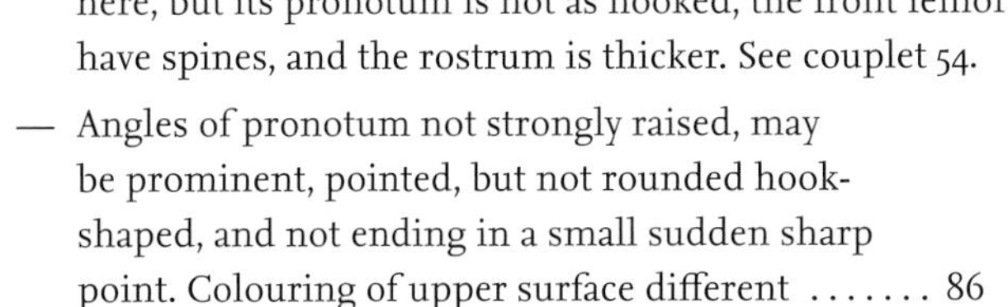

— Angles of pronotum not strongly raised, may be prominent, pointed, but not rounded hook-shaped, and not ending in a small sudden sharp point. Colouring of upper surface different 86

Carpocoris mediterraneus

Carpocoris purpureipennis

Holcogaster fibulata

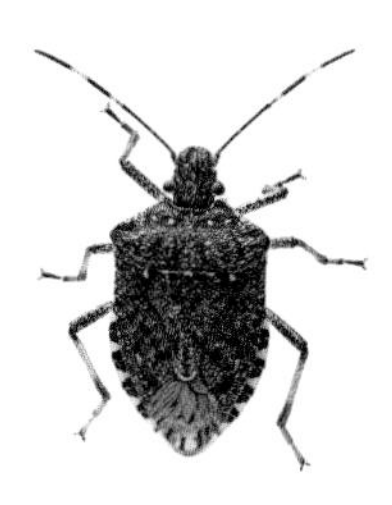

Halyomorpha halys

86 Antennae black but with basal segment contrasting red, pink or orange 87

Carpocoris mediterraneus

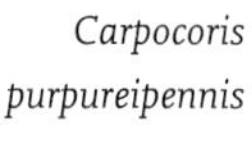
Carpocoris purpureipennis

— Antennae all dark or variously barred pale, but not with contrasting basal segment 88

87 Angle of pronotum sharp and projecting. Upper surface usually brightly coloured, orange-brown strikingly marked with black on head, pronotal corners, connexivum and base of scutellum (see above) .... *Carpocoris mediterraneus* (p. 274)

Known from Channel Islands and mainland Europe, very rare vagrant to Britain.

— Angle of pronotum prominent but not so sharp. Prettily marked, often with purple or pinkish hemelytra but not so strikingly (see above) .... *Carpocoris purpureipennis* (p. 272)

Known from Channel Islands and mainland Europe, irregular vagrant to Britain.

88 Large, 12–17 mm. Minutely mottled brown/grey and cream, tibiae with strongly contrasting broad pale bands. Base of scutellum with five small pale spots along margin with pronotum. Antennal segments 4 and 5 with narrow pale bands .... *Halyomorpha halys* (p. 270)

Halyomorpha halys

Invasive species only recently detected in British Isles. Superficially *Rhaphigaster nebulosa* resembles this species but can easily be distinguished by the sharp forward-projecting spine on the underside of the abdominal base. See couplet 74.

— Smaller, less than 13.5 mm. Differently coloured, tibiae lacking strongly contrasting pale bands 89

89 Smaller, 4.5–8.5 mm, and broader in form. Pronotum with a well-marked groove just behind its front edge. Upper surface coloured with black, grey, pink and/or orange marbling .... *Holcogaster fibulata* (p. 279)

Known from Channel Islands, but not yet from mainland Britain.

Holcogaster fibulata

— Larger, 9.5–13.5 mm. Body form less broad and rounded. Pronotum without a groove just behind front edge. Upper surface more uniformly brownish, although with some mottling because of black punctuation 90

90 Smaller, 9.5–11 mm. Antennae with segments 4 and 5 pale banded. Connexivum strongly marked with alternating pale and black blotches. Tip of scutellum usually pale .... *Peribalus strictus* (p. 280)

Scarce, southern, previously rare vagrant, now small colonies established.

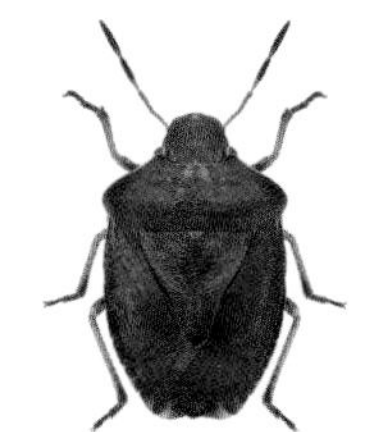

Peribalus strictus

— Larger, 12–13.5 mm. Antennae less obviously banded. Entire body more or less uniformly coloured brownish purple, connexivum feebly marked ....*Palomena prasina* (p. 282)

Winter form of this common species, changes back to brilliant green after emerging from hibernation.

Pyrrhocoridae

Just a single species in the British Isles.

91 9.0–9.5 mm. Distinctively patterned bright red and black. Head lacking ocelli. Hemelytra without a cuneus. Usually brachypterous (short-winged), lacking the membrane ...... *Pyrrhocoris apterus* (p. 322)

Scarce southern species historically known only from Devon, but now established in south-east England, sometimes congregates at base of lime (linden) tree trunks.

Pyrrhocoris apterus

Stenocephalidae

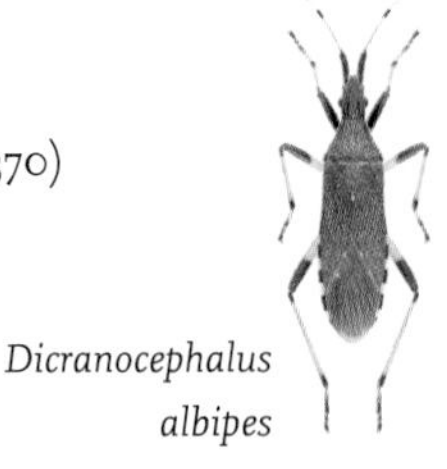

Dicranocephalus albipes

92 Second antennal segment without a dark central ring .... *Dicranocephalus albipes* (p. 370)

Reputed to have been recorded in Britain, but these old records may be misidentifications. Occurs in Europe right up to Normandy.

— Second antennal segment with a dark central ring 93

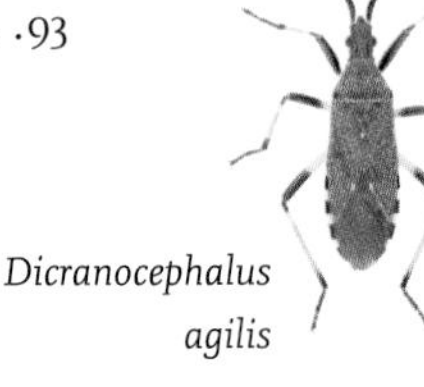

Dicranocephalus agilis

93 Rostrum reaching back to hind coxae, second segment of antennae with hairs only about half as long as the width of the segment .... *Dicranocephalus medius* (p. 367)

Scarce, woodlands, on spurge (*Euphorbia* species).

— Rostrum only reaching to middle coxae, second segment of antennae with hairs about equal to the width of the segment .... *Dicranocephalus agilis* (p. 368)

Rare, southern, dunes and coastal areas on Portland Spurge (*Euphorbia portlandica*) and Sea Spurge (*E. paralias*).

Dicranocephalus medius antenna

Dicranocephalus agilis antenna

Rhopalidae

94 Head as long as broad (or longer), wings shorter than abdomen. Rather narrow and parallel-sided insects. Eyes less globular 95

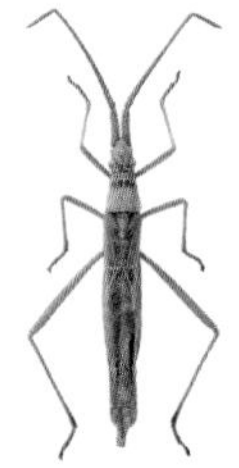

Chorosoma schillingi

Myrmus miriformis female

Myrmus miriformis male

— Head broader than long, wings reaching tip of abdomen. Rather broader insects. Eyes more globular 96

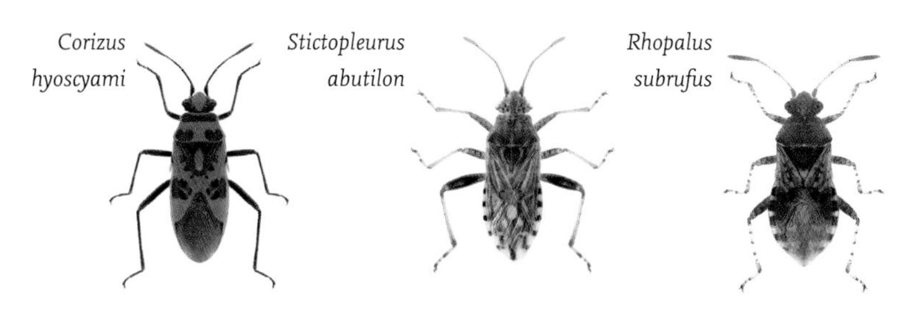

95 Very long and narrow, 12–16 mm. Wings reaching two-thirds length of abdomen. First segment of antennae as long as head. Straw-coloured (see opposite)........................... *Chorosoma schillingi* (p. 363)

Coastal grasslands, sand dunes.

— Shorter (6.5–9 mm), broader, but still rather slim. Wings with red inner veins, often very short, but sometimes full length of abdomen, or intermediate. First segment of antennae shorter than head. Green or brown (see opposite) .... *Myrmus miriformis* (p. 365)

Rough grasslands.

96 Strongly coloured black and red (or orange); corium sclerotised (tough and dark) between the veins (see above) .... *Corizus hyoscyami* (p. 351)

Historically a rare coastal species, but now widespread and common in rough grassy places.

— Coloured otherwise; corium transparent between the veins 97

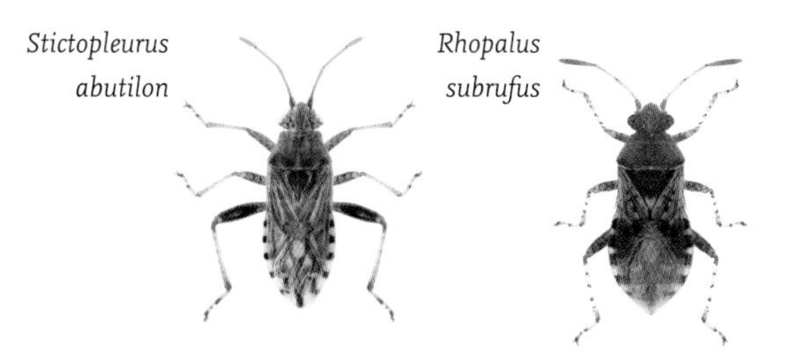

97 Metapleuron (side plate of thorax above the middle and hind coxae) more or less quadrangular, hind margin weakly convex, not obviously divided by a groove, uniformly and coarsely punctured throughout (below left) 98

Stictopleurus metapleuron

Liorhyssus metapleuron

Rhopalus metapleuron

— Metapleuron not quadrangular, hind margin more strongly sinuate, more or less divided by a groove, coarsely punctured in front, finely punctured behind or hardly punctured at all (above centre and right) 101

98 Front of pronotum with a raised ridge (usually pale) and behind it an impressed groove (usually black or dark brown) that broadens and surrounds a raised island (usually pale) on each side 99

— Front of pronotum lacking a raised pale ridge, the dark impressed groove curving round at each side, but not completely enclosing raised islands ....*Stictopleurus punctatonervosus* (p. 361)

Widespread and common.

99 Side lobes of the clypeus nearly as long as central lobe; outline of front of head bluntly stepped. Scutellum slightly constricted, spatulate at tip
.... *Stictopleurus crassicornis* (p. 362)

Not yet recorded from British Isles, but widespread in Europe.

Stictopleurus crassicornis
head scutellum

— Side lobes of clypeus shorter than central lobe; outline of front of head smoothly rounded. Scutellum not constricted 100

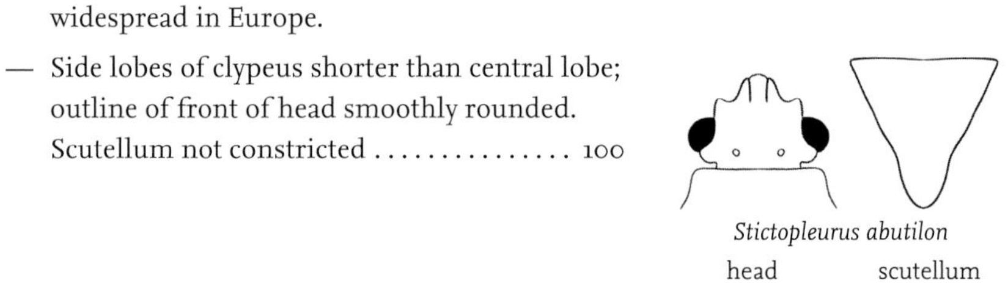

Stictopleurus abutilon
head scutellum

100 Median projection of the posterior edge of the pygophore visible laterally. Paramere sharply bent . . . .*Stictopleurus abutilon* (p. 360)

Widespread and common.

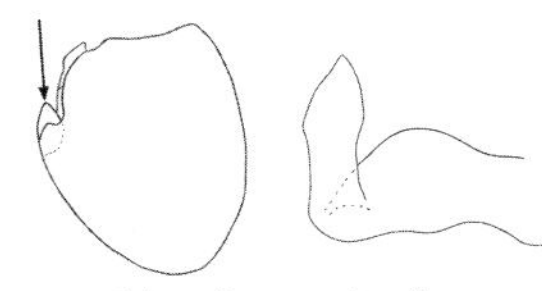

Stictopleurus abutilon
pygophore paramere

— Median projection of the posterior edge of the pygophore barely visible laterally. Paramere less sharply bent . . . . *Stictopleurus pictus* (p. 362)

Uncommon European species. Not yet known from British Isles.

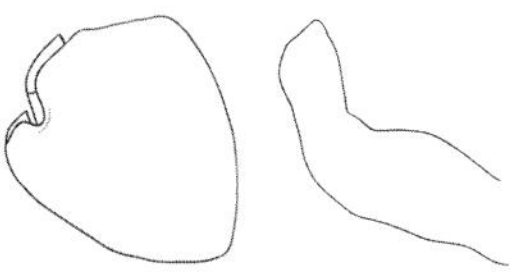

Stictopleurus pictus
pygophore paramere

101 Metapleuron divided by a distinct groove. Membrane usually extending well beyond the end of the abdomen. Upper surface of abdomen (visible through the membrane) mostly black
. . . . *Liorhyssus hyalinus* (p. 353)

Scarce, once considered rare vagrant, but now probably established in southern Britain.

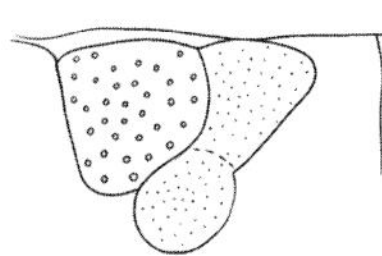

Liorhyssus metapleuron

Liorhyssus hyalinus

— Metapleuron indistinctly divided. Membrane usually reaching just beyond the end of the abdomen 102

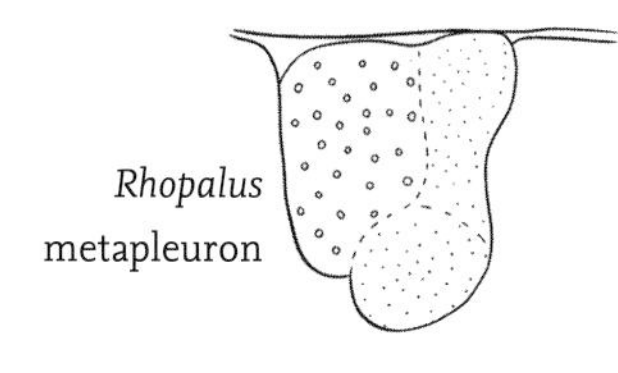

Rhopalus metapleuron

Rhopalus subrufus

102 Head shorter, antennal tubercles not prominent, coloured straw yellow, marked with black
. . . . *Brachycarenus tigrinus* (p. 350)

Scarce, but widespread; recent colonist to Britain.

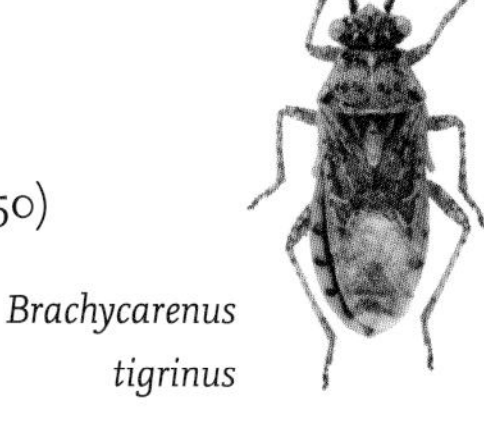

Brachycarenus tigrinus

— Head longer, antennal tubercles prominent, colour brown, red or orange, sometimes speckled or punctured with black 103

Rhopalus subrufus

103 Outer hardened and coloured area of the hemelytra broader, only the wing's inner cells clear and transparent. Upper surface of abdomen (visible through membrane) mostly red but with some dark patches. Underside of abdomen with several rows of black spots .... *Rhopalus maculatus* (p. 354)

Scarce, southern, heathy places, fens.

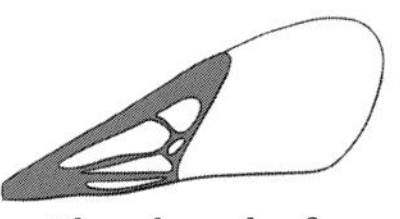
Rhopalus maculatus wing

— Outer hardened and coloured area of the hemelytra narrow, with more clear and transparent cells. Upper surface of abdomen mostly black, but with pale spots 104

Rhopalus subrufus wing

104 Connexivum strongly marked with alternating black and white spots. Tip of scutellum white-marked (see above)........... *Rhopalus subrufus* (p. 357)

Common and widespread.

Rhopalus subrufus

— Connexivum not strongly black-and-white marked, more or less uniformly red, yellowish or greenish 105

105 Hemelytra veins marked with black spots ....*Rhopalus parumpunctatus* (p. 355)

Uncommon, dry sandy places.

— Hemelytra veins not marked with black spots 106

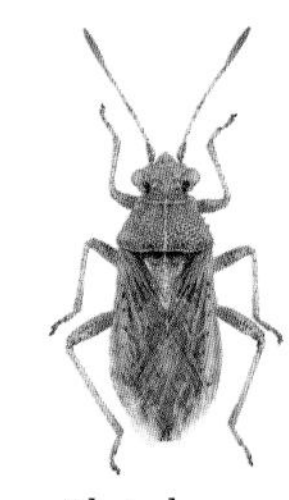
Rhopalus parumpunctatus

106 Connexivum yellow or reddish ..*Rhopalus rufus* (p. 357)

Rare, southern heaths.

— Connexivum green, along with outer edge of hemelytra .... *Rhopalus lepidus* (p. 359)

Not known from mainland Britain yet, but in Channel Islands.

Alydidae

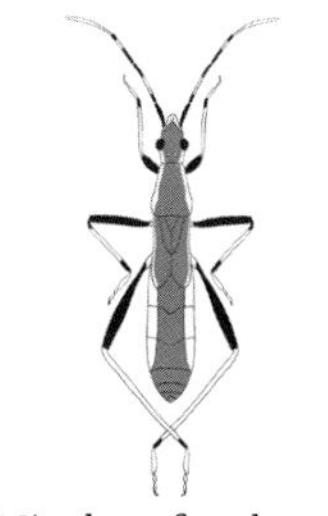

Micrelytra fossularum

107 Form long and slim (although gravid females have swollen abdomen), body more than five times as long as pronotal width. Often brachypterous, with short wings leaving abdomen exposed .... *Micrelytra fossularum* (p. 327)

Not known from mainland Britain, but found in Channel Islands.

— Form broader, body about four times as long as pronotal width. Always fully winged 108

Camptopus lateralis

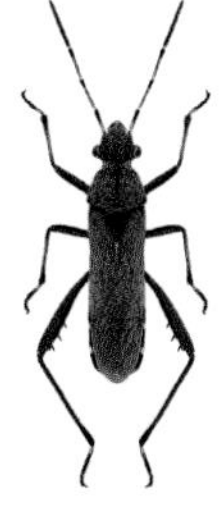

Alydus calcaratus

108 Hind femora strongly thickened in the middle, hind tibiae curved. Brown but with white flecks on connexivum, pale bands on antennae and tibiae, and edges of pronotum and hemelytra narrowly cream (see above) .... *Camptopus lateralis* (p. 326)

Not known from mainland Britain, but in Channel Islands.

— Hind femora not so strongly thickened in the middle, hind tibiae straight. More or less completely brown except for pale flecks on connexivum (see above) .... *Alydus calcaratus* (p. 324)

Dry sandy or chalky places.

Coreidae

Leptoglossus occidentalis

109 Large, 15–20 mm. Hind tibiae spatulate, broadened and flattened on each side. Hind femora with a double series of short sharp spines increasing in size into short but strong curved thorns towards the apex. A distinctive brown bug with chequered connexivum and bright white zigzag line across each corium .... *Leptoglossus occidentalis* (p. 344)

Recent arrival in Britain, now widespread, often comes indoors to hibernate.

— Smaller, up to 14.5 mm. Hind tibiae not spatulate. Hind femora at most with a small group of spines near apex. Corium not marked with strong white zigzag line 110

110 Hind femur with one or more strong spines near apex, as long as width of tibia 111

— Hind femur without spines, or only with small spines not as long as width of tibia 115

111 Rear margin of pronotum with several small spines. Side margins of pronotum with a series of white spines each tipped with a hair 112

Coriomeris denticulatus

Ceraleptus lividus

— Rear margin of pronotum without spines. Side margins of pronotum with teeth, but these are dark. . 114

112 Antennae with recumbent short hairs only .... *Coriomeris scabricornis* (p. 340)

Not known in Britain, but widespread in central Europe to southern Scandinavia.

Coriomeris scabricornis antenna

Coriomeris denticulatus antenna

— Antennae with underlying recumbent short hairs, but also with numerous outstanding long hairs ... 113

113 Side margin of pronotum with 10–12 evenly and closely-spaced teeth. Rear corner of pronotum with 3–4 small teeth. Hind femur with two large teeth, the second with several small teeth behind it at the apex ....*Coriomeris denticulatus* (p. 338)

Widespread, rough grassy places.

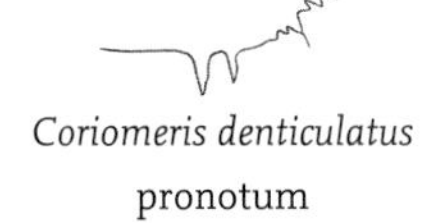

Coriomeris denticulatus pronotum

— Side of pronotum with 5–8 larger and more widely spaced teeth. Rear corner of pronotum with 3–4 large teeth. Hind femur with three large teeth, the second with several small teeth behind it at the apex .... *Coriomeris affinis* (p. 340)

Not yet known in Britain, but widespread in Europe to Brittany.

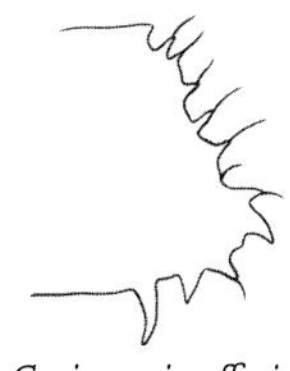

Coriomeris affinis pronotum

114 First segment of antennae strongly spined. Heavier, more strongly built insect. Dark brown, tibiae each with two pale rings .... *Bothrostethus annulipes* (p. 332)

Known from the Channel Islands and Atlantic coast of France.

Bothrostethus annulipes

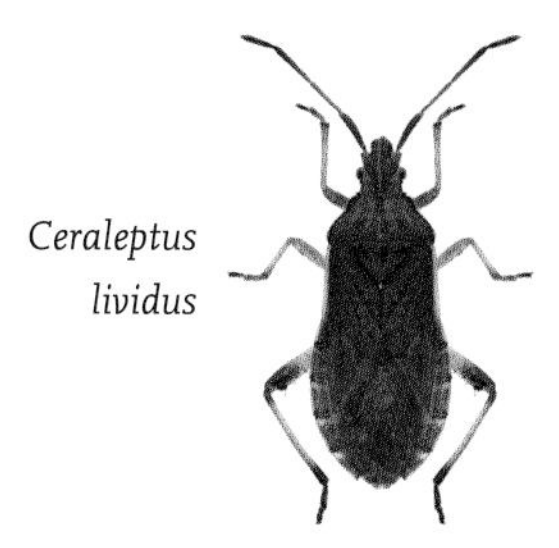

Ceraleptus lividus

— First segment of antennae not strongly spined. Slighter insect. Paler brown, legs not strongly pale-ringed .... *Ceraleptus lividus* (p. 333)

Southern, dry grassy places.

115 Large, more than 9 mm. Rear angles of pronotum prominent, raised; surface of pronotum more or less smooth, but may be punctured, or minutely sculptured 116

Coreus marginatus

Enoplops scapha

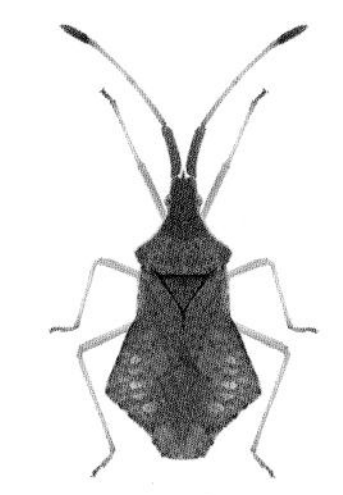

Syromastus rhombeus

Gonocerus acuteangulatus

— Smaller, less than 8 mm. Rear angles of pronotum less prominent, not raised; surface of pronotum may have ridges, spines or tubercles 121

Arenocoris fallenii

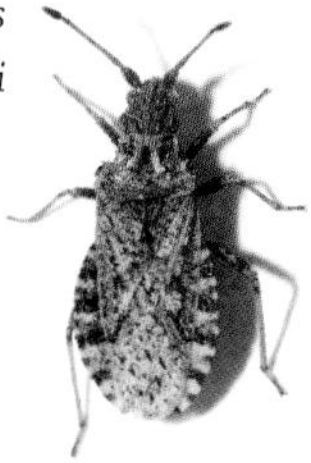

Bathysolen nubilis

Spathocera dalmanii

116 Antennal tubercles produced forwards as a pair of curved, inwardly pointing spines ....*Coreus marginatus* (p. 335)

Common and widespread.

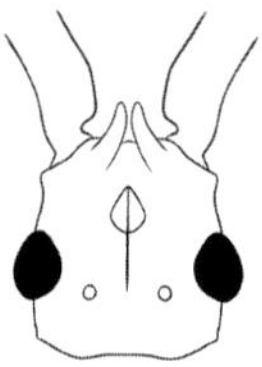

Coreus head

— Without a pair of spines between the antennal bases 117

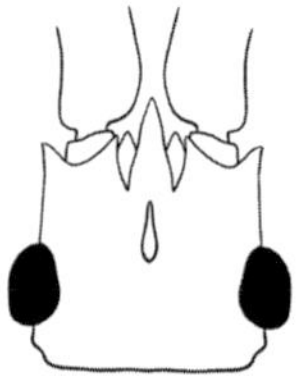

Enoplops head

117 Antennal tubercle with a spine on the outside. A large, strongly built insect. Connexivum broad ....*Enoplops scapha* (p. 341)

Scarce, coastal sandy places.

— Antennal tubercle without a spine 118

Enoplops scapha

118 Connexivum broad, expanded and flattened 119

— Connexivum narrow 120

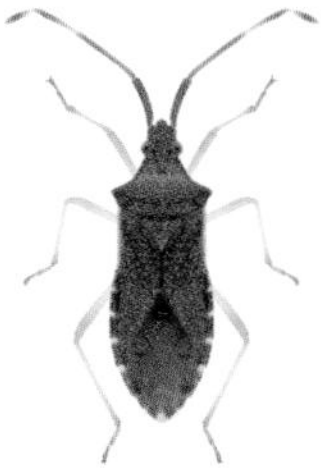

Gonocerus acuteangulatus

Syromastus rhombeus

119 Connexivum distinctly diamond-shaped, angular in middle .... *Syromastus rhombeus* (p. 348)

Local, rough grassy places.

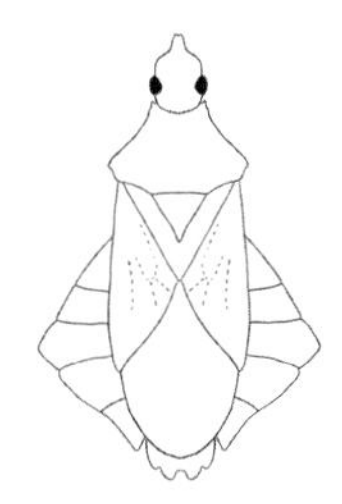

Syromastus rhombeus

— Connexivum rounded, broad, but not angular in the middle ... *Haploprocta sulcicornis* (p. 337)

Not recorded from British Isles, but widespread in France to Brittany and Normandy.

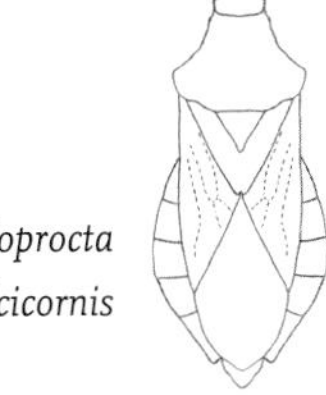

Haploprocta sulcicornis

120 Background colour more or less uniformly mid- to pale brown (see opposite) .... *Gonocerus acuteangulatus* (p. 342)

Widespread, formerly only on Box (*Buxus sempervirens*) trees, now on hawthorn, honeysuckle, bramble etc.

— More prettily marked, body mottled, connexivum pale greenish yellow, often with contrasting dark marks .... *Gonocerus juniperi* (p. 343)

Not recorded from British Isles, but widespread in Europe to Belgium.

121 Margins of pronotum strongly indented, lined with irregular teeth. Third antennal segment long, more than twice as long as fourth segment 122

Arenocoris fallenii

— Margins of pronotum more or less straight, but with a few small teeth. Third antennal segment shorter, about twice as long as fourth 123

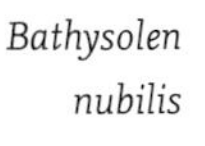

Bathysolen nubilis

Spathocera dalmanii

122 Centre of pronotum with two short and slightly curved arrays of pale rounded warts along with some small dark and pale spines. Third antennal segment about equal diameter throughout .... *Arenocoris fallenii* (p. 329)

Scarce, sandy places, including coastal dunes.

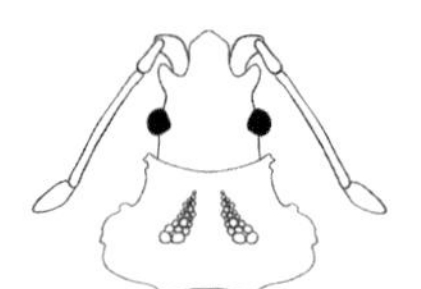

Arenocoris fallenii
head and thorax

— Centre of pronotum without arrays of pale warts, but with some short erect spines. Third antennal segment swollen at tip, where it is as broad as segment 4 ....*Arenocoris waltlii* (p. 330)

Rare, sandy places, mostly in the East Anglian Breckland.

Arenocoris waltlii
head and thorax

123 Larger, 6–7 mm. Pronotum transverse, very short and broad. Third antennal segment longer than first and second together .....*Bathysolen nubilus* (p. 331)

Scarce, rough dry grassy places, mostly Thames estuary.

Bathysolen nubilis

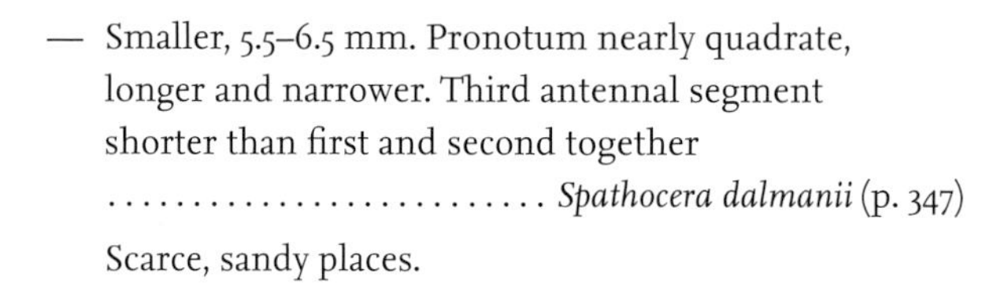

— Smaller, 5.5–6.5 mm. Pronotum nearly quadrate, longer and narrower. Third antennal segment shorter than first and second together .... *Spathocera dalmanii* (p. 347)

Scarce, sandy places.

Spathocera dalmanii

CHAPTER 8

British Shieldbug Species

The following chapter comprises, I admit, a slightly eclectic list of species and, yes, it includes several that are not British – yet. As of January 2023 there are reckoned to be 79 species of shieldbug, squash bug and close relatives currently 'occurring' in Britain, or at least worthy of a mention in the national checklist. This number is unlikely to remain static for long. Some have not been recorded for so long that they are thought to be extinct in the British Isles. Perhaps they were never truly established here in the first place – such is the nature of outlier colonies on islands like ours. Maybe some of these shieldbugs will be rediscovered in the future surviving in secret relict colonies, or they might recolonise from continental Europe. Several new species have been discovered here in recent years, and more are set to establish themselves any time soon. With the huge movement of horticultural and agricultural material around the world, some of these potential new colonists can sweep in from the other side of the planet, not just from the other side of the English Channel.

In addition to the 79 British species, this chapter includes a further 11 known from the Channel Islands, 16 more that may or may not eventually make it here from mainland Europe, and a triad of oddities that have obviously come in from further afield. One species, *Leptoglossus occidentalis*, the Western Conifer Seed Bug, has already made it from North America across the Atlantic, by way of central and southern Europe. *Halyomorpha halys*, the Marmorated Shieldbug, a native of China, Japan and Korea, is already on the cusp, a few specimens having turned up in imported goods or at pheromone lures. I've taken a punt at guessing which European and more distant world species could conceivably turn up here in the near future. I'm quite prepared to be proved wrong. I'll have another go if the book ever gets to a second edition.

Perhaps rather controversially, I've decided to go with my own order of species here, vaguely following the taxonomic schedules of Hawkins (2003), Rintala & Rinne (2011) and Lupolia & Dusoulier (2015), which reflect the recent catalogue of Palaearctic Hemiptera by Aukema & Rieger (2006). Several modern lists place the Acanthosomatidae and Pentatomidae (true shieldbugs) after the Coreidae (squash bugs) and Rhopalidae (sometimes called 'scentless' bugs), but since I've already set the direction of this book by calling it *Shieldbugs*, it's quite clear to me which groups must line up at the front of the grid. I leave arguments over precise checklists to more technical academic publications.

The layouts of the following entries are self-explanatory. I will, however, just explain my particular use of the word 'hemelytron'. Technically this is the entire front wing, membrane included, but here I use it to mean just the hardened, sclerotised, coloured, basal section comprising corium, clavus, embolium and cuneus combined. Lengths are measured from head to tail tip, but exclude antennae.

My intention is not to create a definitive descriptive catalogue; that would be too repetitive and hardly readable. Some entries are long, others short. These decisions hinged around whether I thought a careful text was appropriate, or whether I could just get away with a few brief comparisons. What I have to say about each species turns out to follow a complex algorithm that even I don't really comprehend. If it's a favourite, or I have an anecdote to spin, then off I go.

FAMILY ACANTHOSOMATIDAE

***Acanthosoma haemorrhoidale* (Linnaeus, 1758) – Hawthorn Shieldbug**

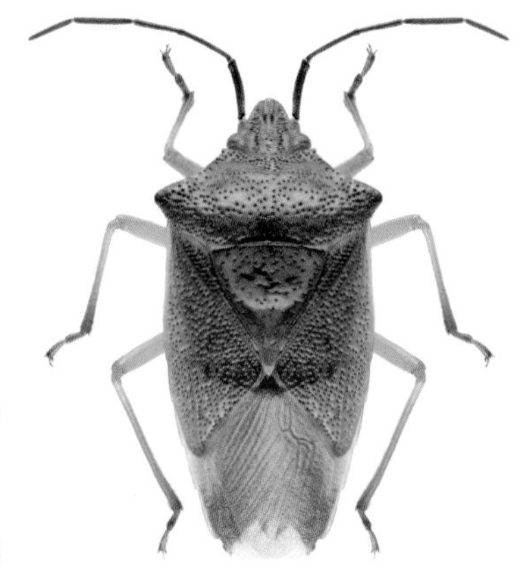

Description. Length 12.5–18.0 mm. This is an elegant, shining, elongate species, with sharply prominent pronotal corners. In life it is mostly bright emerald-green, marked with deep burgundy red across the basal half of the pronotum, down the clavus and corium, onto the smoky membrane, and with the very tip of the abdomen reddish. The scutellum, embolium and cuneus are a contrasting green. The extreme apices of the thoracic protuberances are quite bright red, although often with dark tips. The upper surface is marked with deep black punctures which are larger and more diffuse across the scutellum and pronotal base, smaller and closer across the hemelytra, head and front corners of the pronotum. The orbits and two slightly raised pale islands near the front edge of the pronotum lack punctures. The underside is a yellowish green, without obvious punctures. The antennae are long, very dark green to black, sometimes with the basal segment paler. The head is broad, flat and triangular, with the central lobe narrow, but expanded towards the apex. The scutellum is produced into a narrow unpunctured tongue-shaped process. The legs are long and narrow, green with tibial tips and tarsi reddish brown.

Nymph

Early nymphal instars are mostly red, marked with brown and green; later instars are clearer green, edged with reddish or pink. The stink glands are dark against a reddish blush. The final instar already shows the strong pronotal points.

Similar species. This species is larger and more brightly coloured than even the largest *Elasmostethus interstinctus*, and usually unmistakable, the bright green scutellum contrasting with the reddish corium.

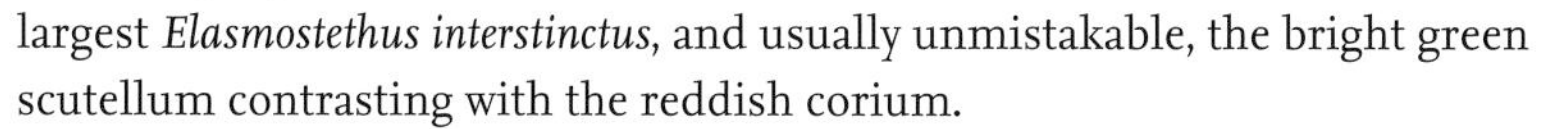

Ecology. Although typically found on hawthorn, where it feeds on the developing fruits, it is nevertheless likely to be found on almost any rosaceous species bearing berries, including rowan, whitebeam, cotoneaster and firethorn (*Pyracantha*). It also occurs on birch, often in company with *Elasmostethus interstinctus*. In parts of Europe it is considered a sometime minor pest of hazel nuts. It has also been

Adult

Fifth-instar nymph

found on apple, pear, cherry, sallow, rowan, elm, elder and rhododendron, though some of these may be mere resting points rather than feeding stations. Typically it is a hedgerow species where hawthorn grows most readily, but it is also found along wood edges and in broad woodland rides, and it is common in parks and gardens where its other foodplants occur. It overwinters as an adult, when it turns a reddish brown, usually reappearing and recolouring in April. Each female lays about 28 eggs. Most eggs are hatched by early July. There is one generation per year. It flies readily and frequently turns up in moth traps.
Distribution and status. *Acanthosoma* is common throughout Britain and Ireland, but is less frequent in Scotland, where it was considered a great rarity until a recent apparent expansion northwards (Ramsay 2014). It is now calculated to be the most widespread shieldbug in the British Isles (Ryan 2019). This is usually the first shieldbug I find out in the countryside in May. Without fail it always surprises me how large it is, and it's always a delight to hold its handsome form in my hand.

Cyphostethus tristriatus **(Fabricius, 1787) – Juniper Shieldbug**

Description. Length 8.5–11.0 mm. This is one of our prettiest and most distinctive shieldbug species, sporting a glossy green and pink livery. It is predominantly a vivid green but with a bright pinkish-orange curved ('boomerang-shaped') streak reaching from each pronotal prominence, across the corium and onto the cuneus. This is sometimes clouded with a pale cream streak. The pronotum has the fore-edges, and sometimes a central line, paler. The scutellum is bright green, produced into a short narrow tongue-shaped process; the lateral corners and apex are marked with creamy yellow. The upper surface has strong punctures; these are variously infilled black in some areas – usually the front and rear edges of the pronotum, the front half of the scutellum, and parts of the corium and cuneus – giving a mottled appearance under the naked eye. The membrane has a small area of smoky cloud, often showing as an X-mark combined with the darkish edge to the abdominal tip. The abdomen is black above, visible through the membrane, but with the apex and connexivum pale. The underside is clear green, with the propleuron (first side plate of underside of the thorax) strongly but diffusely punctured; some punctures are black above the front coxae. The head is long,

ABOVE: Adult

Nymphs feeding on 'cones' of *Thuja occidentalis*.

triangular, flat, with the central lobe broadly expanded at the apex. The antennae are green, often with segments 4 and 5 darkened or reddened. The first antennal segment reaches only to the end of the head – a key diagnostic feature in faded museum specimens. The legs are green with the tarsi slightly dark or reddish.

The nymphs are a bright emerald-green or slightly turquoise, variously marked with black and red, but usually with the lateral margins of the head and pronotum pale. The stink glands are dark against an orange or pinkish patch.

Similar species. The orange-pink markings are more precisely defined than the usually hazy ones in *Elasmostethus interstinctus*. Nymphs have a longer head than those of *E. interstinctus*, and the central lobe is distinctively dark-edged.

Ecology. This was always a species of chalk and limestone hills, where it fed on the berries of Juniper (*Juniperus communis*) bushes. Southwood & Leston (1959) famously wrote: 'there is no evidence to indicate that the bug ever feeds upon any other plant.' However, a few years later Southwood (1963) found it feeding on the unripe cones of Nootka Cypress (*Chamaecyparis nootkatensis*), a widely planted garden conifer. In the ensuing half century it has jumped to several other planted conifers, including *Chamaecyparis lawsoniana* (Lawson's Cypress), *Thuja plicata* (Western Red Cedar) and *T. occidentalis* (Eastern White Cedar) actually both cypresses not true cedars. It has never looked back. Adults overwinter; eggs are laid from late May onwards, with new nymphs achieving adulthood by August.

The jump from juniper to ornamental cypresses has been noted in many other countries – Turkey (Lodos & Önder 1979), France (Reichling 1988), Czech Republic (Stehlik 1988), Denmark (Tolsgaard 2001), Germany (Werner 2002), Portugal (Grosso-Silva 2004), Iran (Linnavuori 2008), Faroe Islands (Tolsgaard & Jensen 2010) where it was imported on Christmas trees, Canary Islands (Aukema *et al.* 2013) and Crete (Heckmann *et al.* 2015). It is now also in North America (Ratzlaff & Scudder 2018). Hawkins (2003) commented that the first anecdotal link to cypresses was of an overwintering adult found on a cypress cone in the Netherlands in 1935 (Aukema 1988) and suggested that since the Netherlands has an extensive history of growing and exporting these and other garden plants, perhaps the switch occurred there and the bug has since been transported about the continent. Incidentally, several other heteropteran bugs have also jumped from declining junipers to burgeoning cypresses, including the lygaeid *Orsillus depressus* (Hawkins 1989) and the mirid *Dychrooscytus gustavi* (Jones 2000). Despite this, there seems to be no appetite to rename the Juniper Shieldbug as the Cypress Shieldbug.

Distribution and status. Though never very rare, in southern England at least, the bug's attachment to juniper meant that it was always restricted to chalk downs. And as chalk downland management moved away from unprofitable sheep- and cattle-grazing regimes in the 1960s, increasingly aggressive ploughing for arable crops, along with scrub encroachment where stock animals were removed, threatened to decimate both juniper and shieldbug populations. By the late 1970s this was a scarce bug in Britain.

A decade and a half later the bug would be more or less dependent on those garden cypresses for its survival. But 'survival' hardly describes the bug's altered fortunes. It was everywhere. Roger Hawkins (2003) writes eloquently about his own discovery of the bug on virtually every Lawson's Cypress tree that he examined. It fired his interest in the bug, and in shieldbugs in general, and his searches for it directly resulted in the shieldbug survey of Surrey and the eventual publication of his Surrey Wildlife Trust atlas. He gives an extensive account of the insect's occurrence in Surrey.

Elsewhere in the country it has spread far beyond its previously constrained range. The doubtful localities in Yorkshire and Northumberland mentioned by Southwood & Leston (1959) have now been overtaken by the bug's northward expansion, well into Wales and with outliers in Scotland, the Isle of Man and Ireland. Perhaps the most startling spread has been into East Anglia, a decidedly juniper-free zone of the country, but one now well populated with garden conifers.

Elasmostethus interstinctus **(Linnaeus, 1758) – Birch Shieldbug**

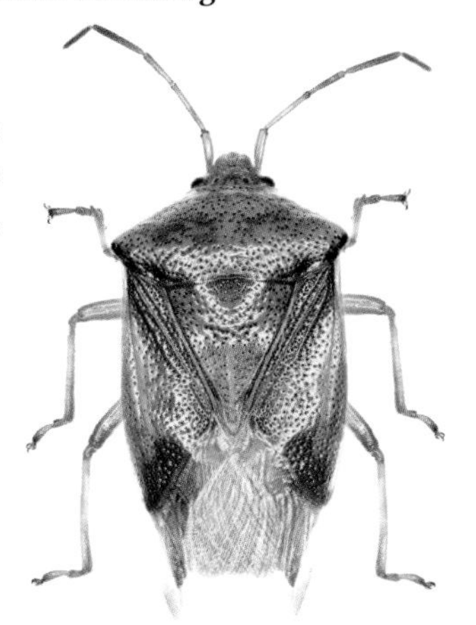

Description. Length 8.0–11.0 mm. This shining brownish-green bug is delicately suffused with red. The upper body is mainly leaf-green, but with a broad smear of ruddy brown along the rear edge of the pronotum, across the base of the scutellum, along the clavus, leaking onto the corium, down to the cuneus edge of the hemelytra. The extreme tips of the pronotal corners are blackened, with this colouring variously seeping along a narrow ribbon across the suture between pronotum and scutellum. The scutellum is extended into a narrow, barely punctured or entirely smooth, tongue-shaped process. The upper surface is punctured throughout; most punctures

Adult

are infilled with black. The punctures on the scutellum and pronotum are larger and sparser than those on the hemelytra. The pronotum has a narrow unpunctured field across its width just behind the front edge, which joins a narrow unpunctured bar inside each lateral margin. The membrane is clear, though with a small dark clouded spot on the outer edge, and showing the abdomen beneath, dark centrally, but with pale lateral edges and red tip. The underside is a clear brownish green, lacking any punctures. The legs are dull green with the tarsi slightly darkened. The head is triangular, slightly narrowed just in front of the eyes, with the central lobe narrow, expanding gently and evenly to the slightly bulbous apex. The antennae are a dirty green with some stronger darkening on the last two segments.

Nymph

The early instars are distinctive by virtue of the dark, almost black, pronotum and head. Later instars have head, pronotum and wing buds a dull green; the abdominal segments are marked above with a reddish V- or Y-mark, and the scent glands are prominently marked with black.

Similar species. This species can resemble a miniature *Acanthosoma haemorrhoidale*, but it is significantly smaller, less smartly marked, and lacks the more prominent pronotal angles. The reddish cloud at the base of the scutellum is variable, but different from the striking green triangle of *A. haemorrhoidale*. The reddish markings follow a similar curve to those in *Cyphostethus tristriatus*, but are much more muted than the bright pinkish orange shown in that species.

Ecology. Not surprisingly, the Birch Shieldbug occurs mostly on birch trees, including various ornamentals such as the very white-trunked *Betula utilis* as well as native *B. pendula* and *B. pubescens*, where it feeds on the catkin-shaped fruits. It is also regularly found on Hazel (*Corylus avellana*) and Aspen (*Populus tremula*) trees. After the adults emerge from hibernation from March to May, the eggs are laid on the underside of a leaf in small clusters. Nymphs emerge about July and migrate to the fruits to feed. They reach maturity as new adults in August. Adults leave the trees at leaf fall and hibernate in evergreen foliage or in log crevices, where they may darken or even become completely black. Hawkins (2003) reports one specimen found in May 1986 being wholly glossy black – 'the only British shieldbug that I ever misidentified' (Roger Hawkins, personal communication).

Distribution and status. This is a widespread and common species throughout almost all of Britain and most of Ireland. It flies readily and frequently appears in moth light traps.

Elasmucha ferrugata **(Fabricius, 1787)**
Description. Length 8.0–11.0 mm. This stout orange-brown shieldbug has sharp pointed thorn-like pronotal corners. The base colour is brown or tawny, with some green hints, offset by the very dark brown to black of the pronotal prominences, dark head, and an oval or diamond-shaped spot in the centre of the scutellum. The long sharp pronotal spines are particularly striking. The scutellum is drawn out into a contrasting pale narrow tongue-like process at the apex. The hemelytra sometimes appear mottled, especially towards the rear edge. The membrane is darkly clouded but with a clear spot at the very tip. The upper body is dinted with deep black punctures. The thoracic pleura and abdominal sternites have sparse black punctures. The connexivum is spotted with black at each abdominal suture. The antennae are pale fawn, but the first and last segments are darkened. The legs are pale brown, with the tarsi slightly darker.

The nymphs are strongly coloured brownish across the head, pronotum, scutellum, developing wing buds, and around the stink glands, contrasting with the bright emerald-green abdomen.

Similar species. No other acanthosomatid has such prominent pronotal spines. *Picromerus bidens* is more sombrely coloured, apart from having the usual pentatomid complement of three tarsal segments rather than two.

Ecology. In mainland Europe this bug normally feeds on Bilberry (*Vaccinium myrtillus*) and Cowberry (Lingonberry) (*V. vitis-idaea*), but has also been recorded on Fly Honeysuckle (*Lonicera xylosteum*), bramble, hawthorn and cherry. In Belgium it has been shown to develop its entire life cycle on Blackcurrant (*Ribes nigrum*) (Devillers & Lupoli 2016), and one of the British records was from cultivated raspberries. Like *E. grisea* it shows maternal behaviour by guarding its egg clutches, which are laid on the underside of the foodplant leaves. It remains with the nymphs during their first- and sometimes their second-instar stages.

Adult

Distribution and status. Possibly now extinct in Britain. Five records from Caernarfonshire, Derbyshire and Yorkshire, between 1889 and 1950. Having said this, at least these localities are uplands, where an insect feeding on Bilberry might be expected to occur. One was in Leeds market where, who knows, maybe it was found in a portion of soft fruit on one of the stalls. This insect is widespread in Europe and appears to be extending its range. It was recently found in southern Belgium (Baugnée 1999), in lowland Montenegro, feeding on brambles (Hemala & Hanzlik 2015), and in Lombardy on Stinging Nettles (Salvetti & Dioli 2015). It might turn up here again yet.

***Elasmucha grisea* (Linnaeus, 1758) – Parent Bug**

Description. Length 6.0–9.0 mm. The Parent Bug is a small, compact slim oval brownish-grey species, with black-flecked abdominal margins. The ground colour is mostly creamy brown but it is clouded with a highly variable pattern of brown, reddish or green marks. The strongest dark blotch is usually across the base of the otherwise paler scutellum, though this is sometimes a dull red. The rear, most domed portion of the pronotum is usually darker than the front slope, with the pronotal prominences often darkest. The extent of dark patches on the hemelytra is very variable, sometimes giving a chequered appearance at first glance. The upper surface is heavily pockmarked with black punctures; those on the scutellum (especially in the basal half) are stronger than those on the pronotum and hemelytra. The lateral margins of the pronotum are clear of punctures, and usually pale-edged. The head is broad triangular, with the central lobe narrow, more or less flat-ended, barely over-reaching the lateral lobes. The connexivum is very distinctly black-spotted, at each abdominal suture, the black spots sometimes bleeding inwards and around, meeting each other to leave the connexivum white-spotted. Underside pale, but thoracic pleura (and sometimes abdominal segments) punctured black. Spiracles black-spotted. The membrane is clouded to a variable degree, but in the darkest specimens there is an apical pale patch. The legs are pale greenish buff, spotted with black, especially in more heavily patterned individuals; the tibiae (towards the apices) and tarsi are darkened. The antennae are pale but with the terminal segments darkened, and often obviously puncture-marked.

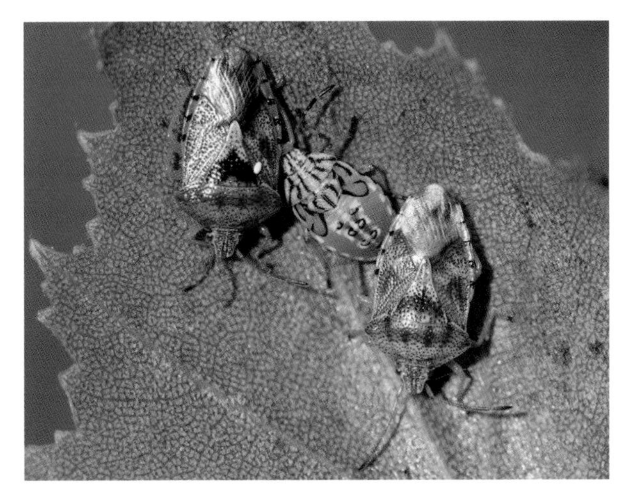

Adults and final-instar nymph.

BELOW: Adult and late-instar nymph cluster

Early-instar nymphs are brownish red across the head and thorax and green across the abdomen, with the stink glands black in a pinkish-green cloud. Later-instar nymphs are highly and distinctly decorated; the head, pronotum, scutellum and wing buds are a dull brownish or pinkish green with strong black humbug stripes. The abdomen is bright emerald-green, but with the stink gland prominences marked black and circled with pale cream; the abdominal margins are also marked with short streaks of pale cream.

Similar species. The black-marked connexivum is a key character for this species, but see *E. fieberi* (below), which might conceivably turn up in the British Isles in the future.

Ecology. The parental behaviour of the female is well known and long celebrated. Douglas & Scott (1865) paraphrase De Geer by likening this to a hen with her chicks. The first detailed modern exploration was given by Jordan (1958). Emerging from hibernation in April and May, the bugs mate and the female lays 30–60 pale green eggs in a roughly diamond-shaped batch on the underside of a birch leaf, sometimes Alder (*Alnus glutinosa*). She then sits over them, guarding them until they hatch two or three weeks later. If a predator or parasitoid approaches, the female will tilt and rotate her body towards the enemy, flicking her wings open and shut. These are mostly ants and ladybird larvae, though egg-parasitoids also attack. Poking the egg or nymph cluster with a finger or a twig will also elicit this defence. The female remains with the first-instar nymphs, protectively clambering over them, almost seeming to draw them to her using her legs and antennae. After they moult, the female accompanies the second-instar nymphs to the catkins to feed. They often reassemble on leaves to shelter, and females have been recorded sitting over or with groups of second and third instars too big to all fit beneath her body. Hawkins (2003) counted instars over several years, and his phenology diagrams suggest that after the female leaves her nearly-grown first brood she may mate again and lay a second batch of eggs. Thus, sitting females peak in June and again in August, final-instar nymphs and adults peak in August and again in September.

Larger females lay more eggs, and this seems to be linked to the maximum number she can successfully defend against invertebrate predators and parasitoids. By experimentally manipulating clutch size in different-sized females, Mappes & Kaitala (1994) showed that small females given too many eggs were apt to lose them, mainly to ants. Throughout the world other members of this genus show similar brood-guarding behaviours. Pairs (sometimes threes) of females guarding their clutches on the same leaf were also better able to defend their offspring against attack than single females (Mappes *et al.* 1995).

Distribution and status. This shieldbug is widespread and common throughout most of Britain and Ireland. It is typically a species of hedges, wood edges and broad woodland rides, but is also common in parks and gardens, heaths and moors, in fact wherever birches grow, and also along damp stream valleys where it feeds on Alders.

Elasmucha fieberi (Jakovlev, 1865)

Description and similar species. Length 6.0–9.0 mm. In general appearance this species is virtually identical to *E. grisea*. The underside of the body is more strongly black-punctured and spotted, particularly the abdominal sternites. The front angles of the pronotum are more pronounced into a sharp reflexed pale denticle easily clearing the edges of the eye. The antennae are wholly black or dark brown.

Ecology. Its life history appears to be essentially the same as that of *E. grisea*, although the maternal behaviour was only confirmed recently (Hanelová & Vilímová 2013). It is known from birch, Alder and Hazel (*Corylus avellana*).

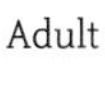

Adult

Distribution and status. Although not recorded from Britain, it is widespread in Europe including (sporadically) right up to north-west France and southern Scandinavia. It may be worth looking out for.

FAMILY CYDNIDAE

Aethus (*Byrsinus*) *flavicornis* (Fabricius, 1794)

Description. Length 2.6–4.0 mm. *Aethus* is a very small, very dark brown bristly oval bug. The head, pronotum and scutellum are black, shining, and sparsely but deeply punctured. The hemelytron is sometimes lighter brown, often with bronze

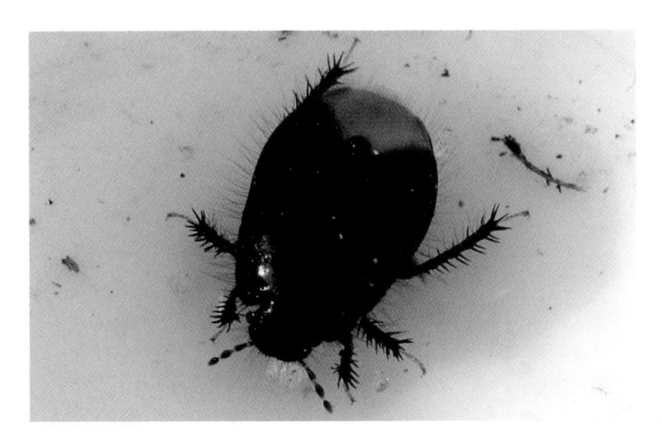

Adult

glints; it is also sparsely but deeply punctured. The junctions of head, pronotum, scutellum and hemelytra are all slightly paler, enhancing the segmented appearance of the insect. The margins of the body have long brownish setae. The front angles of the pronotum have especially long black setae, and those along the front edge of the head arise from small thickened spines, giving a clumped bristly appearance. All legs are usually brown, and covered with long stout spines and setae; the tarsi are pale. The antennae are short; the basal segments are a lighter brown than the darker apical segments.

The nymphs, at least in later instars, are very dark brown to black; the abdomen is a pale cream, with narrow reddish horizontal bars and large black plates around the stink glands.

Similar species. In *Geotomus*, the only other small dark cydnid genus with a bristly head, the setal bases are simple, not thickened.

Ecology. The biology of this species is poorly understood. It is normally found under a variety of plants on coastal dunes, landslips, river banks and other sparsely vegetated sites.

Status and distribution. This species has only a very tenuous claim to be British. It is known, in Britain, from a single specimen found on the Isle of Wight in 1895. It occurs sporadically along the Atlantic coast of France, along the Cherbourg peninsula, on some of the Channel Islands, and just onto the North Sea coast of Belgium. This is a similar French distribution to *Geotomus punctulatus*, so Wight and West Country might be possibilities.

Geotomus punctulatus (Costa, 1847) – Cornish Shieldbug

Description. Length 3.5–5.0 mm. This is a small, oval, glossy black bug. The adult is almost completely black, except for the eyes, and even the membrane is usually darkened. Very occasionally there are reddish or brown marks on the corium and pronotum. The upper surface is strongly punctured, though with a broad shining virtually unpunctured field (or two islands) across the front half of the pronotum. The head and front lateral margins of the pronotum have long outstanding bristles. The legs and antennae are black or very dark brown; the tarsi are slightly paler. The tibiae have rows of long erect spines.

The nymph has the head, pronotum, scutellum and wing buds black; the abdomen is red but with broad black marks surrounding the stink glands. The legs are reddish, but the spines are black.

Similar species. *Geotomus petiti*, recently found in Britain, is externally virtually identical, though maybe a fraction smaller, but distinction can only be confirmed by dissection of the male genitalia. *Aethus flavicornis* (see above) has similar but stouter bristles along the front margin of the head.

Ecology. In Britain this bug is associated with Lady's Bedstraw (*Galium verum*) growing on sand dunes. On the continent it is recorded from under a wide variety of plants including lavender, mullein, poppies, spurges and Sea Kale (*Crambe maritima*), but usually in dry sandy coastal localities.

Distribution and status. The Cornish Shieldbug is critically endangered in the British Isles. It is only recorded from a handful of coastal localities in Britain, and recently only from Sennen Cove in Cornwall, where it was first found in 1864. Thankfully, Alexander (2008) comments that the population has proved to be very resilient, and appears to be in no danger. When I visited that site on holiday on 13 May 1992 I took samples of several small insects crawling amongst the loose sand and Marram Grass (*Ammophila arenaria*), and was pleased when I later identified this species among them.

Geotomus petiti Wagner, 1954

Description and similar species. Length 3.0–4.5 mm. Externally, this very small glossy black shieldbug is virtually identical to *G. punctulatus*, and although it is notionally a fraction smaller, firm identification can really only be achieved by dissection of the male genitalia and examination of the shape of the parameres of the aedeagus. The nymph is also identical to that of *G. punctulatus*.

Ecology. It is a burrowing species, at roots of plants in dry sandy areas, and in France it is associated with acid heathland and coastal dunes. Specific foodplants are unknown.

Distribution and status. A fairly strong colony was unearthed at Dungeness in July 2019 (Walker & Hollamby 2020). Further specimens were found in May 2020, including 45 on the concrete patio of the RSPB Observatory, and three in a moth trap. Hemipterists wait to see if it becomes permanently established. It is uncommon and scattered in France, mostly on the Mediterranean, Atlantic and Channel seaboards.

***Macroscytus subaeneus* (Dallas, 1851)**

Description and similar species. Length 8.0–9.0 mm. This is a small shining, dark brown bug. It is sparsely punctured above, the head being nearly impunctate. The head is smoothly rounded at the front. It has the same approximate body form as *Aethus* and *Geotomus*, but lacks the numerous bristles along the front of the head and edges of the pronotum.

Ecology, distribution and status. This is a rare South-East Asian species, known from the Philippines, Thailand and Indonesia, where it probably lives under low-growing plants. Its place on any British list is highly dubious and rests on just one record from Bath. A single dead specimen was retrieved from a Starling (*Sturnus vulgaris*) nest in the loft of a house in Odd Down, on the south-west outskirts of the city, in May 2015 (Lis & Whitehead 2019). No other information is available. The authors suggest it had arrived from Asia, possibly in or on a shipping container. They do, however, suggest that a Starling would be unlikely to scavenge a dead insect, and that the bug was probably alive when it reached Bath. Members of the Cydnidae are known to be sometimes attracted to lights, and it is tempting to suggest that this behaviour might have lured the insect from its native 'wild' habitat in the Far East to a container port for onward transportation to England.

***Adrisa sepulchralis* (Erichson, 1842)**

Description and similar species. Length 7.0 mm. A small oval blackish-brown cydnid bug. Head and pronotum fringed with several long stout setae. A single stout seta at the very base of the costal edge of the hemelytyra. Upper surface diffusely punctured.

Ecology, distribution and status. This is an Australian species until recently known only from a single female found in Tasmania over 180 years ago. Its ecology was unknown, but a male was found in 2004 associated with an Australian tree fern imported into the British Isles (Lis & Webb 2007).

Like *Macroscytus* (above), this bug is not really worthy of a place on any British list. However, at the time of its discovery here half of the known specimens in the world had been found in Britain, so I include it for its curiosity value.

***Crocistethus waltlianus* (Fieber, 1836)**

Description and similar species. Length 4.0–5.0 mm. Of typical cydnid form, broad, oval, domed, with glossy slightly metallic bronze head, pronotum and scutellum. Hemelytra pale whitish beige with a large dark blotch, and with a network of black veins on the pale membrane.

Ecology. Ground-dwelling in dry sandy places.

Distribution and status. This circum-Mediterranean shieldbug has occasionally been found in imported foodstuffs (Dolling 2008), but there is no suggestion that it might become established in the British Isles – yet. The last sighting was in the cauliflowers at a supermarket in Littlehampton, West Sussex, in March 2022 (Chris Birder, Facebook), but it fell out of view into the vegetable display after a snap taken on a phone camera. Nothing like this has ever turned up in my cauliflower au gratin.

***Adomerus biguttatus* (Linnaeus, 1758) – Cow-wheat Shieldbug**

Description. Length 5.0–6.5 mm. This is a small, elegant, oval, glossy black and white shieldbug. It is shining black, but with the narrow rims of the pronotum and hemelytra white, and with a smart white spot on each corium. The upper surface is strongly punctured, but with a shining unpunctured field running across the front of the pronotum. Very occasionally the two white spots are dark, but they are still indicated by small unpunctured fields on the hemelytra. The membrane is dark, clouded, almost black. The eyes are prominent, rounded. The antennae are slim, black. The legs are short, black, spined.

The nymph is black, but the upper abdominal segments are broadly red-marked on each side.

Adult

Similar species. *Canthophorus impressus* has the pale edges to the pronotum and hemelytra, but lacks the two distinctive white spots on the hemelytra. *Sehirus luctuosus* is entirely black.
Ecology. This species feeds on Cow-wheat (*Melampyrum pratense*) in open broadleaf woodlands, rides and clearings, where it is mostly ground-dwelling, feeding at the roots or low stems.
Distribution and status. This uncommon species occurs mostly in south-east and southern England, from Essex to Cornwall, but with scattered outlier records from Wales, Shropshire and Cumbria. A nationally scarce insect, it has much declined in the last 100 years, with former records from Norfolk, Lancashire and Scotland. Cessation of woodland coppicing is thought to have drastically reduced the habitat for its foodplant, with concomitant loss of the bug. Ramsay (2019) concluded that it was probably extinct in Scotland, so the rediscovery of the insect in Strathspey in the Cairngorms in early 2021, the first Scottish record for 30 years, was enough to make national news headlines. Woodland conservation work in Blean, Kent, and various old woodland localities in Somerset for the benefit of the Heath Fritillary (*Melitaea athalia*), the caterpillars of which likewise feed on Cow-wheat, appears to have also benefited the shieldbug.

Canthophorus (Cydnus) aterrimus (Forster, 1771)

Description. Length 8.0–12.5 mm. This oval glossy black shieldbug is the largest member of the Cydnidae. It is wholly black, except for the contrasting pale creamy grey membrane. The head is broadly rounded, with the front edge slightly indented at the centre and slightly upcurved. The scutellum is an equilateral triangle, the apex just meeting the membrane. The hemelytra have the boundary to the membrane with an almost square castellate incursion, giving the wing cases a very short appearance. The pronotum is heavy and domed, the front face suddenly sloping away, and with a strong transverse impression across the upper disc. The upper surface is densely punctured, being very heavily pockmarked on the pronotum. The lateral pronotal margins have a series of long fine black setae. The hemelytra are evenly and closely punctured, but with a narrow unpunctured band where they meet the membrane. The front legs are broad with a series of close spine-like teeth along the outer edges of the flattened tibiae. The middle and hind tibiae are strongly spined throughout. The antennae are black.

Nymphs have the head, pronotum, scutellum and wing buds black. The abdomen is a deep red with black plates around the stink glands. The legs and antennae are dark brown.

Similar species. This is Europe's largest cydnid shieldbug. All the other shieldbugs in this family have a longer, isosceles-triangle scutellum.

Ecology. This mostly ground-dwelling bug feeds on spurges (Euphorbiaceae), at the roots and in the leaf litter around the base of the plants, although it is sometimes seen crawling over the flowers.

Distribution and status. Although known from the Channel Islands (Jersey), it is not recorded from mainland Britain. In France it is mostly coastal, commonest in the south, but occurs on the dunes south of Boulogne (Roger Hawkins, personal communication); it is fairly widespread in Europe, but not Scandinavia. This might be one to look for under spurges growing on sunny undercliffs or even possibly brownfields in southern England.

Canthophorus impressus Horvath, 1871 (formerly *Sehirus dubius* in Britain) – Down Shieldbug

Description. Length 5.5–7.5 mm. This small shining blue-black bug has clear neat white edges to the pronotum and hemelytra, and white marks on the connexivum. The membrane is a contrasting dull grey. The depth of the blue appears to vary from fairly bright metallic to nearly black depending on the incident light, and can seem greenish or violet. It is strongly punctured all over, with slightly more diffuse fields on the pronotum and the base of the scutellum and head. The strongest punctures occur across the disc of the pronotum, which is impressed into a small wrinkled depression on each side of the middle. The antennae are long and slim, blue-black. The legs are blue-black and spiny.

The nymph is shining black, but with a strongly contrasting red pattern on the abdominal segments. The white edges of the pronotum appear in later instars, but the pale rims of the hemelytra do not become distinct until the final moult to adulthood.

Similar species. *Legnotus* species have a pale hemelytral rim, but have distinctive projecting conical eyes. *Adomerus biguttatus* has white spots on the corium. *Zicrona caerulea* is a similar metallic inky blue, but lacks the pale rims on the body and has no spines on the legs. Until recently this insect was called *Canthophorus/Sehirus dubius* in the British Isles, but all British specimens have so far proved to be *C. impressus*. The two species can only be firmly separated by examination of the male genitalia. The spine of the phallus is fine and strongly curved in *C. impressus*, but short and straight with a basal nodule in *C. dubius*. The true *C. dubius* occurs in scattered localities across Europe, and, in France at least, occurs through the same range as *C. impressus*, as far north as Cherbourg and Calais.

Ecology. In Britain *C. impressus* feeds on Bastard Toadflax (*Thesium humifusum*), but in Europe it also occurs on the more montane *T. alpinum*. This originally gave rise to the idea that *C. impressus* was a high-mountain species, whilst *C. dubius*, which was also recorded from Viper's Bugloss (*Echium vulgaris*), sage (*Salvia*), thyme (*Thymus*) and wormwood (*Artemisia*), was considered the lowland species. Nymphs have been reared using Marjoram (*Origanum majorana*) (Binding & Binding 2015).

Distribution and status. This is a very scarce bug of southern England, occurring only in rough unimproved chalk and limestone grassland where its scarce foodplant occurs. It is now more or less confined to the Isle of Wight, Hampshire, Dorset, Oxfordshire and Wiltshire. It was recently rediscovered in West Sussex, but seems to be lost from Kent, Surrey and East Anglia, where it once occurred.

***Legnotus limbosus* (Geoffroy, 1785) – Bordered Shieldbug**

Description. Length 3.4–4.8 mm. This small shining black bug has distinct contrasting narrow white borders running along the edges of the hemelytra, from the shoulder nearly to the membrane. It is strongly punctured all over, but with finer, more diffuse areas on the front and disc of the pronotum, and with the scutellum sometimes appearing wrinkled. The antennae are black, but with the basal two segments usually brownish. The legs are black, with the tarsi brownish. The strongly projecting conical eyes are highly distinctive. The central lobe of the head is short, and depressed at its tip, giving the front of the head a strongly notched appearance. The membrane is brownish.

Adult

The nymph is shining black, but with the abdominal segments broadly marked with dull beige or pinkish or bright red around the central black spot.

Similar species. *Legnotus picipes* is very similar, but is a fraction smaller, 3.2–4.5 mm, and the white rims along the edges of the hemelytra reach only about two-thirds of the distance to the membrane. The central lobe of the head in *L. picipes* is nearly as long as the side lobes and therefore not indented, giving the front edge of the head a more evenly rounded appearance.

Ecology. This pretty little bug feeds on bedstraws (*Galium*), including the common Goosegrass, *G. aparine*, in rough grassy meadows, verges, arable field edges, woodland rides, parks and gardens. It usually lives in dry sandy or chalky places, where it feeds at the roots and can burrow into loose friable soil. Adults occur all year round, but are most abundant in May and June; they hibernate in dead leaves in hedge bottoms, or in moss. Despite its soil-dwelling habit, adults can often be swept and it flies readily. One landed on my shoulder in the playground of Ivydale School, Nunhead, in south-east London, in June 1999.

Distribution and status. This shieldbug is widespread, but local in England to Yorkshire, and Wales, where it is mostly coastal.

***Legnotus picipes* (Fallén, 1807) – Heath Shieldbug**

Description and similar species. Length 3.2–4.5 mm. This small shining black bug is very similar to *L. limbosus*. The white edging (sometimes said to be slightly more yellowish in this species) to the hemelytra is much shorter, reaching from the shoulder to only about halfway, or perhaps two-thirds. The central lobe of the head is nearly as long as the side lobes and is not depressed, giving a much cleaner smoother outline when viewed from above or behind – appearing mildly impressed rather than sharply notched.

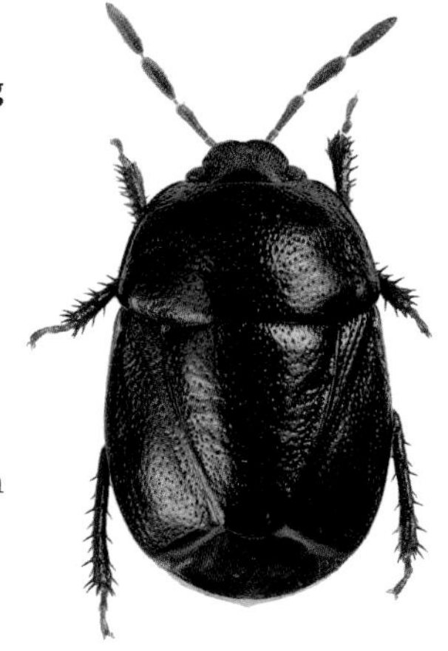

Ecology. Like its congener, *L. picipes* feeds on bedstraws (*Galium*) on sandy soils. It is assumed to be a root-feeder.
Distribution and status. This is a very scarce bug of sandy heaths in Surrey, Dorset and Hampshire, and especially the Breckland of East Anglia; elsewhere it is mainly coastal, from Pembroke to Spurn Point. I've only ever found it once – Deal Sandhills, Kent, 8 September 1976.

Ochetostethus nanus Herrich-Schäffer, 1834

Description. Length 3.0–4.4 mm. This small, neat, narrow oval shieldbug is entirely brown and black. Sometimes the head, pronotum and scutellum are nearly black, but with the hemelytra, legs and antennae slightly paler. The upper surface is punctured evenly and closely. The pronotum is rather long and more parallel-sided than in many other cydnids. The front of the head is smoothly rounded. The scutellum is long and drawn out to a narrowly rounded tip.
Similar species. It is smaller than *Sehirus luctuosus*, and the eyes are less prominently conical than in *Legnotus picipes*, very dark specimens of which might have the usual pale shoulder bar difficult to appreciate.
Ecology. In France it is recorded in uncultivated places, hillsides and sand dunes, under low plants, stones and plant debris. No specific foodplants are recorded.
Distribution and status. This bug is known from the Channel Islands, but nowhere in mainland Britain yet. It occurs along the Atlantic seaboard of France, the Cherbourg peninsula and just into Belgium, and was discovered in the Netherlands in 2017, so its occurrence in the British Isles is not entirely impossible.

Sehirus luctuosus Mulsant & Rey, 1866 – Forget-me-not Shieldbug

Description. Length 5.5–8.3 mm. This is a compact, slightly shining, all-black bug. The body is uniformly black, with some slight bronze reflections, but the second segment of the antennae is pale brown to yellow, the tarsi are pale brown, and the membrane is clouded, greyish to

Adult

brownish yellow. The body is punctured all over, those punctures on the wings being perhaps finer and denser than in close relatives. The underside is much more finely punctured than the upper surface. There is a smooth more-or-less unpunctured field across the front middle of the pronotum. The side lobes of the clypeus meet, enclosing the central lobe.

The nymph is glossy black but with the abdominal segments pale and dull except for the black dorsal spots and connexivum marks.

Similar species. All other British Cydnidae have white markings, except the smaller and much rarer *Geotomus* species. *Ochetostethus nanus*, should it turn up, is smaller and narrower. In Europe, *Sehirus morio* is larger (8.5–11.5 mm), less bronze and more deeply black, with a longer rostrum reaching to the rear of the middle coxae.

Ecology. The Forget-me-not Shieldbug obviously feeds on forget-me-nots (*Myosotis* species), usually on open, dry, well-drained soils. Although adults occur all year round, peak activity seems to be in May and June. On hot days it runs rapidly about on the ground around its foodplant.

Distribution and status. Though never common, this is a widespread species across most of England to Yorkshire, though it is seemingly absent from the West Country, and is scarce in Wales.

Sehirus morio Linnaeus, 1761

Description and similar species. Length 8.5–11.5 mm. This is a medium-sized glossy black shieldbug. Essentially, it is extremely similar to *S. luctuosus*, but is slightly larger – among the European Cydnidae it is second only to *Canthophorus aterrimus* (8.0–12.5 mm) in size. The rostrum is proportionately slightly longer than that of *S. luctuosus*, reaching beyond the middle coxae when at rest, and this is usually partially visible in side view.

Ecology. In Europe *S. morio* is recorded from Alkanet (*Anchusa officinalis*) and Hound's-tongue (*Cynoglossum officinale*); these are in the same family (Boraginaceae) as forget-me-nots.
Distribution and status. Though unknown in the British Isles, this shieldbug occurs across a similar overlapping European range to *S. luctuosus*, and is frequent in France right up to Brittany and Normandy. It is certainly one to keep an eye open for.

***Tritomegas bicolor* (Linnaeus, 1758) – Pied Shieldbug**
Description. Length 5.5–7.9 mm. This is one of Britain's most handsome insects – a glossy, prettily marked broad oval black and white shieldbug. It is such a distinctive species: all shining black, but boldly marked with strongly contrasting white – a large spot on the front angles of the pronotum, a C-shaped blotch on the shoulders of the hemelytra, a jaggedly edged triangle at the end of the corium against the membrane, and spots on the connexivum. The underside of the thorax is minutely punctured, but the underside of the abdomen is more shining. The head is broad and rounded, with the central lobe of the clypeus shorter than the side lobes, which are slightly raised giving a depressed and notched appearance. The insect is punctured all over, with some more diffuse fields on the disc of the pronotum and the scutellum. The wing membrane is a dull grey. This is a very convex species, with a distinct and abrupt angle as the hemelytra and scutellum meet the membrane. The extreme tip of the scutellum is down-turned. The legs are black, but are extensively streaked with white on the outer edges of the tibiae to about two-thirds the length. The antennae are black.

Early-instar nymphs are shining black but with the abdominal segments mostly reddish pink; the colour fades to white in later instars, and there are strong white marks on the head, pronotum, scutellum and wing buds.
Similar species. Until recently this insect was unmistakable, but now great care has to be given to separate it from *T. sexmaculatus*, which has a longer white streak down the side of the pronotum and darker wing membranes. In general appearance it is also slightly broader and rounder than *T. sexmaculatus*. In central-eastern France the very similar *T. rotundipennis* is slightly broader and rounder and has larger white blotches; in particular the mark at the front angle of the

ABOVE: Adult

Nymph

pronotum is white right to the edge along the entire blotch, but in *T. bicolor* the marginal ridge is finely edged with black for the rear half of the spot.

Ecology. The Pied Shieldbug feeds on White Dead-nettle (*Lamium album*) in damp meadows, road verges, woodland rides and rough grassy places, sometimes woundworts (*Stachys* species), and occasionally Black Horehound (*Ballota nigra*) in drier places. Adults overwinter, but appear to die off by June, when nymphs predominate.

Distribution and status. This attractive insect is widespread and fairly common across southern and central England to Yorkshire; it is scarcer in Wales and western England, where it is often more coastal.

Pied is a word I'd like to see brought back into more widespread use when describing black and white things. After wagtail and piper, shieldbug is a good start.

Tritomegas sexmaculatus (Rambur, 1839) – Rambur's Pied Shieldbug

Description and similar species. Length 6.0–8.0 mm. This is a small, round, shining black and white bug very similar in appearance to *T. bicolor*. It appears slightly less rounded than that species, more elegantly shaped with a flatter pronotum, which is more evenly punctured over its rear half. The white marks on the pronotum are longer and narrower, extending nearly the full length of the pronotal edges; in *T. bicolor* they are shorter and rounder, leaving about three-quarters of the margin black. In some lights the black colour of *T. sexmaculatus* can appear slightly inky blue. Its head is longer and flatter and the central lobe of the head is slightly longer than the side lobes, giving a more rounded appearance (the central lobe is shorter in *T. bicolor*, giving an indented or notched appearance). The wing membrane is also very dark, nearly black, giving the insect a crisper, more deeply graphic appearance than *T. bicolor*. Also, the legs are more extensively black, with white marks on the outer edges of the tibiae less than half the length.

The nymphs are similar to those of *T. bicolor*, but often darker, showing less of a pied pattern at this stage. They are also more orangey in *T. sexmaculatus* compared with the paler more yellowish-marked nymphs of *T. bicolor*.

Adult

Nymph

Ecology. This species is usually recorded from Black Horehound, therefore in drier habitats than *T. bicolor*, which occurs more in damp meadows, road verges and woodland edges where its preferred foodplant White Dead-nettle occurs in lusher situations; however, both foodplants are recorded for each species. It often occurs in large numbers on the foodplant, with scores of adults and nymphs clustered into the flower heads.

Distribution and status. First discovered in Britain, in Kent, in 2011 (Bantock 2017), after massive range expansion in Poland, Germany and the Netherlands, this species appears to be a recent colonist from mainland Europe and is now spreading. Its arrival here was predicted by Butler (1923), and Barnard (2016) suggested it had probably been present for a couple of years before it was officially noted. It is currently recorded from numerous sites across Kent, particularly in the north, right into Dartford in south-east London (Barnard in preparation), and in Essex, Middlesex and Surrey, and is set to spread further. I was very pleased to find it myself for the first time shortly after writing this paragraph, swept from Black Horehound on a hot dry derelict brownfield site in Becton on 15 June 2020, and subsequently in several other London localities. This will be an interesting species to monitor and to appraise its spread in any second edition of this book.

FAMILY THYREOCORIDAE

Thyreocoris scarabaeoides Linnaeus, 1758 – Scarab Shieldbug

Description. Length 3.8–4.1 mm. This tiny, glossy, domed, shieldbug is entirely jet black, except the eyes, tarsi and antennae, which can all be reddish brown. The upper surface is very convex, shining, sometimes with slightly metallic bronze reflections. It is punctured all over, strongly on the head and pronotum, less strongly towards a vague nearly unpunctured central line down the scutellum. The under surface is also glossy black, and finely punctured throughout. The head is short and broad, with the eyes subconical; the central lobe is completely contained in the lateral lobes, which meet each other at their tips. The pronotum is short and very broad, and narrowly ridged along the lateral margins, but this narrow line disappears under a domed tumescence at the hind corners. The scutellum is huge, subcircular, broadest behind its middle and sinuously narrowed into the pronotum; the hemelytra are short and narrow, hardly visible from above. The membrane is dark grey, but barely visible, being almost completely covered by the scutellum. The legs are short; all tibiae have rows of short stout spines. The antennae are very short, only slightly longer than the head is wide.

The nymphs are shining black, but the exposed abdominal segments are reddish brown or pale grey around the black scent gland marks and marginal abdominal blobs.

Similar species. Only *Coptosoma scutellatum* has a similar huge scutellum and shiny black form, but it is smaller and even more convex than *Thyreocoris*.

ABOVE: Adult

Nymph

Ecology. This is a root-dweller, seemingly associated with violets (*Viola*). It has been recorded feeding on violet seeds, but has been kept in captivity on other plants. It is usually found by fingertip grubbing at plant roots, or in moss, or when vacuum sampling. It is a regular under squares of roofing felt, carpet tile or corrugated metal left out to monitor reptiles.
Distribution and status. The Scarab Shieldbug is widespread in southern England and south Wales, with outliers to Barrow-in-Furness and Doncaster, but is not common and is easily overlooked. It is more or less limited to chalk downland, sandy soils and coastal dunes.

In some accounts this species has been given the English name Negro Bug, with the plaint that this is simply the Spanish word for black. But this word is now so heavily tainted that it cannot possibly be used.

FAMILY PENTATOMIDAE

Mecidea lindbergi Wagner, 1954

Description. Length 11.5–12.0 mm. This long, narrow, straw-coloured bug is very un-shield-shaped. The head is very long, the central lobe much shorter than the sharply protruding side lobes. The pronotum is a narrow trapezoid, with the hind corners only very slightly protruding. The scutellum is very long and narrow, particularly at the apex. The hemelytra are long and narrow, with the membrane clear, and clearing the end of the abdomen. The colour is more or less uniformly pale brown or beige, but with some vague streaks down the pronotum, and on the abdomen visible through the slightly translucent hemelytra and clearly transparent wing membranes. The corium is sometimes darker. The upper surface

Adult

is marked all over with moderately dense even punctures. The long legs are pale beige. The antennae are long and narrow, beige; the second segment has a variable raised flange or swelling on one side. It really looks nothing like a shieldbug.

The nymphs are equally long and narrow, streaked with straw-yellowish, beige or brownish.

Similar species. No other north European shieldbug looks anything like this, although several Mediterranean and North African species in the genus are separated by very subtle differences in jugal lobe length, pronotal shape and antennal structure. In the field it might easily be overlooked as a mirid grass bug such as *Stenodema* or *Leptopterna*, but the presence of ocelli rules out Miridae.

Ecology. This is a species of warm, dry, well-drained places, where it feeds on grasses, including Marram Grass (*Ammophila arenaria*).

Distribution and status. This is a southern European and North African species, occurring from Cape Verde to Pakistan, with scattered records from Italy and Greece, and a single specimen in southern France. Numerous specimens turned up in moth light traps on the Devon, Dorset and Hampshire coasts, between Exeter and Portsmouth, between 17 and 30 December 2015, and others were found in the garden of an entomologist, suggesting a sudden migratory influx (Bantock 2016b). This coincided with an unprecedented appearance of moths apparently moving north on unseasonably mild airflows from North Africa during this time. To date there have been no further records, but it shows that, really, anything might turn up here.

Arma custos (Fabricius, 1794)

Description. Length 10.0–13.5 mm. *Arma* is a large brownish shieldbug with prominent pronotal shoulders. The long head is bluntly rounded in front, with the central lobe of the clypeus slightly shorter than the side lobes. The pronotum is broad, extended into a short sharp tooth on each side. The scutellum is subtriangular with the apex extended into a long slim lobe. The bug is generally a mid-brown, but has neatly defined dark punctures giving a smudged appearance. In more strikingly marked individuals the pronotal angles are black-edged in front, pale-edged behind. The connexivum has alternating but variable pale and dark squares. The antennae are pale brown with variable contrasting rings of dark brown on all or some segments.

The membrane is darkened, sometimes appearing black. The legs are reddish brown with scattered black punctures, sometimes coalescing into streaks. On the underside the thoracic plates are a similar brownish to the upper surface, but not so densely punctured; the abdominal segments are often slightly olive green.

The nymph is mottled brownish green; earlier instars are more contrasting, with shining black or dark bronze-brown against the pale abdomen.

Similar species. *Arma* is more muted than *Troilus luridus* and has shorter spines than *Picromerus bidens.*

Ecology. This predator of moth and sawfly caterpillars and beetle larvae is predominantly arboreal, on various broadleaved trees; mostly in damp woodlands

Adult

Nymph

and on river banks. In Europe it is often found feeding on the Alder Leaf Beetle (*Agelastica alni*), a species thought to be extinct in Britain until its discovery in Lancashire about 2004, and which is now widespread and common throughout much of England.

Distribution and status. Though not known from the British Isles, this species is widespread and common in mainland Europe to Normandy, Calais and Belgium, and up into Denmark and southern Scandinavia; it is apparently spreading, so might turn up one day.

***Jalla dumosa* (Linnaeus, 1758)**

Description. Length 11.0–16.0 mm. *Jalla* is a large oval brown shieldbug marked with a slightly variable pattern of pale streaks. The flat head projects stoutly in front of the eyes. The broad pronotum has edges straight or only gently curved. The scutellum is broad and long, subtriangular, but with the apex slightly extended, almost parallel-sided. The ground colour of the upper surface is mid- to olive-brown, but with fairly strongly contrasting pale cream or white marks. A prominent streak passes from the clypeus back across the head, onto the pronotum; another, or a continuation of the first, through and down the centre of the scutellum. The edges of the pronotum and the extreme base of the hemelytra are bordered pale. The most striking marks are two teardrop-shaped blotches, one at each side of the base of the scutellum. There are sometimes narrow pale tick marks on the connexivum. Excepting the pale areas, the surface is covered with heavy black punctuation, accentuating the dark streak marks on the front of the pronotum. The antennae are long and slim – dark, but with the tip of each segment sometimes slightly paler. The legs are dark brown or black, but each tibia is broadly ringed with white or cream. The pale markings are sometimes pinkish or red.

The nymph is dark brown, often with blue or brassy reflections, but later instars already show the pale streaks along the edges of the pronotum. The abdomen is broadly red except for the central dark plates around the dorsal glands and the spots on the connexivum.

Similar species. With its banded tibiae, *Jalla* looks a bit like a large *Rhacognathus punctatus* but with the pronotal corners less pronounced. On the whole, the pale marks on the pronotum and scutellum are pretty and distinctive.

Adult

Ecology. In Europe it occurs in rough grassy and flowery places. It is a predatory species, attacking Lepidoptera caterpillars and beetle larvae, and probably whatever else it can find. In Britain, records have been confined to coastal sandy places.
Distribution and status. In Britain this species is known only from a handful of specimens collected in the Sandwich and Deal area of Kent between 1850 and 1885. One example was a nymph, suggesting that it had been breeding in the area. None has been seen since, and it is thought to be extinct in Britain. Douglas & Scott (1865) state: 'Two specimens taken by Mr H. J. Harding, and one by Mr Ernest Adams, but they have no note of the locality', which seems a bit lax of them. It is widespread but uncommon in mainland Europe.

Picromerus bidens (Linnaeus, 1758) – Spiked Shieldbug

Description. Length 10.0–13.5 mm. One of Britain's most distinctive shieldbugs, *Picromerus* is medium-sized, stout, brown and with distinctive sharp thoracic spines. The head is long, flat, and bluntly rounded at the front, neatly and deeply set into the pronotum; it sometimes appears tinged deep blue. The pronotum is broad, also sometimes tinged deep blue along its edges, and is extended at each side into a long sharp thorn with a small nodule on its rear surface. Several small yellowish denticles line the edges of the pronotum on the approach to

Adult with prey

the front angles. The scutellum is long, subtriangular, with the apex extended into a rounded lobe. The entire upper surface is dark brown, heavily and rather densely sculptured with black punctures. There are a few paler flecks, often with two small pale spots on the disc of the pronotum, two spots on the base of the scutellum, and a distinctive curved creamy yellow rim at its tip. The underside is similarly coloured, or slightly paler reddish, black-punctured, but not so strongly or densely as the upper surface. The orifice of the metathoracic stink gland has a broad reddish semi-crescent-shaped area. The legs and antennae are rather bright orange-red, with perhaps the knees and final antennal segment slightly darker. The front femur has a short spine at the apical third.

The nymph is mostly black with a red abdomen, becoming metallic bronze and black-punctured brown in later instars.

Similar species. *Pentatoma rufipes*, *Troilus luridus* and *Carpocoris purpureipennis* have pronounced thoracic extensions, but none matches the distinct sharp spike of *Picromerus*; the only species that comes near is the brightly coloured and highly unlikely *Elasmucha ferrugata*.

Ecology. *Picromerus bidens* is unusual for a British shieldbug in that it overwinters in the egg stage. Laid in batches of about 28 on the leaves of trees, from June to September, the eggs hatch the following May. The nymphs feed up until new

adults start to appear in July, with a peak in September. This is mostly a predatory species, attacking moth and sawfly caterpillars, beetle larvae and other small invertebrates. It usually occurs on shrubs and trees, or tall herbage in a wide variety of habitats including downlands, woods, heaths and marshy meadows, but avoiding built-up areas and farmland according to Hawkins (2003). It has been seen feeding on nectar, particularly thistles, so is not entirely carnivorous.
Distribution and status. *Picromerus* is widespread and fairly common across most of Britain and Ireland, becoming more scattered in Scotland.

***Pinthaeus sanguinipes* (Fabricius, 1781)**
Description. Length 11.5–16.5 mm. This is a large striking brown, yellow and red bug. The large flat head is broadly rounded at the front. The pronotum is broad, drawn out to a round, hooked prominence at each side. The scutellum is long and drawn out to a tongue shape at the apex. The hemelytra are broad and flat, the membrane dark and smoky. The connexivum is brightly marked with alternating black and white (sometimes black and reddish brown) squares. The legs are fairly bright red. The antennae are black, with the fifth segment broadly pale at the base. Apart from the connexivum, the upper surface is predominantly a rich brown, with

Adult

the side margins of the pronotum, a narrow tick on each side of the scutellum base and a broad spot at the scutellum tip pale yellow, sometimes nearly white. It is heavily punctured black, more heavily but less densely on the pronotum and scutellum than on the hemelytra.

The nymph is dark brown, heavily marked with black, and with bright white marks on the edge of the pronotum, the sides of the connexivum, and a broad band on each tibia.

Similar species. At first sight *Pinthaeus* is very similar in appearance to *Pentatoma rufipes* by virtue of its general shape, dark brown colour, red legs, spotted connexivum and contrasting pale scutellar tip. With the rounded pronotal corners, however, the shape is different, with the straighter margins suddenly flexing into the hook where *Pentatoma* has a more broadly rounded transition. There is a small tooth on the front femur; this is lacking in *Pentatoma*. *Pinthaeus* also has the thickened rostrum of the Asopinae.

Ecology. This is apparently mostly a species of woodland edges, where it feeds on arboreal moth caterpillars.

Distribution and status. A single specimen landed in a moth trap in Lowestoft, Suffolk, in September 2021 (Antony Wren on Twitter). It occurs throughout much of France and central Europe, just about into Denmark but not the Scandinavian peninsula, but is generally regarded as a rare insect. This one may be on the cusp of arrival into the British Isles.

Rhacognathus punctatus (Linnaeus, 1758) – Heather Shieldbug

Description. Length 7.5–9.5 mm. This is a medium-sized shining brown shieldbug. The pronotum is broad, with side angles prominent, though not sharply pointed. The scutellum is subtriangular with the apex extended into a broad rounded subparallel lobe. The upper surface is dark shining brown, mostly a combination of uniform dense and evenly spaced black punctures over a mid-brown base colour. It can appear metallic bronze-green in some specimens, or the background colour can take on a reddish blush in others. A central line down the pronotum, often continued onto the scutellum, is pale. Some specimens appear freckled with paler marks because of irregularities in the punctuation. The connexivum is marked with variable pale yellow tick marks to large spots. The membrane is dark. The antennae are long and slim, dark brown but with

Adult

Nymph

the joins often pale-ringed. The legs are stout, dark brown, but the tibiae have a broad pale ring around the middle.

The nymph is shining black with metallic bronze glints contrasting with the pale reddish-pink abdominal colour.

Similar species. The pale-barred tibiae are very distinctive. The very unlikely *Jalla dumosa* is larger and has less pronounced pronotal angles.

Ecology. Adults occur all year, but peak in August; nymphs appear from June to August. This is a species of dry heaths and wet boggy places where heathers grow. It is a predatory species, particularly attacking larvae of the Heather Leaf Beetle (*Lochmaea suturalis*), though other prey species are recorded.

Distribution and status. This species is recorded from England, Wales, Scotland and Ireland, but localities are often widely scattered, and this is a rather local insect. Its main centres appear to be the Lake District, Yorkshire, the lowland heaths of Surrey, Hampshire and Dorset and the Weald of Sussex and Kent.

Troilus luridus (Fabricius, 1775) – Bronze Shieldbug, Banded Shieldbug

Description. Length 10.0–13.0 mm. *Troilus* is a striking shining brown shieldbug with prominent rounded pronotal angles. The long flat head is rounded at the front, but with the central lobe of the clypeus shorter than the side lobes, giving a notched appearance. The pronotum is broad, expanded into a flat, rounded, hooked prominence, ahead of which the sides are lined with a row of small pale denticles

Adult

Nymph

to the front angles, though these can be muted, rounded, blended together or asymmetrical, one side being more jagged than the other. The scutellum is nearly triangular, but with the apex drawn out into a slight lobe. The upper surface is smudged brown by virtue of the dense black punctures over a paler background colour. The punctures on the head, pronotum and scutellum are very strong and deep, those on the endocorium smaller, finer and sparser, and there is usually a small darkened unpunctured triangle at the end of the endocorium. The general appearance is brown, but with bronze glints, sometimes more obviously greenish metallic. The serrated edges of the pronotum, the flecks at the base of the scutellum, a vague cloud at the end of the scutellum, and large squarish spots on the connexivum are yellow. The membrane is cloudy, with a darker smudge at the tip, although this sometimes fades in museum specimens. The underside is a dirty yellow, sparsely stippled with tiny black spots. The antennae are dark, but with the apical half of the penultimate segment brightly marked with a broad yellow ring. The legs are pale to mid-brown, but appear dark or at least dark-clouded because of the many black punctures. The tibiae (especially the front pair) are flattened on the outer surface, and edged with fairly sharp carinae (ridges).

The head, pronotum and developing wing buds of the nymph are shining black, in later instars becoming an attractive metallic coppery bronze, contrasting against the pale reddish, yellow or brown abdominal segments.

Similar species. The hooked shape of the pronotal prominence is distinctive; that of *Pentatoma rufipes* is longer and rounder and more sharply hooked at the apex, that on *Picromerus bidens* straighter and sharper. *Troilus* is shorter and broader than *Pinthaeus*, with darker legs, and it also lacks the pale spot at the tip of the scutellum.

Ecology. Overwintered adults lay eggs in spring and these emerge from June. New adults peak in September. *Troilus* is usually found on trees, where it is a predator of moth and sawfly caterpillars, beetle larvae and other small invertebrates. It is often found hanging from a twig or leaf, with its hapless prey dangling from its proboscis, frequently still wriggling as its innards are slowly sucked out over some time. It mostly occurs in woodlands (both conifer and broadleaf), but also occasionally in hedgerows and parks.
Distribution and status. *Troilus* is widespread across England, Wales and Ireland, with a scatter of records in Scotland. It is generally common.

Zicrona caerulea (Linnaeus, 1758) – Blue Shieldbug

Description. Length 5.5–8.0 mm. This is a small but highly distinctive shining metallic-blue shieldbug. The head is long and rather narrow, exposing the attachment of the antennal base, which has a small but distinct sharp tooth on the outer edge. The pronotum is broad and domed; the front angle has a small but sharp denticle; the outer corner is pronounced but not projecting beyond the shoulders of the hemelytra. The scutellum is long subtriangular, extended at the apex into a

Adult

short broad, nearly parallel-sided lobe. The entire body is a dark inky metallic blue-black, appearing brighter in some specimens, sometimes greenish. The membrane is dark, sometimes appearing black. The upper surface is punctured all over, but with some diffuse fields on the pronotum and scutellum. The under surface is more finely punctured and more shining.

The nymph is also shining metallic blue-black, but with the exposed abdomen reddish apart from the central blotches and connexivum marks.

Similar species. Really, *Zicrona* is unmistakable. Only *Canthophorus impressus* has a similar metallic blue colour, but that species is darker, a different shape and has white edge marks. Fifth-instar nymphs of *Troilus luridus* and *Rhacognathus punctatus* closely resemble fifth-instar nymphs of *Zicrona*, but are generally greenish bronze rather than blue.

Ecology. Adults peak in May, when they emerge from hibernation, and in August after the ensuing generation of summer nymphs has matured. This is another predatory species, usually attacking the larvae (and sometimes adults) of leaf beetles, seemingly metallic blue *Altica* species in particular. It occurs in a wide variety of habitats, but especially rough grassy places, heathland, brownfield land, damp meadows and woodland edges.

Distribution and status. *Zicrona* is a local species, never common, but widespread across most of England, Wales and Ireland, and with a few records in Scotland.

***Aelia acuminata* (Linnaeus, 1758) – Bishop's Mitre**

Description. Length 7.0–10.0 mm. With its slim, elegant, narrow oval, pale-streaked form, and pointed triangular head, there is no mistaking *Aelia acuminata*. The long triangular head is narrowed to a blunt point in front where the side lobes of the clypeus meet well ahead of the very short central lobe, which they envelop; it appears down-turned in side view. The ocelli are small red dots. The broad pronotum is domed and the side angles are prominent but not sharp, though a narrow pale flange running along the margin stops short, leaving a small tooth. The scutellum is long, gradually narrowed to a broadly rounded tip. The abdomen is very narrow for a shieldbug, with the connexivum a mere thin streak. The membrane is transparent, the dark abdomen showing through. The whole upper surface is streaked brown and beige, normally with

Adult

Nymph

a narrow pale stripe extending from the clypeus, over the head and pronotum, to the tip of the scutellum, and this is flanked by two broader variable streaks. The side margins of the pronotum are narrowly pale; a narrow pale raised and unpunctured streak also runs near the edge of the hemelytra, and the narrow connexivum is pale too. The entire upper side is covered with strong even punctures, the colour of these (varying from black through brown to clear) creating or emphasising the dark longitudinal streaks. The underside is more uniformly greenish beige, with scattered black marks. The legs are pale, but with scattered black punctures especially on the front pair. The antennae are pale but with the penultimate segment vaguely orange and the final segment darker.

Aelia nymphs are long, narrow oval, not angular, pale beige, streaked with a variable pattern based around four dark stripes.

Similar species. In Britain nothing else comes close to the narrow streamlined appearance of what must qualify as one of our most elegant shieldbugs. There are, however, several other species in the genus in mainland Europe, some of which occur commonly in France and well up into southern Scandinavia, so might potentially occur here. In *A. acuminata* the middle and hind femora have a pair of small dark spots or streaks on the hind face, whereas other members of the genus have this area of the legs with one mark or none. The third antennal segment of *A. acuminata* is clearly longer than the second (about 1.7–2.0 times), but in others of the genus the two segments are about equal in length. Two possibilities are *A. klugii*, which is smaller (6.0–8.5 mm) and sometimes slightly more strikingly and contrastingly coloured than *A. acuminata*, and *A. rostrata*, which is much brighter and, at 9.5–12.0 mm, also much larger.

Ecology. The adults emerge from hibernation in about April and numbers peak in May, and then again in September after the nymphs (which occur mostly in July and August) of that year's generation achieve maturity. *Aelia* feeds on grass seed heads, against which it is remarkably camouflaged as both nymph and adult. It occurs in rough grassy and flowery places including downlands, hay meadows, verges and brownfield sites. In the eastern Mediterranean this species is regarded as a major pest of wheat.

Distribution and status. In the middle of the twentieth century this was regarded as a widespread but rather local insect of southern chalk downs, or coastal localities, but it has since increased in numbers and in range. It is now common and widespread in south-east England, with scattered records to Yorkshire, but it remains mainly coastal in the West Country and Wales. Roving across the South Downs as a boy, I was always very excited to find this highly distinctive, elegant and then very local species. Now sometimes the net is bustling with dozens of them.

Aelia klugii (Hahn, 1833)

Description and similar species. Length 6.0–8.5 mm. Rather similar to *A. acuminata*, in shape, colours and patterning, though a little smaller. The undersides of the middle and hind femora lack the paired dark marks found in *A. acuminata*. Segment 3 of the antenna is only slightly longer than segment 2, where in *A. acuminata* it is almost twice as long.

The nymph is extremely similar to that of *A. acuminata*.

Ecology. In Europe this bug is recorded feeding on a wide range of grasses in warm, dry, well-drained places.

Distribution and status. Although unrecorded from the British Isles, this species is relatively common in Europe, through to Denmark and southern Scandinavia, and might feasibly turn up.

Adult *Aelia klugii*

Aelia rostrata **Boheman, 1852**

Description and similar species. Length 9.5–12.0 mm. Generally larger and more brightly marked than *A. acuminata*, especially in the pale-edged head, which appears longer and straighter. Like *A. klugii* it lacks marks on the middle and hind femora and has more or less equal second and third antennal segments. On the underside of the head *A. rostrata* has a pronounced ragged thorn on the buccula, visible in side view.

The nymph is very similar to that of *A. acuminata*.

Ecology. Like others in the genus, *A. rostrata* is recorded feeding on various grass species in warm, dry, well-drained places.

Distribution and status. This species is commonest in southern Europe, but it does occur

Adult

sporadically into northern France, Denmark, and just in southern Scandinavia. It might turn up in Britain one day.

Neottiglossa pusilla **(Gmelin, 1789) – Small Grass Shieldbug**

Description. Length 4.0–6.0 mm. This is a small broad, stout, pale-brown shieldbug. The head is large and protruding, rather pointed at the front, but the tip is deflected downwards, appearing strongly curved in side view. The eyes are very prominent, and subconical. The pronotum is broad, with the side angles prominent but not protruding. The scutellum is large, broad, and bluntly pointed at the apex. The hemelytra are short, about as long as the scutellum. The upper surface is brown, but covered all over with black punctures, giving a clouded appearance. Generally the head is black (or at least darker than the rest of the body), the pronotum has darker patches near the front angles, and the scutellum has dark basal patches and a dark tip. The edges of the pronotum have distinct pale cream beading, mirrored on the hind body by pale edges to the hemelytra and connexivum. There is usually a narrow variable pale stripe from the middle of the head, back across the pronotum and three-quarters of the way down the scutellum; the scutellum also has two pale tick marks at the basal corners. The membrane is clear, but the veins are darkened. The underside (especially the abdomen) is dark, sometimes

Adult

appearing bronze, with heavy black punctuation more or less throughout, but leaving the underside of the connexivum clear yellowish. The antennae are pale at the base, but darkened in the two apical segments. The legs are pale, but with dark shadows around the femur/tibia and tibia/tarsus joints.

The nymph is a variable mottled or streaked brown or olive, with the abdomen pale, except for marks around the dorsal glands and connexivum spots. Even in the earlier stages, the downward-curved head is very obvious.

Similar species. There is nothing else like it in Britain and Ireland, and similarly sized *Aelia* nymphs do not have the head so strongly curved over. Several similar species occur in mainland Europe, including the widespread *N. leporina*, which occurs right up to the Cherbourg peninsula and the Channel Islands, sometimes even more commonly than *N. pusilla*. *N. leporina* has the underside of the connexivum more broadly pale, and the scutellum is obviously longer, by about a quarter, than the hemelytra.

Ecology. Adults peak in June, July and September, with nymphs appearing in July and August. This rather secretive species feeds on various grasses in rough places like damp meadows, verges, open woodland rides, hillsides, heaths and downland.

Distribution and status. This shieldbug is widespread in southern England, through East Anglia to Yorkshire, and just into Wales. It is a secretive and rather local species, and is never common.

Neottiglossa leporina (Herrich-Schäffer, 1830)

Description and similar species. Length 5.0–6.5 mm. This shieldbug is extremely similar to *N. pusilla*. It is perhaps a fraction larger, but should immediately be identifiable by virtue of its longer scutellum, which projects beyond the tips of the corium by about a quarter of its length. In side view the pale under-edge of the connexivum is broader and more pronounced.

The nymph is extremely similar to that of *N. pusilla*.

Ecology. Like its congener, *N. leporina* feeds on grasses in rough, lush habitats like meadows, and verges.

Distribution. Although not known from Britain or Ireland, this species occurs in the Channel Islands and is widespread in France.

Adult Nymph

Halyomorpha halys Stål, 1885 – Marmorated Shieldbug, Brown Marmorated Shieldbug

Description. Length 12.0–17.0 mm. This is a large mottled brown shieldbug. The head is long in front of the eyes, broadly rounded at the front. The pronotum is short and broad, with the corners prominent, but not pointed. The scutellum is subtriangular, extended at the apex into a short broad rounded tongue. The overall ground colour is usually pale brown, but the entire upper surface is covered with black punctures. The head is dark; the pronotum is dark except for a pale blotch on the side margins and four small slightly raised pale fields near the

Adult

front margin. The scutellum has 2–5 small pale spots along its basal margin, with the side ones usually the largest. The connexivum is strikingly pale-and-dark marked. The antennae are mid-brown but segments 3–5 have broad strong dark bands across each, leaving pale bases and tips. The legs are pale brown variously darkened by black punctures, especially on the hind femora. The underside is mostly pale, with some areas of vague darkening because of punctuation.

The nymph is dark brown; early instars are distinctively spiny along the sides of the fore-body and later instars have echoes of the pale spots on the pronotum and scutellum to come. The legs are black but with a broad pale band around the middle of the mid and hind tibiae.

Similar species. In general appearance adults are extremely similar to *Rhaphigaster nebulosa*, which is also large and has the same mottled coloration, though it lacks the small yellow spots on the pronotum and scutellum. Despite their approximate similarities they are in completely different shieldbug subfamilies. The key distinction is that, on the underside, the third sternite of the abdomen (the second visible segment) is simple in *Halyomorpha* but has a long sharply pointed projection reaching forwards between the hind coxae in *Rhaphigaster*. In late-instar nymphs this spine becomes evident as a wrinkled protuberance. Nymphs are easier to distinguish because the pronotal edges of *Halyomorpha* are spiny, and there are additional spines on the sides of the first abdominal segments, and in front of the eyes; nymphs of *Rhaphigaster* are smoothly edged.

Ecology. This is a very general plant-feeder on a wide variety of species (over 300 recorded) in woods, gardens, parks, farmland, indeed anywhere. Its attacks on fruit trees, causing cat-facing on peaches and apples for example, have raised it to pest status in many countries. Adults have a propensity to come into buildings to overwinter, sometimes invading houses or offices in plague proportions. The scent they give off is notoriously powerful and pungent, and large accumulations of many thousands cause great consternation to the property owners, who then have to spend time and money clearing them out and cleaning the buildings.
Distribution and status. This shieldbug is not yet (April 2023) definitely confirmed as fully established in the British Isles, but confirmation is due any time now. This species is native to China, Taiwan, Korea and Japan, but has been accidentally transported around the world where it has become widely established. It is now regarded as a notorious pest species in North America since its arrival in 1996, and has also been discovered in Liechtenstein (2004), Switzerland (2007), Germany (2011), Italy (2012), France (2013), Hungary (2013), Greece (2014), Corsica (2018) and Malta (2019). Specimens have been found in passenger luggage and imported goods, and two males were found in a pheromone trap (part of an interception and monitoring study) at the RSPB's Rainham Marshes in Essex in 2020. Another turned up in the wildlife garden of London's Natural History Museum that same year. Although it is not technically a legally notifiable pest, Defra has already produced a fact-sheet for this species and any sightings should be reported. Across the world it is usually called the Brown Marmorated Stink Bug (rather than Shieldbug), since it first acquired an English name in the USA.

***Carpocoris purpureipennis* (De Geer, 1773)**

Description. Length 10.5–13.5 mm. This is a large, bright, strikingly marked shieldbug. The head is long triangular, flatly rounded in front. The pronotum is broad, extended into a prominent curved angular process at each side. The subtriangular scutellum is rounded at the tip. Though rather variable in colour, it is generally a muddy green, with the pronotal corners darker, the front edges of the pronotum and the scutellum tip lighter green or yellowish. The head and hemelytra are pinkish, sometimes also with a reddish cloud near the pronotal

Adult

Nymph

angles. The connexivum is marked with black and yellow alternating squares. The legs are pale, yellowish, the femora greenish, the tarsi and tibial apices pinkish. Apart from the first segment, the antennae are more or less uniformly dark. This, at least, is the typical colour pattern, but this species is very variable and can be almost entirely orange-brown, with lighter or heavier dark marks, or the reddish clouding can be much more extensive.

The nymphs are also very striking (later instars anyway), being a deep purple-red with creamy yellow stripes down the head, pronotum, scutellum and wing buds; the abdomen is pale pinkish with dark blotches around the dorsal glands and on the connexivum, giving a highly stylish graphic appearance.

Similar species. The typical colour form might be mistaken for *Dolycoris baccarum*, but that species is distinctively hairy and the antennae are strikingly bicoloured. There are several *Carpocoris* species known in Europe and there has been much taxonomic debate to organise them into acceptable taxa. The pronotal angles of *C. mediterraneus* are more pronounced than those of *C. purpureipennis*, with the black marking more concentrated to the angle point where in *C. purpureipennis* the black mark is more diffused along the pronotal edge.

Ecology. This species feeds mostly on composites in grassy flowery places such as rough meadows, field edges, verges, woodland rides, parks and gardens.
Distribution and status. It is probably just a vagrant so far to the British Isles, with irregular scattered records. It turned up in Gloucestershire in 1995 (Barclay & Nau 2001), and at Portland in Dorset in 2005, and one was photographed in East Sussex on 18 May 2017. It is widespread and common in France. It is, perhaps, on the cusp of invasion and colonisation here.

***Carpocoris mediterraneus* Tamanini, 1958**

Description. Length 10.0–14.5 mm. This is another large striking shieldbug marked variously with orange, green and/or pink. The broad pronotum is extended into sharp points at the side angles. The scutellum is subtriangular, and extended into a narrow nearly parallel-sided lobe at the apex. The upper surface is often broadly orange-brown, although it is sometimes olive-green with the front of the head, pronotal angles and hemelytra variously clouded or coloured with purple-pink. The scutellum often has three black spots arranged in a triangle near the base; these marks are sometimes doubled up. The pronotal angles are smartly black-tipped. The connexivum has black tick marks or large black square blotches. The membrane is brown to black. The legs

Adult

Adult *C. fuscispinus* for comparison

are more or less clear orange-brown. The antennae are mostly black, apart from the pale first segment.

The nymphs are strikingly marked, and very similar to those of *C. purpureipennis*.

Similar species. As with *C. purpureipennis*, green and pink colour forms might be mistaken for *Dolycoris baccarum*. Several similar species occur across Europe, including *C. fuscispinus* throughout much of France, and until recently these were thought to be one variable species. *C. fuscispinus* has a shorter pronotum, such that a line drawn through the spines is closer to the front edge than to the rear. In *C. mediterraneus* the line through the spines more or less equally divides front and back halves of the pronotum, which is slightly more produced forwards, and is more angular on the front edges.

Ecology. It occurs in various habitats, on a wide variety of plants including many composites, umbels and grasses.

Distribution and status. Once thought to have been a rare native in south-west Britain during the late nineteenth century (recorded as *C. fuscispinus*), this species is now deemed to be extinct here. It is, however, known from the Channel Islands and much of Atlantic France.

Chlorochroa juniperina **(Linnaeus, 1758) (formerly** ***Pitedia juniperina*****) – Juniper Shieldbug**

Description. Length 10.5–13.0 mm. This is a large broad green shieldbug edged neatly with yellow or cream. The broad head is flat, and slightly notched where the central lobe of the clypeus is shorter than the side lobes. The pronotum is broad with the front edges straight, gently domed across the rear. The scutellum is flat, broad and triangular, and slightly extended into a lobe at the apex. The upper surface is generally smooth and shining, gently and evenly punctured all over. The bug is a deep emerald-green, fading to a slight yellowish with age, but with the edges of pronotum, hemelytra and connexivum brightly contrasting yellow or pale cream. The tip of the scutellum is this same pale colour. The membrane is dark, nearly black. The legs and antennae are green, often becoming darker, with antennal segments 3 to 5 often nearly black.

The nymph is a bronzy grey-black; the abdominal segments are sometimes purplish mottled, but with the pronotum and wing buds already showing the pale margins.

Nymph

Adult

Similar species. *Chlorochroa* is generally similar in shape and green colour to the common *Palomena prasina* and *Nezara viridula*, but immediately distinguished by the bright yellow piping around the edges of the body. *Piezodorus lituratus* has a vague yellow beading, but this is never so pronounced as in *C. juniperina*. A pale-edged form of *N. viridula* is very rare but does not have the pale scutellum tip. The similar but more olive/brownish-green *C. pinicola* occurs on pine trees in Europe.

Ecology. This shieldbug feeds on the berries and developing fruits of Juniper (*Juniperus communis*), but also apparently Crowberry (*Empetrum nigrum*) in parts of northern Europe. In Britain it was only ever known on juniper on chalk downland or limestone hills.

Distribution and status. This species was always considered very rare; there are historical records from Surrey, Kent, Derbyshire and Lancashire (where it was last seen in 1925), but this species is now thought to be extinct in Britain. Juniper has declined on chalk and limestone soils because of drastic changes in land use as grazing became commercially unviable. It is also rather scarce in mainland Europe.

Dolycoris baccarum **(Linnaeus, 1758) – Hairy Shieldbug, Sloe Bug**

Description. Length 9.5–12.5 mm. This medium-sized purple and olive shieldbug has dapper black and white marks on connexivum and antennae. The following description is for a well-marked live specimen, but this species can be uniform mid-brown to olive-green, and colours also fade to brown after death, although the extreme end of the scutellum usually remains paler. The head is long, flat, and bluntly triangular in front, though with the central lobe of the clypeus shorter than the side lobes, giving a notched appearance; it is dark, heavily black-punctured though with a variable central stripe paler. The pronotum is broad, deeply indented in front to take the head, and with the sides pronounced, but not protruding; the side margins are raised into a fine narrow keel, with a minute denticle at the front corners, just behind

Adult

Nymph

the eyes; the front half is approximately olive, the hind half pinkish purple; it is heavily black-punctured throughout, but with a narrow pale bar of conjoined unpunctured flecks across the disc. The broad scutellum is subtriangular, olive green, black-punctured, densely near the base, and extended into a tongue-like process, and narrowly rounded at the slightly raised apex which is very finely and diffusely punctured. The hemelytra are pinkish purple, heavily and densely black-punctured throughout, but with some pale flecks where gaps between the punctures are large. The connexivum is prominent, wrinkle-punctured, black, but each segment is marked with a large white spot and the segmental boundaries are very narrowly pale. The membrane is pale, brownish. The underside is a pale olive-beige, stippled lightly with small black marks. The antennae are black, but strikingly white-marked at all the joints. The legs are pale, dirty white, but stippled with tiny black marks, and segments 1 and 3 of the tarsi are dark-marked. The entire body is covered with long pale hairs, though these are shorter and less obvious on the underside and antennae. The purple tints that often make this such a pretty species are not present in freshly moulted individuals in autumn (Roger Hawkins, personal communication). This species becomes noticeably darker and browner in winter.

Nymphs are mottled brownish, sometimes with abdominal segments pinkish. Whatever their instar, though, they are immediately distinctive because of their hairiness.

Similar species. The *Carpocoris* species described above can have similar purple and olive colour forms, but all lack the all-over body hairs of *Dolycoris* and the prettily marked antennae.

Ecology. Adults emerge from hibernation in April and peak in May, then again in August and September as the summer nymphs reach maturity. Although one of its English names is Sloe Bug (or Sloebug), and it sometimes occurs on related hawthorn, damson and wild roses, Hawkins (2003) points out that this insect was never recorded on sloe (Blackthorn, *Prunus spinosa*) during the long Surrey shieldbug survey. Nevertheless he also comments (personal communication) that this was a well-established name given by country people living in Sussex (Butler 1923), and by the fruit-growers of Kent (Massee 1963). He sees no reason to doubt reports of the bug on sloes, and suggests that these orchard workers were looking at plums, damsons and sloes for 100% of their time, whilst he was beating Blackthorn less than 1% of his survey. Indeed, this species has very broad foodplant tastes including Stinging Nettles, figwort, birch, knapweed, thistles, dead-nettles and Ribwort Plantain, *Plantago lanceolata*. Consequently it occurs in a wide variety of habitats including rough flowery places like verges and brownfields, woodland rides, parks and gardens.

Distribution and status. This is a very common and widespread species across England, Wales and much of Ireland. It seems to be increasing in the north of England, and with a few records into Scotland, where it has a tendency to be coastal.

***Holcogaster fibulata* (Germar, 1831)**

Description. Length 4.5–8.5 mm. This small-to-medium broad oval, mottled or marbled pinkish-grey shieldbug is highly variable in its colour and pattern. There is a transverse furrow close to the front edge of the pronotum. The rostrum is very long, reaching right underneath the thorax and resting in a depression in the abdomen. The third antennal segment is longer than the second. The upper surface is often reddish orange and grey/black, forming a pretty marbled pattern from quite bright red to dark grey and black; there is usually a reddish streak on the claval fold, emphasised by a dark shadow on its inside. The pronotum also appears

Adult

Nymph

streaked, lengthways, with about four dark shadows. The head is also streaked, usually with the side lobes paler than the darker central lobe of the clypeus. The connexivum is spotted with cream marks against the dark background. The whole surface is relatively shiny, but covered all over with wrinkled punctuation.

The nymph is similarly marbled, but with slightly more contrast between the reddish and black marks often creating a striking harlequin pattern – easily one of the most attractive and recognisable of all the northern European shieldbugs.

Similar species. Despite the vagueness of that description, this is a highly distinctive insect, unlikely to be mistaken for anything else, although there are others in the genus elsewhere in Europe.

Ecology. It feeds on the developing fruits and cones of various conifers including juniper, garden cypress species, and pines, thus also making confusion with other species unlikely.

Distribution and status. This shieldbug is not known from Britain or Ireland, but occurs on the Channel Islands, and through much of Mediterranean and Atlantic France. Given the wide movement of garden trees in the horticultural trade, this seems like a likely contender for colonisation sometime soon.

Peribalus strictus **Fabricius, 1803, (formerly *Holcostethus strictus/vernalis*) – Vernal Shieldbug**

Description. Length 8.0–10.5 mm. This is a neat, elegant, brown shieldbug. Indeed, Nau (1996) calls it 'the epitome of a shieldbug yet mainly recognisable by a lack of really striking features'. The head is very flat, long, rounded and triangular, with the side lobes of the clypeus meeting, or indeed crossing, in front of the central lobe. The pronotum is broad, with leading edges slightly concave, and corners produced – prominent but not sharp. The scutellum is long triangular, narrowly rounded at the tip. The entire upper surface is evenly and finely black-punctured. The membrane is slightly darkened. The ground colour is dirty yellowish to reddish brown, sometimes bronze, but with yellow or cream markings on the leading edges of the pronotum and the tip of the scutellum, and regular marks on the connexivum; the head and the front half of the pronotal disc are often darker. The legs are pale, but stippled

ABOVE: Adult

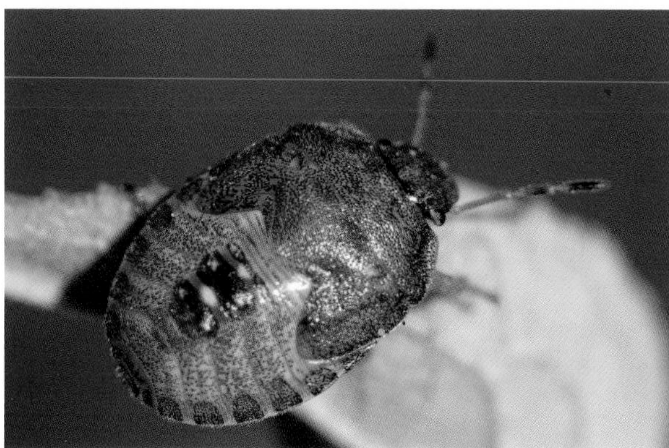

Nymph

with black punctures. The antennae are pale, sometimes reddish, but with segments 4 and 5 strongly blackened. The underside is more uniformly pale dirty yellow or straw, with some of the punctures black.

The nymph is brown or bronze but with contrasting pale abdominal segments.

Similar species. This is a very evenly coloured shieldbug, and might be confused with the brown overwintering form of *Palomena prasina. Peribalus strictus vernalis* is the northern subspecies known in the British Isles, but the more Mediterranean *P. strictus strictus* has antennae more uniformly coloured, and the pale scutellum tip is much less pronounced.

Ecology. This shieldbug species is known from a wide range of foodplants, usually in rough grassy and flowery places like woodland clearings, verges, meadows, field edges and hedgerows.

Distribution and status. Historically it is recorded from a few widely scattered localities in southern England, many of them coastal, from Suffolk to Bristol. It appeared to have fairly long-lived colonies in Sussex and Kent during the 1950s, but then died out. In the last 10 years, however, it seems to have recolonised successfully and is well established in the Bournemouth area of Hampshire. I found a good colony in West Sussex in 2019 and there have been many more scattered records from other sites in the area, suggesting that it is, again, established here.

Palomena prasina (Linnaeus, 1761) – Green Shieldbug

Description. Length 11.5–15.5 mm. This is perhaps the most familiar and friendly of Britain's shieldbugs – large and emerald-green, chunky, walking with a reassuring clockwork gait. It is a distinctive broad oval species, almost entirely bright leaf-green, contrasting with the dark, almost black, membrane. The eyes, tarsi, and borders of the pronotum and connexivum can appear narrowly tinged with yellowish or orange red. The antennae are pale, with segments 4 and 5 slightly darkened. It is fairly evenly punctured all over with small neat black

Adult

Adult dark winter colour

punctures. The underside is paler, yellowish green, sometimes with a pinkish tint. In autumn it changes colour to an olive-brown or sometimes purple, with reddish or brown legs and antennae. This seems to be a physiological change connected to hibernation. After emergence the next spring it reverts to its original bright green.

The nymphs are bright green, black-punctured and sometimes pale-edged, but early instars are strongly marked black on the pronotum and around the abdominal glands, and have legs and antennae dark green to black. The tiny, nearly hemispherical black and green spotted first- and second-instar nymphs are frequently confusing to budding naturalists, who often ask if they are some sort of ladybird.

Nymph

Empty eggshells

Similar species. *Nezara viridula* is slightly narrower, often with two or three small pale spots along the base of the scutellum, and has the membrane transparent, showing the green abdomen beneath, not contrasting dark. And its nymphs are prettily marked with pink and white spots. The extinct (in Britain) *Chlorochroa juniperina* is broader and bordered with a pale cream beading. Subtle differences from the very similar European *Palomena viridissima* are discussed below.
Ecology. Adults emerge from hibernation almost any time from February onwards, and eggs are laid in May or June. Nymphs predominate through to the end of August and into September as the overwintering adults die off, but the new generation of adults appears from late August with a peak in late September. This shieldbug occurs everywhere, in fields, verges, woodland rides, parks and gardens. It feeds on a huge range of wild and domesticated plants, and is sometimes regarded as a minor garden nuisance on runner and broad beans.
Distribution and status. *Palomena prasina* is common throughout England, Wales and Ireland, but seemingly absent or rare in Scotland. This is often *the* shieldbug as far as non-entomologists are concerned, and they know it from their gardens, where they see it flying about or crawling across the patio.

***Palomena viridissima* (Poda, 1761)**

Description and similar species. Length 11.0–15.0 mm. Like its congener, this is a large bright green shieldbug. It is extremely similar to *P. prasina* and distinguished by some subtle differences. The leading edges of the pronotum are very slightly convex, where in *P. prasina* they are very slightly concave. The second antennal segment is 1.5–1.8 times the length of segment 3, where in *P. prasina* the segments are about equal. Generally, it is slightly more green than *P. prasina*, lacking the red/pink tinge around the edges of the head, pronotum and connexivum.

The nymph is also very similar to that of *P. prasina*.
Ecology. Its life history is less well known than that of *P. prasina*, but it is apparently also broad in its habitat and foodplant choices. It has been recorded on sallow, oak, bramble, privet and juniper.
Distribution and status. This species is not known from Britain, and although very rare in France there are recent records from the Channel coast in Normandy; it is also known from Denmark and Estonia. It is more common in southern and eastern Europe. Perhaps this is a potential arrival in the future.

Eysarcoris aeneus **(Scopoli, 1763) – New Forest Shieldbug**

Description. Length 5.0–6.0 mm. This is a small broad oval brownish shieldbug marked with two bright pale spots on the scutellum. The head is long, with the central lobe of the clypeus shorter than the lateral lobes, giving a notched appearance; the head is slightly down-turned in side view. The scutellum is very broad, distinctly concave along the lateral edges and produced into a prominent, but not particularly sharp, point at the tip; it nearly covers the abdomen. The hemelytra are narrow, about as long as the scutellum. The general background colour is a pale brown, but it is marked all over with black punctures, though these are more diffuse on the centre of the pronotum and scutellum. The pronotum is edged pale, but with a large dark blob covering the front angles behind the eyes. The scutellum has a bright clear cream or white spot near each front corner. The connexivum is narrowly pale with a tiny black tick mark at each segment junction. The underside is bronze-black down the centre, but paler towards the connexivum. The legs are pale brown, with varying black punctures. The antennae have the basal segments light brown, and the apical two segments dark brown to black.

The nymph is pale brown, heavily punctured black, the pronotal edges broadly cream and with the shadows of the dark blotches starting to appear, along with two vague white patches on the scutellum.

Adult

Nymph

Similar species. *Eysarcoris venustissimus* is immediately distinguished by the large bronze blotch at the base of the scutellum. In Europe the similar *E. ventralis* has less concave pronotal edges and smaller less prominent scutellum flecks.
Ecology. In Britain this bug is associated with Slender St John's-wort (*Hypericum pulchrum*) on damp acid heathland, although in Europe it is reported from hemp nettle (*Galeopsis*), woundwort (*Stachys*) and other labiates.
Distribution and status. The New Forest Shieldbug lives up to its name – it is very rare and only known from the New Forest area and the Isle of Wight, although a specimen was found at Sennen Cove, Cornwall, on 12 July 2016. There are, however, scattered older records from Norfolk, Bedfordshire, Sussex, Kent, Cornwall and Wales.

Eysarcoris ventralis (Westwood, 1837)

Description. Length 4.5–6.0 mm. This is an attractive small oval brownish shieldbug. The head is long, and broadly rounded at the front. The pronotum is broad, with the leading edges straight, and corner angles distinct but not overly protruding. The scutellum is large, triangular, and broadly rounded at the tip. The hemelytra are flat, about as long as the scutellum. The membrane is clear, showing the dark abdomen beneath. The upper surface is generally mid-brown, but is covered all over with black punctures. The pronotum has the rear half concolorous with the scutellum and hemelytra, but the front half is pale and with a large contrasting dark blotch at each front corner, behind the eyes. The head is dark with a narrow pale streak on the rear half. There is a small pale spot on each basal corner of the scutellum. The underside is bronze-black down the centre, but paler at the edges. The legs and antennae are brown.

The nymph is brown with pale edges and dark blotches on the head and pronotum, and around the abdominal glands.
Similar species. Like *E. aeneus* this species lacks the bronze blotch on the scutellum so obvious in *E. venustissimus*. The pronotal corners are not as prominent as in *E. aeneus* and the sides of the pronotum are straight, not concave. The pale scutellar spots are also smaller than in *E. aeneus*.
Ecology. This species occurs mostly in herb-rich flowery and rough grassy places. It is found on various grasses and sedges and is apparently a rice pest in parts of Asia.

Adult

Distribution and status. This shieldbug is not known from mainland British Isles, but is found in the Channel Islands, and is widespread in Atlantic and Mediterranean France.

***Eysarcoris venustissimus* (Schrank, 1776) (has also been called *E. fabricii* Kirkaldy, 1904) – Woundwort Shieldbug**

Description. Length 5.0–6.5 mm. This bright button of a bug is small, round, brown and metallic bronze. The head is broad and flat, nearly quadrangular and parallel-sided, but bluntly rounded at the tip. The pronotum is broad and domed with the hind angles prominent, but not sharp. The scutellum is large and broad, nearly parallel-sided, and broadly rounded at the tip. The hemelytra are flat, slightly longer than the scutellum. The membrane is clear, showing the dark abdomen beneath. The upper surface is mostly pale grey and brown/beige, marked all over with black punctures. The head, two large blotches on the pronotum behind eyes, and a large triangular mark on the base of the scutellum are shining metallic coppery bronze, although this appears

ABOVE: Nymph

Adult

metallic green in some lights. The connexivum is marked with small white spots. The underside is almost entirely dark metallic bronze or coppery green. The pronotum and the base of the hemelytra are narrowly white-streaked; this is particularly noticeable in side view. The legs are pale brown but with dark stipples and shadows underneath near the tip of the femora. The antennae are pale at the base but with segments 4 and 5 darkened.

The nymph is shining black with the abdomen pale.

Similar species. This species is immediately distinguished from others in the genus by the large triangular coppery bronze mark on the base of the scutellum, and the entirely bronze-black underside.

Ecology. Mating pairs are a frequent sight in May, after this bug emerges from hibernation. Eggs laid in June hatch into nymphs in July and the new generation of adults starts to appear in August, with a peak in September. It is usually found on Hedge Woundwort, *Stachys sylvatica*, but also sometimes on White Dead-nettle, *Lamium album*, in damp hedgerows, marshy field edges, stream banks and woodland rides. In drier localities it is sometimes found on Black Horehound (*Ballota nigra*).

Distribution and status. This bug is generally common and widespread through most of England to Yorkshire, although it is scarce in the West Country and in Wales. Hawkins (2003) describes how this species was once uncommon in Surrey, and according to Butler (1923) England in general, but increased from the middle of the twentieth century. It also appears to be increasing in Nottinghamshire and Yorkshire. My impression is that it is now declining again, at least in south-east London, and I rarely see it.

Stagonomus bipunctatus (Linnaeus, 1758) subspecies *pusillus* (Herrich-Schäffer, 1830)

Description. Length 4.5–6.2 mm. *Stagonomus* is a small, broad, roundish, pinky-brown shieldbug. The head is rather long, broadly flattened along the front edge. The pronotum is broad and domed. The scutellum is large, broadly rounded at the apex. The hemelytra are broad and flat and the membrane is pale and clear. The bug is generally a pinkish brown, but there is a strong pale spot on either side at the base of the pronotum and the insect is marked all over with strong black punctures. The underside is similarly pale beige-brown with black punctures. The legs and antennae are pale straw, with small black puncture marks. The typical form, subspecies *bipunctatus*, is more greenish brown, has the tip of the scutellum darkened with a strong but vaguely defined black mark, and has the black marks on the connexivum more precisely delineated. Subspecies *pusillus* has a more rosy pink tint; any darkening to the tip of the scutellum is very obscure and the black marks of the connexivum are less clearly defined because of heavy black punctures in the paler areas.

The short squat nymph is mottled brown (sometimes pink) on the fore-parts, with the abdominal segments a dirty greenish.

Similar species. The strong pale marks on the scutellum are similar to those of *Eysarcoris aeneus*, but that genus has the front edges of the pronotum

Adult

slightly concave where those of *Stagonomus* are slightly convex. The underside of *Eysarcoris* shows a strong bronze patch along the centre of the abdominal segments, but in *Stagonomus* the tergites are uniformly brownish.
Ecology. Seemingly on various foodplants including bugle (*Ajuga*), thyme (*Thymus*), mullein (*Verbascum*) and figwort (*Scrophularia*). Subspecies *bipunctatus* is a thermophile, occurring in dry sandy and chalky places, particularly the garrigues of France. Meanwhile subspecies *pusillus* occurs in woodlands, parks and gardens.
Distribution and status. This species is not known from the British Isles, but subspecies *pusillus* occurs throughout a large part of Europe, well into southern Scandinavia and the Baltic. It is found sporadically through France and was noted as spreading through the Netherlands by Aukema (2001).

Nezara viridula (Linnaeus, 1758) – Southern Green Shieldbug

Description. Length 11.5–18.0 mm. This is a large slim parallel-sided green shieldbug. The head is long, flat and triangular. The pronotum is broad, the side angles prominent but not sharp, the front margins very slightly concave. The scutellum is large, broad, flat and triangular, but drawn out to a narrow parallel-sided lobe at the apex. The hemelytra are broad and flat. The body is almost entirely a rich leaf-green, and is concolorously punctured all over. The edges of the pronotum are sometimes very narrowly edged with cream or yellow. The scutellum has three or five very small white spots along the basal margin, and a tiny black spot at each basal corner. The membrane is clear, pale, showing the green abdomen through it. The underside is pale yellowish green. The legs are green. The antennae are also green, but slightly darkened on the apical segments. Further south in its range a form with broadly cream/yellow head and pronotum (form *torquata*) occurs, though this appears to be very uncommon in the British Isles.

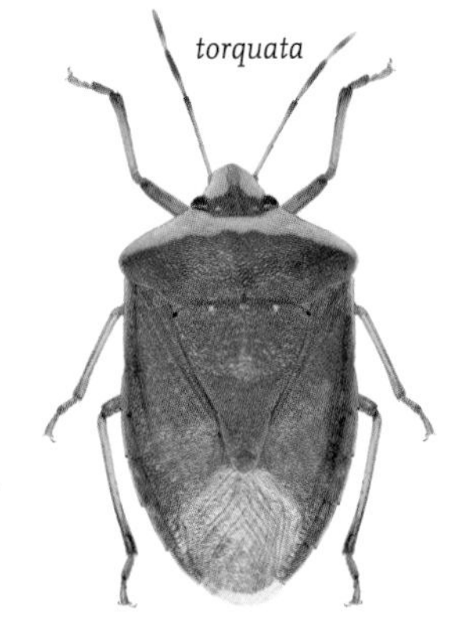

torquata

In contrast to the uniformly green adult, the nymph is very brightly patterned and unmistakable. The early instars are black with strong white and red spots. Later instars are bright green with the edges of the pronotum, wing buds and connexivum reddish pink. The abdominal glands are circled with pink.

Adult

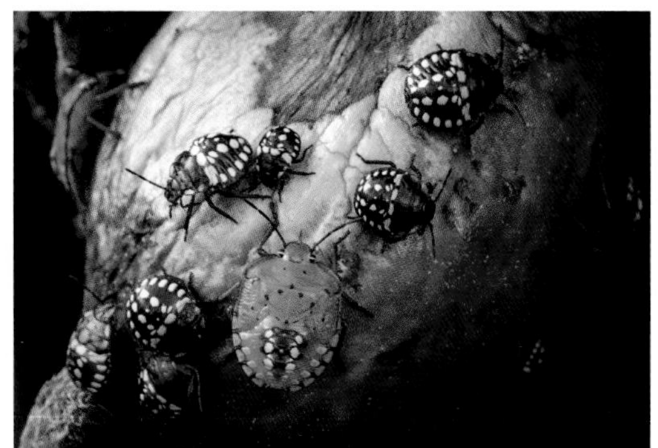

BELOW LEFT: *Torquata* form

BELOW: Nymphs

The sides and central line of the abdominal segments are strongly white-spotted.
Similar species. This shieldbug is slightly longer, narrower and paler than the rather similar Green Shieldbug (*Palomena prasina*), and lacks the black punctures and black membrane so distinctive in that species. The brightly coloured nymphs are unmistakable and very striking.
Ecology. Like *Palomena prasina, Nezara* changes colour during the late autumn to a dull olive-brown to hibernate. On re-emerging the following May it starts to green up again. Eggs appear to be laid from May right the way through to September, with the nymphs often still present into late October. Whether this is because there is an extra generation, or simply because adults are able to lay multiple egg batches, still needs to be ascertained. This bug is found on a wide variety of foodplants, particularly on beans and tomatoes in gardens and allotments. In some parts of its international range it is considered an agricultural pest.

Distribution and status. This shieldbug was originally a native of North Africa, but since at least the 1930s it has often been found on imported foodstuffs. This species has spread throughout the world and is now considered a major pest of broad beans, peas and other crops. Southwood & Leston (1959) considered it 'unlikely to become established', but it was first recorded breeding in the British Isles in 2003 in the London area (Shardlow & Taylor 2004). It is now widespread and common around the capital and Home Counties, and is still spreading. Salisbury *et al.* (2009) review its British status and distribution at the time. Early in 2021 it was photographed in Suffolk (Stuart Reed, Twitter), and it was also recorded from Lincolnshire, Yorkshire, Scotland and Northern Ireland. I was recently sent a photo of a dead specimen of form *torquata* found wedged into a leaf joint of an imported *Trachycarpus* palm being planted on the Dorset coast, suggesting that it is still being moved about in horticultural material, some of these perhaps being fresh importations from elsewhere in the world.

Pentatoma rufipes **(Linnaeus, 1758) – Red-legged Shieldbug, Forest Bug**

Description. Length 12.0–16.0 mm. This large bronze-brown shieldbug with distinctive curved hook-shaped pronotal corners and bright red legs is one of our most familiar species. The head is long, parallel-sided at the base but rounded at the tip. The pronotum is broad with the corners extended into strong hooks, rounded in front and concave behind. The scutellum is large, triangular, extended into a narrower tongue behind. The hemelytra are broad and long, strongly ridged along the endo-/exocorium fold. The upper surface is puncture-wrinkled all over. It is mostly a glossy bronze-brown colour, but the front edge of the pronotum is vaguely yellow-margined and the prominent angles are nearly black; the tip of the scutellum has a large bright yellow or pale cream spot, and the connexivum is strongly spotted with square yellow and black chequer marks. The membrane is dark, brown-tinted. The underside is mostly dark brown. The legs are reddish brown to bright reddish yellow. The antennae are brown to black.

The nymph starts out greenish and brown/bronze, but soon becomes strikingly bronze-marked against the pale pinkish or beige abdominal segments. The dark pronotal hooks appear in the final instar.

Adult

Nymph

Similar species. The predatory *Pinthaeus sanguinipes*, which has turned up once in Britain, looks remarkably similar to the unaided eye because of its overall colour, red legs and pale scutellum tip, but has the thick rostrum of the Asopinae, much less rounded and hooked pronotal corners, and a small tooth on the front femur. The pronotal corners of *Troilus luridus* are also less sharply hooked at the corners than those of *Pentatoma*, those of *Picromerus bidens* more pointed. The contrasting red legs and pale scutellum tip are usually highly characteristic. Nymphs might be confused with those of *Rhaphigaster nebulosa*.

Ecology. The Forest Bug is unusual amongst British shieldbugs in that it overwinters as a second-instar nymph. It starts feeding again in April or May and final instars peak in June. Adults occur from July to early October, with a peak in August. It occurs mostly on broadleaved trees, feeding on the fruits, and sucking sap from the twigs, with oak, alder, hazel, maple, sycamore, birch, apple and cherry often quoted. It must, however, be slightly omnivorous because it is often recorded feeding on honeydew, and on living and dead insects including aphids and moth caterpillars.

Distribution and status. This is a common species, and recorded throughout Britain and Ireland; it is seemingly our most widespread species, occurring across the British Isles right through Wales and Scotland, although avoiding highland areas.

Rhaphigaster nebulosa (Poda, 1761) – Mottled Shieldbug

Description. Length 13.5–17.0 mm. *Rhaphigaster* is a large, narrow, mottled grey shieldbug, arguably our largest pentatomid species. The head is long, parallel-sided, neatly rounded at the front, and with the sides slightly twisted and flexed upwards. The pronotum is broad and domed, with corners prominent and relatively sharp. The scutellum is large and triangular, but drawn out to a narrow tongue at the tip. This shieldbug generally has a pale grey-brown background mottled with dark grey to black and is covered all over with black punctures, some of which coalesce into dark blotches. The connexivum varies from vaguely to strongly black and yellow squares chequer-marked. The membrane is pale grey but uniquely (in British species) black-spotted. The underside is pale yellowish beige spotted with black. The legs are pale, with dark shadows around knees and tarsi. The antennae are long, black, but with the basal half of segments 5, 4 and sometimes 3 yellowish cream.

The nymph is black and cream on hatching from the egg, becoming pale mottled, stippled with black punctures and with dark connexivum marks. Later instars have the head, pronotum and wing buds darker, often bronzy.

Similar species. Worryingly, *Rhaphigaster* is very similar in general appearance to the potential pest species *Halyomorpha halys*, but it lacks the small yellow marks

Adult

Nymph

on the pronotum and scutellum, and has broader white bands on the antennae. In *Halyomorpha* the underside of the third sternite of the abdomen (the second visible segment) is simple, but in *Rhaphigaster* there is a long sharply pointed projection reaching forwards between the hind coxae and nearly to the front coxae (see diagram in Chapter 7, couplet 73), easily visible in side view. The outline and chequered connexivum might give the impression of a faded *Dolycoris baccarum*. The nymphs of *Halyomorpha* are spiny along the pronotal edges and first abdominal segments, with another spine in front of the eyes, but nymphs of *Rhaphigaster* are smooth-edged. Nymphs look very like those of *Pentatoma rufipes*, although the pronotal hooks so distinctive of that species start to show in later instars.

Ecology. Precise life-cycle details have still to be worked out in the British colonies, but nymphs are recorded from July to September, with adults more or less all year round. *Rhaphigaster* is recorded from a wide range of trees including oak, beech, birch, hazel, poplar, sallow, lilac and apple. It has also been noted feeding on leaf-beetle larvae so is probably partly predatory. It has been found in a wide range of habitats including hedgerows, woods, parks, gardens and brownfield sites.

Distribution and status. This is a recent arrival into the British Isles. It had been noted spreading north through Europe and had appeared in the Netherlands in 2001 (Aukema 2001). It got to the Channel Islands in 2009 and was first found on mainland Britain in the London area in 2010. It was found breeding in a south London park in 2011 (Bantock *et al.* 2011), and is now known from several localities in and around the metropolis (for example, crawling over a desk in a Covent Garden office on 3 March 2022), and also from the Sussex coast.

***Piezodorus lituratus* (Fabricius, 1794) – Gorse Shieldbug**

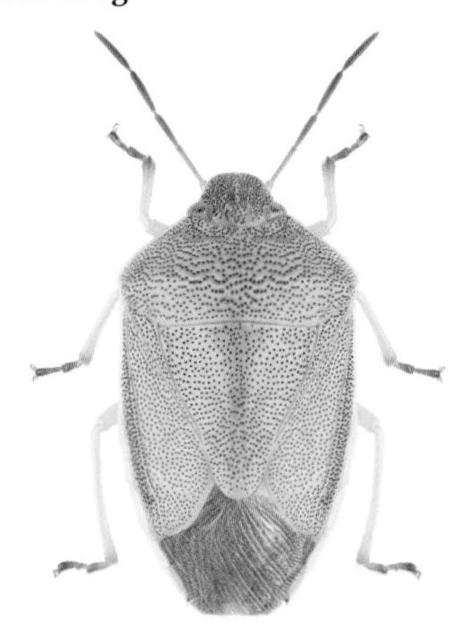

Description. Length 10.0–12.5 mm. *Piezodorus* is a large elegant greenish shieldbug delicately pale-edged. The head is flat, broad, bluntly rounded at the front. The pronotum is large, domed, fairly deeply indented to take the head, with the outer corners prominent but rounded. The scutellum is large and triangular. The hemelytra are broad and flat. The colour varies from bright apple- to olive-green, sometimes with the rear half of the pronotum and inner portions of the hemelytra with a reddish blush. It is finely but densely and evenly black-punctured all over. The sides of the pronotum (occasionally also the hind margin slightly) and edges of the connexivum are contrastingly bright smooth yellow, particularly where the hemelytra sometimes take on a slight turquoise tinge. The membrane is clear, showing the black abdomen beneath. The underside is pale yellowish green. The legs are a dirty yellow. The antennae are reddish orange. There is a rare melanic form, with the upper surface nearly entirely black (see Fig. 87).

The nymph is a dirty yellowish green but stippled with black, or with vague black streaks. Later instars have the pronotum and wing buds sometimes blackish. The pale rim to the pronotum and abdominal segments is variable, but usually discernible.

Adult

Adult head

Nymph

Similar species. The yellow edging to the pronotum and connexivum and pinkish antennae are very distinctive. Only *Chlorochroa juniperina*, no longer found here, has similar pale body margins. The other green shieldbugs, *Palomena prasina* and *Nezara viridula*, are a deeper, cleaner, brighter green.
Ecology. Like several other greenish shieldbugs, the adults of *Piezodorus* turn brownish during hibernation. They emerge in March or April and turn greenish again. The pretty banded eggs are laid in small batches, usually 14, on the pods of the foodplant, and these hatch in about May. Nymphs peak in August and the new generation of adults appear in August and September. Not surprisingly the Gorse Shieldbug occurs on gorse (*Ulex*), but also broom (*Cytisus*), and Dyer's Greenweed (*Genista tinctoria*), or sometimes on *Laburnum* in parks and gardens. It may even breed on herbaceous plants like Red Clover (*Trifolium pratense*) (Roger Hawkins, personal communication).
Distribution and status. *Piezodorus* is widespread and common throughout most of Britain and Ireland, though becoming more coastal in Scotland.

Dyroderes umbraculatus (Fabricius, 1775) – White-shouldered Shieldbug

Description. Length 7.0–9.0 mm. This is a distinctive broad, almost square, brown and grey shieldbug. The head is flat, broadly oval, neatly rounded at the front, with the side lobes of the clypeus meeting ahead of the central lobe. The pronotum is very broad, with rounded and expanded front corners giving a swollen appearance. The scutellum is broad and triangular, bluntly rounded at its apex. The connexivum is

Adult

rounded, giving a broad swollen appearance. The upper surface is brownish, with each front angle of the pronotum marked with a large pale grey blotch, and smaller grey spots on the connexivum. The scutellum has the tip bright white or at least pale grey. The upper surface is black-punctured all over against the brown, but only sparsely on the pale pronotal patches. The legs are pale brown to yellowish white with conspicuous black spotting. The antennae are yellowish, though the last two antennal segments are darker, and highlighted by pale rings at the joints.

The nymph is mottled brown, with abdominal segments pale; later instars have the pale pronotal patches already developing.

Similar species. This is a highly distinctive species, immediately identified by the broad square shape and pale scutellum tip and pronotal corners.

Ecology. Little is yet known about this bug's precise life history in the British Isles, but nymphs have been recorded in August. It feeds on bedstraws (*Galium*) in warm open grassy sites; it is probably mainly ground-dwelling but can sometimes be swept.

Distribution and status. This insect has been known from the Channel Islands since the 1960s, but is a new arrival into the British Isles. It appears to have been spreading through Europe and was first found in Belgium in 2004. In Britain it was first found in south-west London in 2013 (Nau *et al.* 2014). More specimens were found in Southampton in July and September 2015, and it is now recorded from several localities in southern England. I found a nymph in west London in 2021. It has a scattered Mediterranean and Atlantic distribution in France.

***Menaccarus arenicola* (Scholtz, 1847)**

Description. Length 4.8–6.5 mm. This is a small neat oval pale sandy-coloured shieldbug. The head is very broad and widely rounded at the front. The pronotum is broad, almost rectangular. The scutellum is broad, hemi-elliptical, completely rounded at the apex. The hemelytra are short, square-ended, at most about as long as the scutellum. The membrane is pale grey. The connexivum is broadly rounded. The insect is a pale sandy grey all over, but marked with dark streaks and blotches, especially around the front edge of the head and along the clypeal sutures, on the pronotum directly behind the eyes, on the basal corners of the scutellum, and on each segment of the connexivum. It is punctured all over, some in-filled black. The legs and antennae are sandy grey with dark shadows and streaks.

The nymph is sandy grey with the black marks more contrasting.

Similar species. *Menaccarus* is broader and paler than *Sciocoris cursitans*. The front edge of the head is much broader and rounder, and edged with a row of short but distinctive bristles lacking in *Sciocoris*.

Ecology. It is found on sand dunes, usually associated with Marram (*Ammophila arenaria*) or other grasses.

Distribution and status. This species is not known in mainland Britain, but occurs on the Channel Islands, and on the Atlantic coast of France to Brittany. It is an outside possibility for British colonisation.

Nymph

Sciocoris cursitans (Fabricius, 1794) – Sand-runner Shieldbug

Description. Length 4.3–6.2 mm. *Sciocoris* is a small broad flat big-headed sandy-brown shieldbug. The head is large, broad and flat, smoothly rounded at front, though with a small indent where the side lobes envelop the central lobe of the clypeus. The pronotum is broad, nearly rectangular; flat, broadened into a wide flange along each edge. The scutellum is large, broad and rounded at the tip. The hemelytra are short, pointed at the tip, barely as long as the scutellum. The connexivum is very broad and rounded, but indented at the apex in the male. The bug is pale brown with dense black punctures and some dark brownish shadows and streaks, particularly down the centre of the head, across the pronotum, in the front corners of the scutellum and on the connexivum segments. Sometimes it has an all-over mottled appearance. The legs are pale brown. The antennae are pale to dirty brown, the last segment usually darker.

The nymph is pale brown, mottled and streaked with vague darker patterns.

Similar species. No other British shieldbug genus has such a broadly expanded and rounded connexivum. The recently discovered (in Britain)

Adult

S. homalonotus is almost identical, but is slightly larger (5.9–8.5 mm) and has a small notch in front of the eye and a minute denticle in front of this. *S. sideritidis* is also very similar, but has a pale blotch at each front angle of the pronotum, and the central area of the underside of the abdominal segments is entirely dark, whereas it is pale edged with black in *S. cursitans*.
Ecology. Nymphs are usually about from June until August, adults August to June. This bug is strongly ground-dwelling on warm, well-drained sandy or chalky soils. In Britain its foodplant is usually quoted as Mouse-ear Hawkweed (*Pilosella officinarum*), but in Europe many other plants are suggested, including Ling (*Calluna vulgaris*) and sage (*Salvia*).
Distribution and status. This is a scarce species of southern England. Most localities are coastal from south Wales, Somerset, Devon and Cornwall to Kent and Essex, but it is widely recorded on brownfield sites along the Thames estuary and occurs inland in Surrey and East Anglia.

Sciocoris homalonotus **Fieber, 1851**

Description and similar species. Length 5.9–8.5 mm. This is another small broad flat sandy-coloured shieldbug. It is extremely similar to *S. cursitans* in shape and colour, but generally slightly larger: male 5.9–7.2 mm, female 7.2–8.5 mm. A small denticle and indented notch in front of the eye (making it appear pedunculate – stalked) is more prominent than in *S. cursitans*, which has at most a vague rounded nodule and very slight angle. Two other European species with equally or even more prominent eyes, *S. macrocephalus* and *S. microphthalmus*, can be

Adult

distinguished by very subtle comparative differences; these are tabulated by Bantock & Kenward (2017).

The nymph is mottled brown and extremely similar to that of *S. cursitans.*

Ecology. This species is poorly understood, but like others in the genus is strongly ground-dwelling on warm, dry, well-drained soils.

Distribution and status. It appears to have been spreading north through Europe and was found new to Sweden in 1976 (Coulianos 1976). It was discovered, new to Britain, from Chatham, Kent, in 2016 (Bantock & Kenward 2017), and recently on Chipstead Downs, Surrey (Graeme Lyons, via Twitter), suggesting that perhaps all old records for this genus should be re-examined.

Sciocoris sideritidis Wollaston, 1858

Description and similar species. Length 4.6–5.8 mm. This small broad flat brown shieldbug is very similar to *S. cursitans* and *S. homalonotus*, but has the head slightly longer, appearing vaguely triangular rather than semicircular. It is similarly coloured to *S. cursitans*, but has the front angles of the pronotum with a large paler blotch, and the underside of the abdomen has an entirely dark central stripe, or at least a darkened patch.

The nymph is mottled sandy coloured, the long head already evident by the fifth instar.

Ecology. Like others in the genus it is strongly ground-dwelling in hot dry places under various plant species, notably Asteraceae.

Distribution and status. This is a western Mediterranean species known from Spain, southern France, Italy, Slovenia and North Africa. Adults and nymphs

Adult

Adult genital area

were recorded from Purfleet, Essex, in September 2018 (Bantock 2019). There is a large container port nearby and the implication is that this was an accidental introduction, brought in by shipping. Specimens were also found in 2019 and 2020, indicating that the colony had survived the winter and seems to be established here, at least in the short term.

Eurydema oleracea (Linnaeus, 1758) – Crucifer Shieldbug, Brassica Bug, Cabbage Bug

Description. Length 5.5–7.5 mm. This small, elegant, narrow-oval black shieldbug is highly variable, marked with red, orange, yellow or cream. The head is broad, finely and densely punctured, rounded, sinuously curved in front of prominent eyes, the side lobes of the clypeus meeting ahead of the much shorter central lobe. The pronotum is short and broad, gently indented behind the head, and raised into a collar, shining with sparse punctures arranged in approximately transverse lines, particularly deep and strong across the centre. The pronotal

Adult

ABOVE: Adult

Nymph

edges are raised into narrow beads, with a minute denticle at the front angle just behind the eye. The scutellum is large, broad, triangular, with strong punctuation tending to form transverse wrinkles. The hemelytra are flat, and strongly punctured throughout. The overall colour is shining black with bluish or greenish metallic tints, but it is strongly marked with red, yellow or cream (or rarely orange). There is a central line down the pronotum, widening behind, a spot at the tip of the scutellum (sometimes two extra spots in front), and similar marks at the tip of each hemelytron. The edges of the head, pronotum and hemelytra are narrowly coloured too. The membrane is dark, nearly black. The underside is mostly shining black, and heavily punctured. The legs are black but there is a strong broad coloured ring around the centre of each tibia and usually a very narrow ring around each femur at the knee. The antennae are black.

The nymph is shining black with extensive pale cream marks on the head, pronotum, wing buds and abdomen.

Similar species. All the other species in this genus are more extensively marked with bright, usually red, coloured patches, although the nymphs can all look very similar.

Ecology. Mating takes place in May and June, and nymphs are about from then until July. *E. oleracea* occurs on a wide variety of plants in the cabbage family (Cruciferae/Brassicaceae) including garden and crop cultivars. It is often on escaped Horse-radish (*Armoracia rusticana*) on brownfield sites, or Garlic Mustard (*Alliaria petiolata*) in hedgerows, woodland edges, field margins and verges.

Distribution and status. This attractive insect is generally common and widespread in central and southern England, mostly in the triangle between the Wash, the Severn and Portland, but there are plenty of outliers to at least Leicester, Birmingham and Stoke-on-Trent. It appears to be spreading north, and there are now numerous records up into North Yorkshire and west into Wales. It certainly seems much more common now than when I first started noticing shieldbugs when I lived in Sussex in the 1960s and 1970s.

Eurydema dominulus (Scopoli, 1763) – Scarlet Shieldbug

Description. Length 5.0–7.5 mm. This small bright shining scarlet and black shieldbug is very attractive. The head is short, bluntly rounded at the front, with the side lobes of the clypeus meeting ahead of the shorter central lobe giving a notched appearance. The pronotum is short and broad, and gently indented to

Adult

receive the head. The scutellum is large and triangular. The hemelytra are broad and flat. The entire body is punctured all over, most densely and regularly across the hemelytra. Rather than a black bug with red marks, this species is red with black. The head is black with the front edge narrowly beaded red. The pronotum is red with six large oval black spots. The scutellum is red with a large black blob at the base and two small black semicircles near the tip. The hemelytra are red, each with a large black B-shaped mark and smaller black spot. The connexivum is red with small black tick marks. The legs and antennae are black, very narrowly marked pale at the joints.

The nymph is black with contrasting pale creamish spots and abdominal segments.

Similar species. *Eurydema dominulus* is much more red-marked than *E. oleracea*. It is distinguished from other similarly patterned *Eurydema* species by virtue of its entirely red exocorium, the leading edge of the hemelytra, which has black marks in all the other species.

Ecology. It occurs on wild crucifers like Lady's Smock (*Cardamine pratensis*) in woodland rides, clearings and hedgerows.

Distribution and status. This bug is very rare in Britain. There are scattered historical records from Devon, Hampshire, Surrey, Gloucestershire and Norfolk, but all recent records are from a small area of Sussex and Kent.

Eurydema ornata (Linnaeus, 1758) – Ornate Shieldbug

Description. Length 7.0–9.2 mm. This is a small long-oval harlequin-patterned black and red (or cream) shieldbug. The head is long, triangular, but bluntly rounded at the front. The side lobes of the clypeus meet ahead of the short central lobe. The pronotum is short and broad. The scutellum is large and triangular. The hemelytra are flat. The upper surface is punctured all over, but the punctuation is densest and most even on the hemelytra. It is rather variable in its colours and markings, but generally has the red background colour marked with black. The head is black or with variable pale marks. The pronotum is red with six large sometimes squarish black blobs. The scutellum is red but with a large basal blotch and two smaller black marks near the tip. Each hemelytron is red with

TOP: Adult. ABOVE LEFT: Nymph. ABOVE RIGHT: Mating pair

a large black L-shaped mark following the edge of the scutellum, and with a long black streak on the outer margin near the tip. The connexivum is red, with variable black spotting. The membrane is dark, appearing black. Part or some of the red marks can be bright cream, giving a three-coloured harlequin appearance, or they can be tinged creamish. The underside is largely red or cream. The legs are mostly black, but variably marked with the same red, cream or orange. The antennae are mostly black.

The nymph is black and white, often suffused with red or orange.

Similar species. *Eurydema ornata* is distinguished from *E. dominulus* by its larger size and the black mark on the exocorium, the outer edge of the hemelytra.

E. herbacea and *E. ventralis* have more extensive black markings. The dorsum of the abdomen is completely black in this species, but almost entirely red in *E. dominulus*.

Ecology. This species is poorly understood in the British Isles; it is known from Sea Radish (*Raphanus raphanistrum* subsp. *maritimus*), but recorded from a variety of plants throughout Europe, including various wild crucifers likely to be suitable here.

Distribution and status. This bug has recently been discovered in the Channel Islands. It was first found on mainland Britain in 1997, though at first it was misidentified as *E. dominulus*. It is now recorded from various coastal sites in Dorset, Hampshire and Sussex (Slade *et al.* 2005) and very recently inland in landlocked Surrey.

Eurydema herbacea (Herrich-Schäffer, 1833)

Description and similar species. Length 7.2–8.5 mm. This is another prettily marked bright black and red shining shieldbug. It is more marked than *E. oleracea*, but less than *E. dominulus* and *E. ornata*. In particular, the black pronotal marks are usually partly united, and the black mark at the base of the scutellum extends and merges with the two smaller marks near the tip. The edges of the hemelytra are also more black-marked.

Ecology. This bug is seemingly confined to Sea Rocket (*Cakile maritima*) on coastal shingle and sand dunes.

Distribution and status. In France this is an Atlantic species, but it is also known from the Channel Islands. Its foodplant occurs widely around British coasts, so the bug might feasibly turn up.

Eurydema ventralis Kolenati, 1846

Description and similar species. Length 7.5–11.0 mm. This is yet another smart shieldbug, very strongly patterned in black and red (or creamish orange). The pronotum has squarish black spots, leaving a grid of red lines around them. The hemelytra have a large diagonal black mark following the edge of the scutellum and finishing with a broad rectangular transverse bar, and a small round spot behind it. The legs and antennae are black. It is distinguished from both *E. dominulus* and *E. ornata* by the four large square black chequer marks along the connexivum.

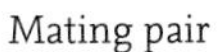

Mating pair

The nymph is black and red, and similar to other *Eurydema* species.

Ecology. In Europe it is known from various crucifers in hedgerows, verges, field edges, parks and gardens.

Distribution and status. This species is widespread in France, right up to the Cherbourg peninsula. It is not yet in the Channel Islands, but is another one to look out for maybe.

Graphosoma italicum **(Müller, 1766)**

Description. 8.5–11.0 mm. *Graphosoma* is perhaps the most unmistakable shieldbug; it is broad, bright red- and black-striped. Its ground colour is a brilliant scarlet, sometimes edging to orange, with strong black stripes: two on the head, six on the pronotum and four on the broad scutellum, which almost completely covers the hemelytra. The head is long and pointed. The pronotum is broad and domed. The prominent swollen connexivum is red with a squarish black blotch or spot on each segment. The underside is red, harlequin-spotted with many large black spots and blobs. The legs are mostly black with some vague reddish streaks on the tibiae.

The nymphs are vaguely streak-mottled pale and dark brown, the only strong contrast being from a single pale line from the head to the end of the developing scutellum bud. The fifth-instar nymph develops a reddish colour, but the strong clear stripes do not appear until the final moult to adulthood.

ABOVE LEFT: Adult

ABOVE RIGHT: Nymph

Mating pair

Similar species. There is nothing else like it in northern Europe. Specimens with mostly clear red legs, from around the Mediterranean, have usually been assigned to *G. lineatum*, although some authors consider them merely subspecies.

Ecology. Mating takes place in June and July, and nymphs occur in July to September. *Graphosoma* feeds on umbellifers, particularly Hogweed (*Heracleum sphondylium*), Wild Carrot (*Daucus carota*) and Hemlock (*Conium maculatum*), and through Europe is a frequent denizen of rough field edges, verges, gardens and waste ground. In Swedish populations, at least, new adults are a dull brown and black, and they do not achieve the final bright red and black pattern until they emerge from hibernation the following year. This is suggested to be camouflage, helping them blend in to the fading flower heads and seed pods of late summer and autumn.

Distribution and status. This bug is widespread and common in Europe right into southern Finland, and has been a potential new arrival here for decades. Hawkins (2003) noted its spread north through France during the 1990s and predicted its imminent establishment in the British Isles. It is now breeding in the Channel Islands. Odd specimens, thought to be accidental imports with agricultural or horticultural goods, have long been recorded from mainland Britain (for example Norfolk in the early nineteenth century right through to Leeds in 2016). Several were found feeding on Hemlock in South Essex on 13 May 2020, and another colony was discovered on Banstead Common, Surrey, on 14 June 2020, so it may finally be established here.

Podops inuncta **(Fabricius, 1775) – Turtle Shieldbug, Rough Shieldbug**

Description. Length 5.0–6.5 mm. The most obvious feature of this small, round, brown shieldbug is its huge scutellum. The long head is notched just in front of the eyes and is rounded at the front, rather domed with a blunt longitudinal keel down the centre. The broad pronotum has distinctive axe-shaped lobes at the front corners, just behind the eyes, and a small central keel towards the front. The side corners are produced into a small prominence, notched just behind to leave a small right-angled node. The scutellum is huge, parallel-sided, and about three-fifths the width of the abdomen. The hemelytra are narrow, and the membrane is just visible beyond the tip of the scutellum. It is all over a dirty brown, darker on the head, pronotum and base of the scutellum, which may have two (sometimes three) small raised pale marks along the base.

Adult

The upper surface is punctured to varying degrees: the punctures are small and even on the scutellum, deep and sparse on the disc of the pronotum, small and dense on the rest of the pronotum and head. The underside is dark brown. The legs are pale brown with dark shadows behind the knees and on the tarsi. The antennae are dark brown, ringed paler at the joints.

The nymph is pale, beige, marked with black punctures and dark streaks on the head, wing buds and around the abdominal glands.

Similar species. *Podops* is unmistakable in Britain, and unique for the sharp knobbly hook-shaped extension on the front angles of the pronotum.

Ecology. Adults can be found almost all year round, nymphs in July and August. Mostly ground-dwelling, *Podops* occurs at roots, usually in dry sandy or chalky places, but can occasionally be found crawling up grass stems or walls. Its foodplants are uncertain, but grasses, thyme (*Thymus*), wormwood (*Artemisia*), mouse-ear (*Cerastium*) and buttercup (*Ranunculus*) are suggested in the European literature. It is often found resting under logs, pieces of wood, planks, and other debris dumped in the countryside, although it is also a regular under metal sheets and roofing felt squares used on development sites for reptile monitoring. Denton (2022) wondered if this shieldbug had some sort of association with ants, since they too are often found under these same logs, stones and roofing squares. His most significant observation was finding a specimen of *Podops* under loose bark of a willow tree surrounded by workers of *Lasius platythorax*, about 40 cm up from the ground, and quite an unusual situation for a shieldbug.

Distribution and status. This is a widespread and common species in southern England, with scattered records to Yorkshire (discovered there in 2014), but mostly south of the Severn–Wash line. It is mainly coastal in the West Country and south Wales.

FAMILY PLATASPIDAE

Coptosoma scutellatum (Geoffroy, 1785) – Trapezium Shieldbug

Description. Length 3.5–4.6 mm. This is an unmistakable bug – tiny, shining black, globose. The body form is broad, domed, subspherical. It is almost entirely glossy black, but appears to have greenish or brassy tints in some lights. The head is broad, smoothly rounded in front of bright red eyes. The pronotum is very broad, appearing

Adult

Mating pair

almost continuous with the huge scutellum in side view. The scutellum is broader than long, and entirely covers the abdomen. The upper surface is finely punctured, but shining. There are some small pale or reddish marks on the connexivum, just about visible in side view. The underside is also shining black, except the underside of the thorax, above the front and middle legs, which is a contrasting matt grey-brown. The legs and antennae are black.

The nymph is reddish, with some greenish hues on the abdomen, and is covered all over with long pale bristles, quite contrasting to the smooth shining adult.

Similar species. There is nothing else like it in Europe. Only *Thyreocoris scarabaeoides* is so shiny and black with a very large scutellum, but it is flatter and has thicker spinier legs. *Coptosoma* might perhaps be mistaken for a small leaf beetle because of its shining domed appearance.

Ecology. *Coptosoma* feeds on various legumes, and in France has been recorded from a wide variety of genera including *Lathyrus, Medicago, Melilotus, Vicia,*

Ononis and *Spartium*. In Britain it is recorded from Dyer's Greenweed (*Genista tinctoria*). The first British specimens were found by suction sampling and by sweeping.

Distribution and status. This wonderful insect was recorded, new to Britain, two specimens, by Graeme Lyons, on 10 July 2019, from Marline Valley, a Sussex Wildlife Trust Nature Reserve near Hastings. It is fairly common and widespread in France, and known across Europe and Asia to China and Japan, and in North Africa. The English name was coined by Graeme for the beast's trapezoidal form – its blunt tail end being wider than its equally blunt head end. Incidentally, the name *Coptosoma globus* appears in a checklist of 'British' species given by Stainton (1861), but this appears to have been wishful thinking, and it was deleted from subsequent lists.

FAMILY SCUTELLERIDAE

Eurygaster testudinaria (Geoffroy, 1785) – Common Tortoise Bug

Description. Length 9.0–11.0 mm. This is a large, round, domed shieldbug, with a broad hexagonal pronotum and massive scutellum that leaves just the shoulders of the hemelytra visible. It comes in a range of colours and patterns from entirely plain or slightly mottled grey/brown to extensively blotched purple/pink and beige. One of the standard forms has the body almost wholly pale brown, marked only by the extensive black punctures and by a vague suffusion of deeper brown on the head, the front edge of the pronotum and the connexivum. In this morph the two pale unpunctured streaks at the base of the pronotum are particularly prominent. The most attractive colour forms have large pale areas either side of the pronotal disc and two pointed patches on the sides of the scutellum, which also has a narrow dorsal line that expands into a long wedge, like an inverted champagne flute or slim tulip, at the apex. Many of these brightly coloured forms fade on death, and museum collections fail to capture their vibrancy. The underside is more or less uniform pale brown, with punctures darker and more obvious under the thorax than under the abdomen. The antennae are slim, and pale, with the last two segments usually

Adult

Nymph

vaguely darker and less shiny by virtue of a covering of short pubescence. The legs are short and stout, pale, but with numerous small dark spines and with many black punctures, usually coalescing into a streak on the upper surface of the tibia.

Early-instar nymphs have a black head and thorax, and a pale abdomen with a series of dark bars and blobs down the centre line. Later-instar nymphs are more streaked and camouflaged with browns, creams and beiges, and other brighter colours arriving when wing buds become prominent.

Similar species. This species is extremely similar in size and colour patterns to the much scarcer *E. maura*. It is, on average, a fraction larger, but there is a wide overlap between the two species, and between the two sexes. Hawkins (2003) examined large numbers of specimens for his Surrey atlas and came to the conclusion that most females fall into the bracket 9.8–11.0 mm (compared to 8.6–9.9 mm in *E. maura*); males were 9.0–10.6 mm (versus 8.8 and 9.7 mm in two *E. maura*). The depressed central lobe of the head, and the slightly shorter second antennal segment (at most 1.5 times the length of the third) are useful characters in the field, but sometimes difficult to appreciate in an active specimen on a hot day. I made this mistake again recently. Dissection of the male aedeagus or examination of the female genital lobes (separated from the edge of the neighbouring segment in *E. testudinaria*, but touching in *E. maura*) is sometimes necessary for complete confidence.

Ecology. The adults overwinter, and eggs are laid in the spring. Nymphs emerge from May onwards, and as they mature there is a peak of adults in August. This is a frequent denizen of rough grassland, often occurring in damp meadows, verges, fens, woodland rides, clearings and edges. A seed-feeder, it usually occurs on grasses, sedges, or rushes, but other plants such as knapweeds (*Centaurea*) are also recorded.

Distribution and status. This species is widespread in England, and also in Wales (Gower and coastal) and Ireland. Historically it was regarded as rather local, occurring in marshy places, but it appears to have spread widely. Douglas and Scott (1865) only knew it from Folkestone (though it was not then distinguished from *E. maura*). Hawkins' survey work in Surrey caught the spread from the southern quarter of that county in 1996 to its entirety by 2003. It is now rather common, at least in the areas of Surrey, Kent and Sussex that I regularly haunt, and new records have recently been made in Nottinghamshire, Yorkshire and Staffordshire.

Eurygaster maura (Linnaeus, 1758) – Scarce Tortoise Bug

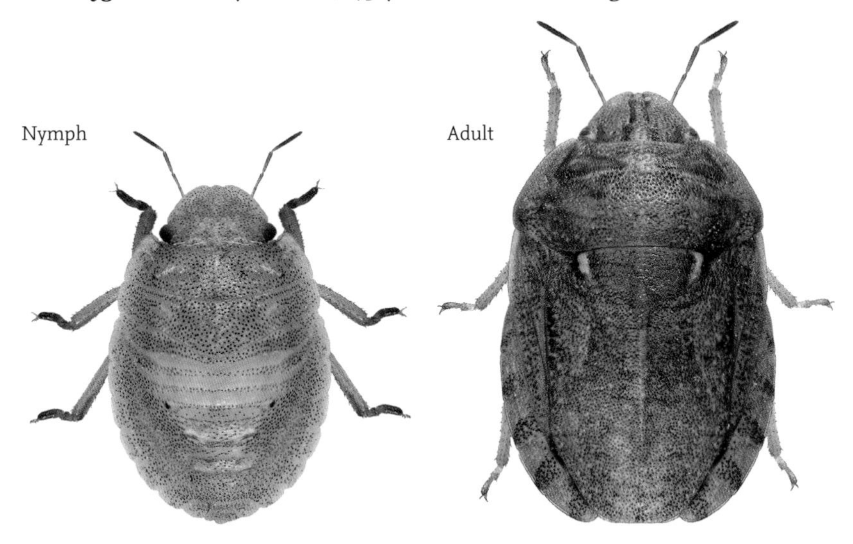

Description and similar species. Length 8.5–10.0 mm. This large broad, domed, heavily punctured shieldbug is so similar in general appearance to *E. testudinaria* that any further description is needless repetition. *E. maura* has less protruding pronotal prominences; side-by-side this is obvious, but this difference is difficult to appreciate without comparison of reference specimens. The flat front edge of the head is usually a good field character. Several key works give varying lengths for the second antennal segment; a threshold of at least 1.8 times the length of segment 3 should be used.

Ecology. This is a species of dry grassland, where it feeds on grass seeds. It has a propensity for chalk downland, but also occurs on sparsely vegetated brownfield sites in the London area and the Thames estuary. In the eastern Mediterranean this species is regarded (along with *E. austriaca*) as a major pest of wheat.

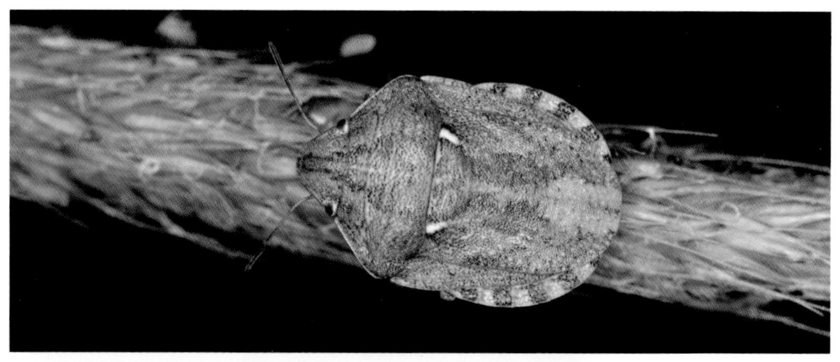

ABOVE: Adult

Nymph

Distribution and status. Although only 'recently' distinguished from *E. testudinaria* by China (1927), this seems to have always been a genuinely much rarer insect in Britain. It is mostly known from the chalk North Downs of Kent (Barnard in preparation) and Surrey (Hawkins 2003), with outlier records in Essex, Sussex, Hampshire, Dorset, the Severn estuary and the East Anglian Breckland.

***Eurygaster austriaca* (Schrank, 1776)**

Description and similar species. Length 11.0–13.0 mm. This is another large domed shieldbug, extremely similar to the two previous species, varying from lighter to darker brown, with a distinct narrow pale streak along the crest of the scutellum, but not appearing to have the more brightly coloured or mottle-patterned forms. It is distinguished from *E. testudinaria* and *E. maura* by its larger size, and by the two side lobes of the head touching each other in front of the shorter central lobe.

Ecology. It feeds on the seeds of various grasses.

Distribution and status. This bug was apparently established along the east Kent coast during the nineteenth century, but has not been recorded since 1885, and is now thought to be extinct in Britain. It has a scattered distribution in France, to Normandy and Boulogne (Roger Hawkins, personal communication), but has

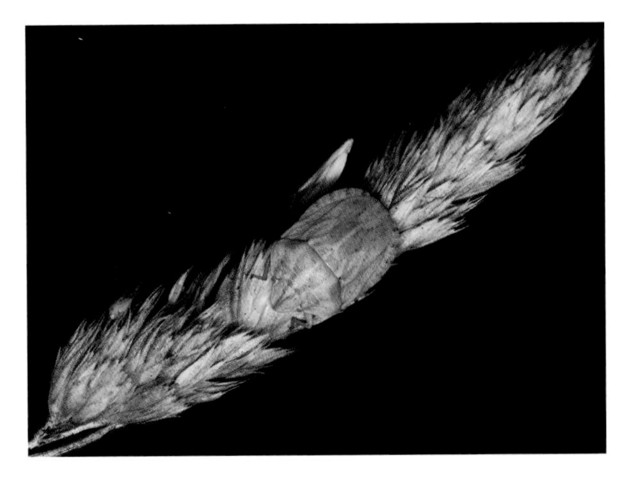

Adult

declined in Belgium and Denmark where it used to occur (Nielsen & Skipper 2015) and has not been found in the Netherlands since 1935 (Aukema 2001). Douglas & Scott (1865) include it as the rather insensitively named *E. hottentottus* (actually a similar Mediterranean species), and although they give no locality they charmingly state 'a single specimen, captured casually by Mr Ernest Adams', as if he were prancing about in his shirtsleeves rather than being properly dressed in suit, wing collars and bow tie as any respectable entomologist ought to have been.

***Odontoscelis fuliginosa* (Linnaeus, 1761) – Greater Streaked Shieldbug**

Description. Length 6.0–8.5 mm. This is a small oval, brown, domed, hairy shieldbug. The head is broad and flat, and usually entirely dark brown to black. The broad pronotum has at least the front half dark brown or black, but is variously mottled with paler brown markings. The broad pronotal margins are marked by narrow grooves running from behind the eyes. There are two short transverse grooves or unpunctured lines on each side of the pronotum, just in front of the middle. The abdomen is completely covered by the huge scutellum, variously light or dark brown, or mottled, but often with a pale central line and/or pale streak on each side, emphasised on the inside by

Adult

a black shadow, and in even the palest specimens these black streaks persist. It is densely punctured all over, and in the pale areas these punctures are infilled black. The entire upper surface is covered with short dark brownish hairs, and the side margins are lined with longer paler brown hairs. The legs and antennae are mostly dark brown or black, but the tarsi are often paler.

The nymph has head, pronotum and wing buds dark brown, more shining than in the adult. The abdomen is slightly paler, marked with dark blotches around the scent glands and along the edges. It is generally covered with longer but less dense pale hairs.

Similar species. The slightly smaller and generally commoner *O. lineola* is very similar but is marked on its upper surface by several dense streaks of short silvery hairs.

Ecology. This genus is strongly ground-dwelling, and this species feeds on stork's-bills (*Erodium* species) on sparsely vegetated sandy coastal dune sites.

Distribution and status. This is a very rare species, apparently much declined in recent decades. Although historically recorded from widely spread coastal localities from Norfolk to Birkenhead, this shieldbug has recently only been found in the Deal/Sandwich area of Kent (my single specimen came from here, 31 July 1977) and Pembrokeshire (Judd 2006). Elsewhere in Europe it is known from inland sandy sites, but former records from the East Anglian Breckland are now reckoned to be identification errors.

Odontoscelis lineola Rambur, 1839 (formerly _O. dorsalis_ in Britain) – Lesser Streaked Shieldbug, Heath Streaked Shieldbug

Description and similar species. Length 4.0–6.0 mm. This small oval, mottled brown, domed and hairy shieldbug is very similar to *O. fuliginosa* in shape, colour patterns and hairiness. However, it is smaller, and also has several streaks of distinctive silvery hairs, which catch the light and give it a much greyer appearance. The first two antennal segments are paler than those of *O. fuliginosa*.

The nymph is very similar to that of *O. fuliginosa*, but the pale silvery hairs are present in larger or lesser degree.

Ecology. This is another strongly ground-dwelling species, feeding on stork's-bills (*Erodium*) in sandy places.

Distribution and status. This is a rather scarce shieldbug, occurring in sandy places where its foodplant grows. Records are mostly coastal, from Norfolk to Cornwall, but also on the East Anglian Breckland and the heaths of Surrey and Hampshire.

Adult

Odontotarsus purpureolineatus (Rossi, 1790)

Description. Length 8.2–11.4 mm. This domed, prettily streak-marked bug has a broad pronotum and large but elegantly shaped scutellum. The pronotal angles are strongly produced into prominent epaulette-like angles. It is a pale brownish beige, marked with flowing almost flame-shaped streaks of pinkish, orange, brown or purple, these markings usually edged with a darker brown to black, giving a highly contrasting attractive graphic appearance. A central pale line usually extends all the way from the clypeus, across the head, pronotum and scutellum to meet two other shorter pale gashes to create an arrowhead-shaped mark at the tail end, emphasised by a blackish shadow on each side. It is more or less strongly punctured all over, with some of the largest punctures on the front of the pronotum infilled black. The underside has the same pale base colour as the upper surface, but is not patterned except for some scattered darker punctures. The legs and antennae are variable, pale beige to dark brown or purple.

Adult

The nymph is broad oval, mottled brown with pale edges to the pronotum and a streak extending from the clypeus to the tip of the developing scutellum. Nymphs are nowhere near as strongly marked as the adults, and might easily be overlooked as *Eurygaster* or *Aelia* nymphs.

Similar species. There is nothing else quite like it in northern Europe, but several others in the genus occur around the Mediterranean, though they are unlikely to ever occur in the British Isles.

Ecology. In Europe this species is recorded feeding on Salad Burnet (*Sanguisorba minor*), knapweeds (*Centaurea*) and thistles (*Cirsium* and *Carduus*) in rough flowery grassy places, including road verges and railway embankments. The striking pattern, so strong and eye-catching in an isolated specimen, is remarkably camouflaged against the mottled streaks of the developing flower heads.

Distribution and status. It is not recorded from the British Isles, Denmark or Scandinavia, but is common and widespread in France right up to Normandy. This, surely, is one to hope for sometime soon.

FAMILY PYRRHOCORIDAE

Pyrrhocoris apterus (Linnaeus, 1758) – Firebug

Description. Length 8.5–10.0 mm. There is no mistaking this bright, narrow-oval, black and red bug. The black head is short and triangular, with the central lobe of the clypeus prominent. The pronotum is trapezoidal, very gently concave behind the head, with the side margins more or less straight, but slightly indented where a transverse groove crosses the middle of the disc; it is mostly red, but with a central rectangular or trapezoidal black mark. The scutellum is black, small and triangular. The hemelytra are flat and broad, short, ending about halfway and leaving the abdominal segments exposed; they are mostly red but with the clavus black, also a large round spot and a small triangle on the corium black. In most specimens (all known British specimens until very recently) the hemelytra are short, the membrane is lacking, and the hind wings are absent. Indeed, the scientific name *apterus* means wingless. The truncated hemelytra are sometimes darkened or black-edged. The hemelytra are slightly longer in the rare winged specimens, terminating in the rounded black membrane. The underside is black, marked with red. The legs and antennae are black.

The nymph is a striking black on the fore-parts and red on the abdominal segments.

Similar species. *Pyrrhocoris* might be confused with *Corizus hyoscyami*, which has similar black and red coloration, but that species is longer and narrower with a clearly different pattern. The black and red *Eurydema* species are much broader and rounder. Similarly bright black and red ground bugs, family Lygaeidae, also have different patterns.

Ecology. Adults occur all year round, nymphs from June to September. *Pyrrhocoris* feeds on the fruits of mallows (*Malva* species; especially Tree Mallow, *Malva arborea*, in Europe) and limes (linden trees, *Tilia* species) in warm, dry, open

Adult

Adults and nymphs

habitats. In continental Europe this insect can be found in large huddled knots of many dozens of adults and nymphs, around the base of street trees in many towns and cities.

Distribution and status. *Pyrrhocoris* is very rare and scattered in the British Isles, where it is right on the northern and western edge of its European range. There are records of individual specimens from Wales, Yorkshire and Norfolk. A strong Devon colony has been known since the nineteenth century – Westwood (1839–40) reported 'some years ago, on some little islands at Torquay in Devonshire, and also on a rock in the sea at Teignmouth ... looked quite red with them'. More recently, colonies have been found in Bedfordshire, Kent, Surrey, Sussex and Essex. For many years only the short-winged form was known in Britain, suggesting it had poor dispersal capabilities and that it had been accidentally brought in on horticultural material from the continent. All of the 30 Surrey specimens examined by Ashwell & Denton (1998) were short-winged and unable to fly, leading the authors to wonder how there came to be a breeding colony on street lime trees and mallow plants in an industrial estate in Epsom. However, macropters (long-winged forms) were recently observed in a strong colony of several hundred individuals in Dovercourt, Essex, in July 2020, possibly the prelude to a much wider spread (see Fig. 92).

FAMILY ALYDIDAE

Alydus calcaratus (Linnaeus, 1758)

Description. Length 10.5–13.0 mm. *Alydus* is an elegant parallel-sided blackish-brown bug. The large head is triangular in front of bulbous eyes. The pronotum is trapezium-shaped, with sides straight and hind angles slightly raised. It is densely punctured, except in the centre just behind the eyes, where a deep dimple is surrounded by smooth integument. The scutellum is long triangular. The hemelytra are flat and long, and the membrane just reaches the tip of the abdomen. The legs are long, especially the hind femora, which are armed with a series of small but sharp spines beneath. The antennae are long, especially the fourth segment which is slightly curved. The insect is entirely dark chocolate-brown, except for a small spot at the tip of the scutellum, rings around the antennal segments, and small tick marks on the connexivum, which are white or cream; the tibiae and some antennal segments are sometimes more reddish brown. The body is covered with a short dense pilosity giving it a rather flat matt appearance. In life this is an active and rapidly moving insect, flitting between grass stems and over patches of bare soil; as it opens its wings its bright red abdomen appears in a sudden flash of colour. This, and its rapid and rather jerky movements, makes it look very like a pompilid spider-hunting wasp.

The nymph is remarkably ant-like. The early instars exactly resemble Black Pavement Ants, *Lasius niger*, in size, colour and activity. Later instars resemble Wood Ants, *Formica rufa*, with dark head, dark bulbous abdomen, and narrow reddish-brown pronotum and wing buds. The deception is quite remarkable and I often have to stop and stare intently at the creature running about in the sweep net before my mind settles on the identification as a het bug nymph.

Adult

Nymph

Similar species. There is nothing else like it in the British Isles. The not-quite-British *Camptopus lateralis* is larger with bolder markings (especially the pale margin of pronotum, hemelytra and connexivum), and *Micrelytra fossularum* is smaller and more slender.

Ecology. Adults overwinter, and nymphs start to appear in June. New adults occur in August and September, often into late October and early November in the right weather. For such a large and obvious insect, details of its life history are surprisingly sparse. It has been found on a variety of plants, including gorse (*Ulex*), broom (*Cytisus*), restharrow (*Ononis*) and medicks (*Medicago*), and it is probably a phloem- and/or seed-feeder. It is particularly associated with dry sandy heathy places including brownfield sites. There has been speculation that it may be a predator of ants, and its ant-mimic nymphs are often found running with them, but there are no first-hand reports of any predation that I can find. Various authors (often quoting each other) claim it has been found feeding on carrion and dung, but I surmise this may simply be for fluid intake in hot dry weather rather than nutrition. Likewise several reports (again often quoting each other) suggest it *may* develop in ant nests. The ant-like appearance of the nymphs is, I think, likely to be a red herring. Although absence of evidence is not evidence of absence, I suggest that the deep, focused and tenacious work by Horace Donisthorpe (1927) and others in the nineteenth century would have found the necessary regular evidence of this insect in the very many ant nests that they dissected, collected, reared and studied in the laboratory, had such a deep myrmecophile behaviour been genuine. Instead, records from ant nests are very occasional and highly anecdotal. True, it has been recorded 'in association'

with various *Formica, Lasius* and *Myrmica* species in the sandy places where the bug is found and where ants actively run about the dry sparsely vegetated terrain, but even a century after Butler (1923) listed these vague relationships, we are no nearer to finding anything other than an approximate coexistence rather than any real ecological interaction.
Distribution and status. This is a very local species of hot dry places, mainly the band of sandy heaths running through East Anglia, Surrey, Hampshire and Dorset, but also on brownfield sites in south Essex and north Kent, and numerous coastal localities in the West Country and Wales. It is very rare in Ireland.

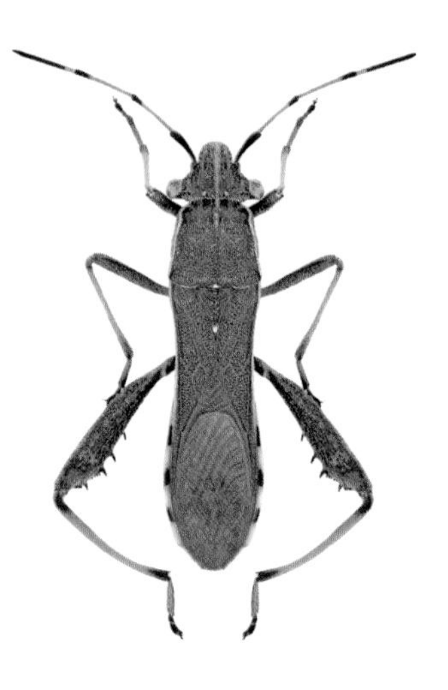

Camptopus lateralis **(Germar, 1817)**
Description. Length 11.0–13.5 mm. *Camptopus* is a large, long-legged, parallel-sided brown bug. The head is large and triangular with bulbous eyes. The pronotum is nearly square. The scutellum is triangular. The hemelytra are flat, long and slim. The legs are long, especially the rear femora, which are armed with a series of small but sharp teeth on the undersides, and rear tibiae, which are rather curved. The antennae are long and slim. This is mostly a deep dark chocolate-brown insect, but with narrow yellowish cream margins to the pronotum and hemelytra, and with pale marks narrow but clear on the connexivum, broad pale rings on antennal segments 2 and 3, and with the tibiae mainly pale, along with the first segment of the tarsi. The tip of the scutellum has a small white spot. The underside is brown, though perhaps not as deep a colour as the upper surface. The whole body is covered with a short but dense pilosity.

The nymph is very ant-like, and rather similar to that of *Alydus calcaratus*.
Similar species. *Camptopus* is larger and broader than *Alydus*, with longer and more enlarged rear femora, white markings on the edges of the pronotum, hemelytra and connexivum, and with pale bands on the legs and antennae.
Ecology. It is recorded from various putative foodplants including bird's-foot trefoils (*Lotus*), restharrow (*Ononis*), gorse (*Ulex*) and medicks (*Medicago*), usually in warm, dry, well-drained sites.
Distribution and status. Although not really recorded from mainland Britain, it is found on the Channel Islands, and much of the Atlantic coast of France. Maria

Adult

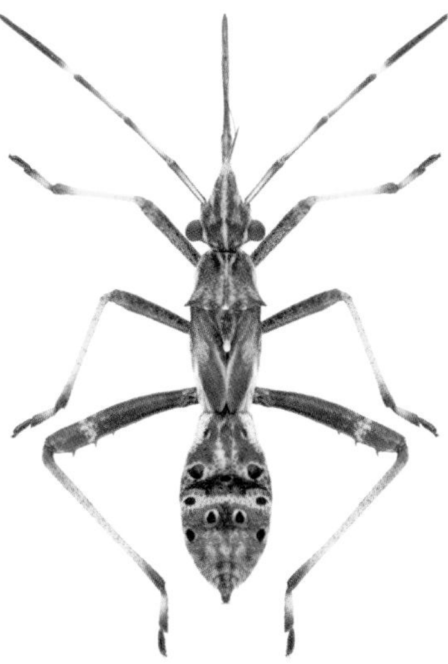
Nymph

Justamond photographed one in Kent, in December 2019, found inside a car that had recently come back from France (see Fig. 139).

Micrelytra fossularum **(Rossi, 1790)**

Description. Length 9.0–12.0 mm. *Micrelytra* is a long, narrow, long-legged, elegant brown and yellow bug. The head is long and narrow, with the clypeus produced into a triangle; the eyes are set in the middle of the side margins. The

Adult

Mating pair

pronotum is long and slim, nearly parallel-sided. The scutellum is narrowly triangular. The hemelytra are long and slim, those of the short-winged form (brachypter) covering about half of the abdomen, those of the long-winged form (macropter) reaching nearly to the tip. The legs are very long and slim. The antennae are long and narrow. The bug is mostly dark brown, but with the margins of the head, pronotum and hemelytra narrowly beaded with pale cream. The connexivum is edged with cream; this is especially prominent in the more swollen abdomen of the female. The tibiae and tarsi are mostly pale cream or beige, contrasting with the dark femora. The antennae have the second and third segments bicoloured brown and cream; the fourth segment is reddish brown.

The nymph is long and narrow, almost entirely dark brown, but with the rear margins and corners of the pronotum cream-edged.

Similar species. *Micrelytra* is much slimmer and slighter than *Alydus* or *Camptopus*, but might be mistaken for a mirid leaf bug (*Stenodema* or *Notostira*) except for the fairly obvious ocelli.

Ecology. It is recorded by Moulet (1995) on Cock's-foot (*Dactylis glomerata*) in Brittany, and it probably occurs on other grass species too.

Distribution and status. This bug has not been found on the mainland British Isles, but is recorded in the Channel Islands. This is a western Mediterranean species, mainly known from Greece, Italy, North Africa, the Iberian peninsula and the Atlantic coast of France. Jersey seems to mark the very northern outlier point of its range, and all records there have so far been of the brachypterous (short-winged) and therefore non-flying form.

FAMILY COREIDAE

Arenocoris fallenii (Schilling, 1829) – Fallén's Leatherbug

Description. Length 6.2–7.5 mm. This is a small, flat, rough-looking squash bug. The head is square or rectangular, with the central lobe of the clypeus standing out spine-like in the middle. The pronotum is short and broad, roughly granular, and with two rows of small pale spines forming a backwards V-shape in the centre. The margins are slightly concave, roughly edged with a row of small sharp granular spines. The scutellum is triangular, ending with a slightly raised double tubercle. The hemelytra are broad and flat, and very heavily punctured between the veins. The abdomen is rounded, with the connexivum broad and strongly granularly punctured. The bug is generally grey/brown but mottled darker, with especially evident blotches on the connexivum and base of the scutellum, and with narrow black streaks on the veins of the hemelytra. Sometimes the entire insect appears dark charcoal-grey, although it also sometimes looks reddish brown or yellowish. The antennae are brown, with the first segment broader and more strongly granular than the others; the last segment is darkest, often nearly black. The legs are light or dark brown, with the femora strongly granular, and tarsi dark.

The nymph is yellowish grey, with the edges of the pronotum and wing buds edged pale, but not yet developed with any granularity.

Adult

Similar species. The similar but much rarer *A. waltlii* lacks the regular V-formation of outstanding pale spines on the pronotum; its penultimate antennal segment is thickened slightly near the end, and the hind femur has a strong spine near the apex. *Bathysolen nubilus* and *Spathocera dalmanii* are both similar in vague shape and size but smoother, less granular.

Ecology. Nymphs occur in June and July, with adults the remainder of the year. It occurs in dry sandy places under its foodplant, stork's-bills (*Erodium* species), under stones or in grass tussocks.
Distribution and status. This is a scarce bug mostly found in the East Anglian Breckland and coastal sites in Suffolk, Cornwall and south Wales. It often occurs on mature coastal sand-dune systems, but also in sand pits and the occasional brownfield site. There is evidence that this species is expanding its geographic range, with modern records from Bedfordshire, Northamptonshire and Essex. Together with the next species, it is top of my 'to find' list.

Arenocoris waltlii **(Herrich-Schäffer, 1835) – Breckland Leatherbug**
Description and similar species. Length 7.5–8.0 mm. This is a small flat grey/ brown granulose squash bug. It is very like *A. fallenii* in general shape and colour, but rather than the granular spines on the pronotum forming an inverted V-shape, they are more randomly spread. The third antennal segment is expanded at its apex in *A. waltlii*, appearing gnarled and bristly, whereas that of *A. fallenii* is evenly narrow. *A. waltlii* has a spine behind, near the end of the hind femur, which is lacking in *A. fallenii*.

The nymph is extremely similar to that of *A. fallenii*.
Ecology. The life cycle is probably similar to the previous species. *Arenocoris waltlii* occurs in dry sandy places under its foodplant the Common Stork's-bill (*Erodium cicutarium*).
Distribution and status. This is a very rare insect, and reliably recorded only from the Breckland of East Anglia, and formerly from the Suffolk and Norfolk coast. It was thought possibly extinct in Britain following no recent records since 1964,

Adult

Nymph

but was rediscovered in 2011. It is reputed to have been found in the Sandwich area of Kent, but this has not been confirmed.

Bathysolen nubilus **(Fallén, 1807) – Cryptic Leatherbug**

Description. Length 6.0–7.0 mm. This is a small, broad, flat, brown squash bug. The head is rather broad, rectangular, warty, with the eyes bulbous, in the middle of the sides. The pronotum is broad and warty, with the lateral angles prominent, but not pointed, and the side margins with a series of small pale tubercles/denticles towards a small projecting prominence at the front angles. The scutellum is small, triangular, drawn out to a slight narrow tongue at the tip. The hemelytra are broad and flat. The membrane is wrinkled, dark. The abdomen is gently oval. Generally the bug is a dark brown, but the front corners of the pronotum, the tip of the scutellum and patches on the connexivum are pale brown to beige, even white. The upper surface often

Adult

Nymph

appears vaguely mottled, and perfectly camouflaged against the soil. The underside is generally mid- to dark brown. The legs are dark brown but the tibiae are obscurely lighter. There are small sharp teeth on the underside at the end of the rear femora. The antennae have the first segment dark brown, large, swollen and tuberous; segments 2 and 3 narrow, paler; segment 4 short ovoid, dark, nearly black.

The nymph is brown, covered all over with paler warts; the sides of the head are often black, matching the dark antennae.

Similar species. This species is generally darker than *Arenocoris* and lacks any pronotal spines. It is broader than *Spathocera*, especially the pronotum. The not-quite-British *Bothrostethus annulipes* is larger, and more spiny on the pronotum, legs and antennae.

Ecology. Nymphs occur in June and July, with adults occurring the rest of the year. This species is strongly ground-dwelling, in hot, dry, well-drained situations, and is usually found under its foodplant Black Medick (*Medicago lupulina*). It is slow-moving and renowned for being almost invisible even when right under the nose of the entomologist.

Distribution and status. This is a very local insect known mainly from the Thames estuary, Essex, Kent and Surrey, but with outlier records from East Anglia, Cambridgeshire and Oxfordshire. At the beginning of the twentieth century it was only known from Deal in Kent, and there is good evidence that it is increasing and spreading.

***Bothrostethus annulipes* (Herrich-Schäffer, 1835)**

Description. Length 9.5–11.0 mm. This is a stout brown rather spiny squash bug. The head is long, the antennal mounds projecting prominently in front of the eyes, and the clypeus projecting into a triangular point in front. The pronotum is trapezium-shaped with a series of small teeth along the side margins and a small hook-shaped prominence at the broadest point. The scutellum is equilateral triangular. The hemelytra are broad and flat. The membrane is dark but with the veins showing prominently. The abdomen is gently ovoid. The insect is generally dark brown, but often rather mottled; the tip of the scutellum is minutely bright white, and the pronotum has a strong black mark in the centre; the connexivum is pale-marked. The underside is generally dark brown.

Adult

The legs are dark, but the tibiae usually have two pale rings. The hind femora are long and dark, with a series of small but sharp spines. The antennae are bristly/spiny, brown, with the last segment and a half often darker.

The nymph is brown with pale wart-like spines all over, the femora and antennae contrastingly dark and bristly.

Similar species. *Bothrostethus* resembles *Bathysolen nubilus*, but is larger and spinier. The third antennal segment is much shorter in *Bothrostethus* than in *Bathysolen* or *Arenocoris*.

Ecology. This bug is mostly found on Fabaceae, sainfoin (*Onobrychis*), brooms (*Cytisus* and *Genista*) and milk vetch (*Astragalus*).

Distribution and status. It is not known from the British Isles, yet. This is a scarce Euro-Mediterranean species, but it occurs up the Atlantic coast of France and is recorded from the Channel Islands.

Ceraleptus lividus Stein, 1858 – Slender-horned Leatherbug

Description. Length 9.5–11.5 mm. *Ceraleptus* is a slender elegant nearly parallel-sided squash bug. The head is long, with the antennal mounds projecting and triangular, the clypeus long and projecting. The pronotum is broad, strongly narrowed towards the head, with the sides straight and only minutely tuberculate in front of the projecting but not overly prominent hind angles. The scutellum is triangular. The hemelytra are broad and flat. The membrane is wrinkled and the veins are prominent. The abdomen is gently and

Adult

narrowly rounded. Overall the colour is mid-brown, but the insect is covered all over with dark punctures; the rear edge and side margins of the pronotum are often clouded darker; the border of the corium and connexivum are narrowly but distinctly pale, and the head has a narrow pale streak back from the clypeus emphasised by a darker streak on each side passing over the eye. The underside is strikingly paler than the upper surface. The legs are pale with some darker spots; the apex of the hind femur is darkened, and bears a large spine and some smaller ones. The antennae are reddish brown, with the apical segment and a half, and sometimes the basal segment, darker.

The nymph is mid- to pale brown, often appearing streaked because of the chocolate colour of the antennae, pronotal edges and wing buds.

Similar species. *Ceraleptus* is larger and paler than *Coriomeris denticulatus* and lacks the distinctive row of pale hair-tipped spines along the pronotal margins of that species. It is also narrower and less granular than *Bathysolen* and *Arenocoris* species. There are several others in this genus occurring in Europe, but the only vaguely likely one is the central and southern *C. gracilicornis*, which has the front corner of the pronotum extended into a sharp forward-projecting hooked prominence.

Ecology. Nymphs can be found from June to August, with adults in almost any month of the year. *Ceraleptus* is usually found on clover (*Trifolium*) and bird's-foot trefoil (*Lotus*), mostly in warm, well-drained places like chalk downs and sandy heaths, but also brownfield sites, gravel pits and sand pits. However, there are increasing records just from rough grassy places – suggesting that its narrow habitat requirements are perhaps broadening.

Distribution and status. This was once considered a relatively scarce species, mostly occurring in the south-east from the East Anglian Breckland, through Essex, Kent, Surrey, Sussex, Hampshire and Dorset. There are now a few outlying records from the coast of Devon, and also some into more central England to Lincoln, Birmingham and Coventry. A Welsh record is thought to be an identification error. This species is frequent on brownfield sites in London and the Thames estuary, and I was very pleased to find it on an arable farmland verge in Burgess Hill, Sussex, and a golf course in Bicester, Oxfordshire, in 2019, two very unprepossessing localities.

Coreus marginatus **(Linnaeus, 1758) – Dock Bug**

Description. Length 10.3–14.2 mm. This is a most gothic, dead-leaf-mimicking, brown squash bug. The head is nearly square, down-turned at the tip, with the side lobes of the clypeus longer than the central lobe, but this is only visible in front view. The eyes are particularly bulbous. The antennal bases expand on the inner side into two long forward-projecting curved spines, which nearly meet at the tips. The ocelli are small. The broad pronotum is depressed at the front, rising to the rear and projecting into a broad prominent triangular lobe at the side angles. The scutellum is an equilateral triangle, sharp at the tip. The hemelytra are flat, though with a

Adult

FAR LEFT: Early-instar nymph

LEFT: Late-instar nymph

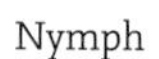

Nymph

slight ridge separating endo- and exocorium. The abdomen is expanded, broadly rounded. The membrane is ovoid, and very wrinkled. The legs are long and thin. The antennae are long, with the first segment stout, and the last segment long and ovoid. The entire body is a deep orange-brown, covered all over with black punctures, these interspersed with small white tubercles on the head and pronotum, side margins of hemelytra, connexivum, femora and first antennal segment. The femora have a series of small tooth-like tubercles underneath.

The nymph is dark brown, spiny with relatively huge thick bristly antennae, body flat, broadly oval and later diamond-shaped. This species appears rather ant-like in early instars. The spines on the antennal bases start to appear in the third instar. By this stage the basal segment of the antennae is strongly triangular in cross-section with sharp edges, segments 2 and 3 being slightly flattened with sharp keeled edges. Along with the early instars of the Green Shieldbug, *Palomena prasina*, which are commonly mistaken for 'some sort of ladybird', the small nymphs of *Coreus marginatus* are perhaps the insect most often found confusing to novice naturalists who cannot quite relate the narrow spiny lanky creature with large bristly antennae to the broad brown adult that they know so well. I include myself here too, and remember being completely confounded when I first found these creatures.

Similar species. The much less common *Enoplops scapha* is of a similar shape and size, but is darker, narrower across the pronotum, with bright white spots on the connexivum, and lacks the spines on the antennal bases. *Syromastus rhombeus* is altogether more delicate and angular. British identifications are usually quite clear, but on the continent the shape of the pronotal angles and the size, length and shape of the sharp antennal prominences vary greatly, leading to the naming of numerous forms and subspecies, several of which are mentioned and illustrated by Moulet (1995).

Ecology. The eggs are usually laid on dock plants, although almost any species in this family has been recorded for the bug, including Broad-leaved Dock (*Rumex obtusifolius*), Sorrel (*R. acetosa*), Rhubarb (*Rheum hybridum*), Redshank (*Persicaria maculosa*) and Knotgrass (*Polygonum aviculare*). Many other species are listed by Hawkins (2003). The nymphs occur from July to September, adults in almost any month of the year. The association with dock has given rise to the bug's common name, but I find it most commonly on brambles. When young, the bug feeds on the fruits of the dock, but it seems that blackberries are just as attractive, particularly when the bug is adult. At rest this insect is flat dark brown, seemingly a good leaf mimic as both adult and late-instar nymph, but when it takes flight its startling bright orange/red abdomen is revealed, perhaps making it appear wasp-like. When it lands, the sudden disappearance of the red flash probably helps it blend back into the foliage, confusing the eye of any would-be predator following it.

Distribution and status. This bug is frequent throughout most of southern and central England, and much of Wales. It is the only coreid native to Ireland. Generally it is one of the commonest shieldbugs and appears to be spreading; it was recorded in Yorkshire in 2015. It is more or less ubiquitous south of the Severn–Wash line, and pretty common to the north of it. This is often the 'brown shieldbug' that non-entomologists know, along with the 'green' *Palomena prasina.*

Haploprocta sulcicornis (Fabricius, 1794)

Description. Length 11.0–13.0 mm. This is an elegant shapely light-brown squash bug. The head is long, with the projecting clypeus triangular, but the eyes are not very prominent. The pronotum is narrow, slightly angular at the rear corners, having the sides in-curved and with a series of minute denticles approaching the front angles. The scutellum is neat, and triangular. The hemelytra are broad and flat, and very parallel-sided. The abdomen is broad, flattened and rounded. The membrane is pale, and wrinkled. Overall the colour is mid- to light brown, but the insect is covered all over with small black punctures. The legs are pale. The antennae are reddish brown, with the first joint broad, and the last joint sometimes darker. The underside is brown.

Adult

The nymph is brownish red or reddish brown, with prominent reddish antennae, and an abdomen with undulating margins.
Similar species. The connexivum is broader than in *Gonocerus*, but not as angular as in *Syromastus*.
Ecology. This delicate insect is mostly found on Sheep's Sorrel (*Rumex acetosella*) in well-drained localities. This was one of several insects considered for possible biological control against *Rumex spinosus*, an Old World species now reaching invasive pest proportions in Australia (Scott & Yeo 1996).
Distribution and status. This species is not recorded from the British Isles, but is frequent in France to Brittany and Normandy.

Coriomeris denticulatus **(Scopoli, 1763) – Denticulate Leatherbug, Trefoil Bug**

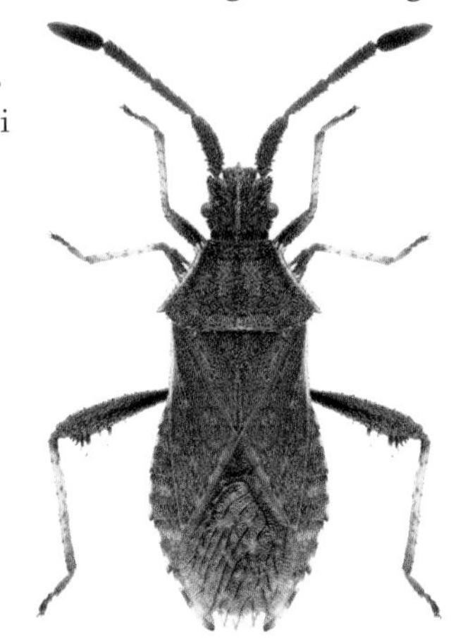

Description. Length 8.5–10.0 mm. This is a small narrow brown squash bug. The head is warty, long, rectangular, with eyes small but bulbous, and ocelli prominent. The pronotum is trapezium-shaped, the rear angles prominent but not sharp; the side margins are nearly straight and armed along their lengths with an irregular series of short but prominent sharp white spines each tipped with a hair. Two or three similar grey hair-tipped spines occur on the rear edge of the pronotum, near each hind corner. The hemelytra are subparallel, flat, toothed along the basal edge with a series of tiny blunt white spines. The scutellum is small, triangular, grey, edged with a series of globular warty teeth. The membrane is wrinkled and darkened, with the veins prominent dark brown. The upper body is more or less uniform mid- or darker brown but punctured all over black, and with the strongly contrasting white spiny margin to the pronotum. The underside is generally slightly paler than the upper surface. The orifice of the scent gland is

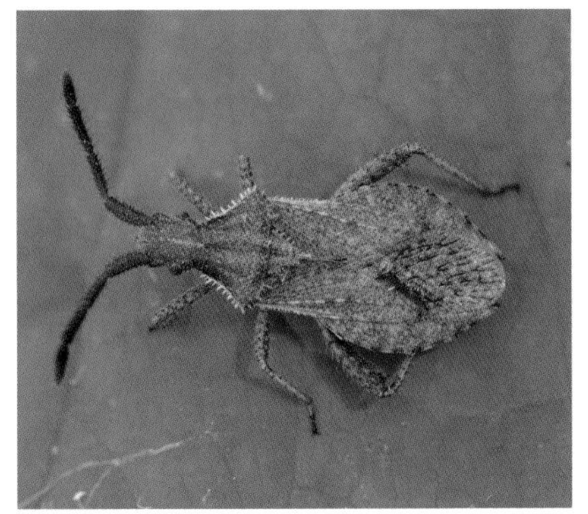

Adult

Nymph

strongly produced into a pale cream ear-like lobe. Legs brown, femora warty, usually darker than tibiae. The tip of each rear femur has two long and several shorter spines. Antennae brown, first segment warty, last segment usually darkest. Head, pronotum, legs and antennae with prominent hairs.

The nymph is mottled brown, warty, with the sides of the wing buds, pronotum and antennae bristly black.

Similar species. This insect is easily distinguished from other native British coreids of similar size and shape (*Gonocerus* and *Ceraleptus*) by the row of strong, contrasting, white spines along the sides of the pronotum. There are, however, several other European species in the genus distinguished by subtle differences in pronotal spines and antennal shape. Two, which might conceivably turn up here, are discussed below.

Ecology. Nymphs occur from late June through to August, and adults almost all year long. This bug occurs on dry sandy or chalky soils where its favoured foodplants are medicks (especially Black Medick, *Medicago lupulina*) and other legume species. It is usually ground-dwelling, but is frequently found in the sweep net.

Distribution and status. This species is local but widespread across most of England and Wales, although it appears to be mainly coastal in the west. It is commonest in south-east England, but there are outlier records to Newcastle-upon-Tyne and Barrow-in-Furness.

***Coriomeris scabricornis* (Panzer, 1809)**

Description and similar species. Length 8.0–9.0 mm. This bug is very similar to *C. denticulatus*, but the antennae are less bristly, covered with much shorter hairs, or with only a few longer hairs projecting. The white spines along the edge of the pronotum are also shorter and more even, tipped with only very short hairs.
Ecology. According to Moulet (1995) it is found on various clovers, trefoils and medicks in warm dry situations.
Distribution and status. It is not known from the British Isles, but is widespread in central Europe, Italy and Greece, and also into Denmark and Sweden, though not much of France. It has been recorded in the Netherlands since about 1987 (Aukema 2001). It's an outside possibility perhaps.

Adult *Coriomeris scabricornis*

***Coriomeris affinis* (Herrich-Schäffer, 1839)**

Description and similar species. Length 7.5–9.5 mm. This species is extremely similar to *C. denticulatus* but has only 5–8 rather widely spaced spines along the pronotal edge, compared to 10–12 in *C. denticulatus*. The two spines on each side of the rear edge of the pronotum are slightly longer and stronger.
Ecology. Although the host plants are uncertain, calamint, thyme, fenugreek and mallow are listed by Moulet (1995) and an association with pine trees is suggested as being for shelter rather than feeding.
Distribution and status. This species is not known from the British Isles, but has an Atlantic and Mediterranean range in Europe with records through France to Saint Malo in Brittany.

Enoplops scapha **(Fabricius, 1794) – Boat Bug**

Description. Length 11.0–13.0 mm. This is a large stout greyish-brown squash bug. The head is square, with the short triangular clypeus slightly protruding, and the eyes hemispherical. The pronotum is broad, strongly indented behind the front angles, then produced into very broad rounded hind angles. The hemelytra are broad and flat. The abdomen is broad, flattened and splayed. The membrane is dark brown, wrinkled. The general background colour is a dark greyish brown, but with the sides of the head, pronotum, hemelytra and connexivum narrowly but brightly pale. The connexivum is pale-marked at each segmental intersection. The underside is this same pale cream. The legs are brown, though the tibiae are paler. The antennae are brown, but segments 2 and 3 are narrow, usually more reddish, and segment 4 is swollen, dark, sometimes nearly black.

The nymph is brown, prickly with large spiny antennae. Early instars have the abdomen pale.

Similar species. This large bug is very similar in shape and size to *Coreus marginatus*, but in that species the pronotum is more sharply produced and has no white edging; *Coreus* also lacks the pale marks on the connexivum, and has two distinct curved and pointed processes projecting forwards from the antennal tubercles at the front of the head.

Adult

Ecology. Nymphs occur from June until August, with adults the rest of the year. This bug is associated with mayweeds (*Matricaria* and *Tripleurospermum*), especially *T. maritimum*, but is also recorded on ragworts (*Senecio*), chamomiles and other composites, usually in sandy coastal localities like cliffs, undercliffs and dunes.

Distribution and status. This is a scarce bug, being almost entirely coastal in the British Isles, with scattered localities from Kent to Anglesey, but with some historical records from Yorkshire.

This is my favourite shieldbug name. A scapha was, apparently, a small oar-powered Roman galley-type ship (a skiff?), so perhaps Fabricius thought the rounded and elevated connexivum was boat-shaped.

Gonocerus acuteangulatus (Goeze, 1778) – Box Bug

Description. Length 11.5–13.5 mm. This is a large narrow and elegant brown squash bug with prominent sharp pronotal angles. The head is long, with the projecting clypeus triangular, and the eyes bulbous. The pronotum is gently curved to the sharp, raised, rectangular side angles. The scutellum is triangular. The hemelytra are narrow, flat and parallel-sided. The abdomen is gently rounded. The membrane is dark, and wrinkled. Above, this insect is almost entirely a rich brown colour, sometimes with olive tones, but covered all over with black punctures; the connexivum has intersegmental junctions narrowly pale. The underside is a

Adult

Nymph

warmer orange-brown. The legs are long and narrow – a pale orange-brown. The antennae are long, slender, brown, but with the apical portion of each segment often darkened. In contrast to other British genera in the family, *Gonocerus* has a small nodular sub-segment between the normal third and fourth segments.

The nymph has the abdomen bright green, contrasting with the brown prickly fore-body, and later with the developing brown wing buds. Early instars have the antennae disproportionately large and stout.

Similar species. *Gonocerus* is intermediate in size between *Coreus*/*Enoplops* and *Coriomeris*/*Ceraleptus* and is altogether a more narrow, smooth and elegant creature than most other coreids, particularly in the male; the parallel-sided body contrasting with the sharp pronotal angles makes it a highly distinctive species. There are several other European species in the genus, one of which is discussed below.

Ecology. Nymphs start to appear in July and peak in late August; new adults appear in September and October. Although formerly seemingly only known to feed on the fruits of Box (*Buxus sempervirens*), it has recently been recorded from hawthorn, honeysuckle, rose, yew, cypress, and many other trees and shrubs.

Distribution and status. For many years this insect was known in the British Isles only from Box Hill, Surrey, where it was said to feed on the fruits of Box trees. However, in January 1990 it was found a few kilometres away at Bookham Common. Later that year it was found in several parts of the common, mostly beaten from hawthorn, including nymphs, and by 1995 it was found regularly throughout this area of central Surrey. This was the beginning of a great expansion in the bug's range, still continuing today. By the time of Hawkins' atlas (2003), the Box Bug was widespread in Surrey, but barely known beyond its borders. However, it has continued to spread. Presently it is widespread throughout England to Hull, Nottingham, Birmingham, Bristol and Exeter, and with outliers to Manchester and North Yorkshire. It will probably go further.

Gonocerus juniperi Herrich-Schäffer, 1839

Description and similar species. Length 11.5–13.5 mm. This species is very similar in general appearance to *G. acuteangulatus*, but is often less uniformly coloured and more brightly patterned reddish, brown and grey, with an inverted dark Y-shaped mark on the pronotum, mottled chevrons on the hemelytra, the connexivum often with a dark and light pattern, and legs and connexivum sometimes yellow or greenish. The underside is also more yellow or green. The head is a little longer than that of *G. acuteangulatus*, and the pronotal angles are slightly sharper, although these differences are almost impossible to appreciate until both species are examined together down the microscope.

Ecology. This bug is recorded on junipers and cypresses.

Adult

Adult

Distribution and status. Although not known from the British Isles, it is widespread in Europe to France and Belgium and could easily be introduced here with the large movement of garden cypresses in the international horticultural trade. On 28 July 2003 I beat a nymph of what I took to be *G. acuteangulatus* from a cypress tree in Battersea Park; it most likely was that species, but there remains a tantalising doubt in my mind that it could just conceivably have been this one. I'm still keeping my eyes peeled.

Leptoglossus occidentalis **Heidemann, 1910 – Western Conifer Seed Bug**

Description. Length 15–20 mm. There is no mistaking this large, striking, distinctively patterned squash bug with its swollen back legs. The head is long, with the clypeus well extended, triangular, and eyes moderately bulbous. The pronotum is broad, with the hind angles prominent, each ridged into a dark carina, though it is strongly narrowed towards the head. The scutellum is triangular, pointed, shining at the base, but with a narrow stripe of glistening golden-silver hairs down the centre. The hemelytra are broad and flat, but parallel-sided, and covered all over with short glistening golden-silver hairs. The membrane veins appear as a dense wrinkled network. The connexivum is broad, and gently rounded. This large insect is predominantly a rich chestnut-brown, but it has a distinctive bright white zigzag line following some of the veins, across the middle of the hemelytra, although this can be faint, or even absent. The connexivum is strongly marked black and white. The hemelytra are sometimes pale at the base, and the head and pronotal shoulders are often darker, nearly black. The scutellum is often edged white and divided by a pale median line. The underside is mostly a mid-brown. The legs are brown, but the femora are darkened. The front and middle femora have small spines near the apex; the hind femora have a double row of small spines beneath, larger, stouter and more curved towards the apex. The tibiae have a broad pale ring around the middle. Among the most distinctive features are the tibiae of the back legs, which are expanded and flattened into leaf-like blades, pale brown at the base but mottled with dark brown or black. The antennae are brown with the middle segments paler, and the final segment dark, sometimes nearly black; the basal antennal segment is stout, curved, and black above.

The nymph is prickly, with an orange abdomen contrasting with the dark brown head and thorax, and later with the developing wing buds.

Similar species. This large bug is unmistakable. Its size, rich colours, white zigzag marking on the hemelytra and flattened hind tibiae immediately identify it. When my father sent me an out-of-focus photograph of a vaguely insect-shaped

ABOVE LEFT AND RIGHT: Adults

Nymph

blob sitting on a trowel handle, there was still no mistaking it. The only other European member of this genus is the large, dark brown, pan-tropical *L. gonagra*, which occurs sporadically around the Mediterranean and which is very unlikely to turn up here. For what it's worth, McPherson *et al.* (1990) give a key to the *Leptoglossus* species of North America.

Ecology. The nymphs start to appear in July and feed through to September, with new adults appearing from late August. This bug feeds on the seeds in the cones of various conifer trees – many different species of pine, fir, cedar and spruce have been recorded. In captivity it has been reared on the fruits of pistachio. It overwinters as an adult, usually under pieces of loose bark, or in bird or rodent nests, but it also has a propensity to enter houses – its large size causing consternation to the householders. It flies well, and loudly, and its long legs allow it to run fast too.

Distribution and status. This large and distinctive insect is a native of western North America (hence its common English name). Originally recorded from California, Colorado and Vancouver, it started to spread east during the second half of the twentieth century (McPherson *et al.* 1990), reaching New York in 1992. From the USA it was accidentally transported to Europe and was first found here, in Italy, in 1999. It quickly spread to Switzerland (2002), Spain (2003/04), Hungary (2004), Austria (2005), Germany, France, Serbia and the Czech Republic (2006), Belgium, Slovakia and Britain (2007), Poland (2008), East Asia (2010), North Africa (2013), Turkey (2018) and South America (2017); no doubt it is still spreading. Its first appearance in the British Isles was in Weymouth, Dorset, in 2007, but the following year many specimens were found from various sites along the south coast of England (Malumphy *et al.* 2008). This was thought to have been a natural invasion from the continent, but there were also numerous inland records to Lancashire and Cumbria. Movement of ornamental conifers through the horticultural trade is likely to have been responsible for transatlantic and cross-Britain movements. By the time of the provisional shieldbug atlas 10 years later (Bantock 2018), it had reached Newcastle, Sunderland, the Stranraer peninsula and the Isle of Man, and was obviously common across most of England and Wales southwards. The latest map from the National Biodiversity Network database shows it well into Ireland, Scotland (the only coreid that makes it north of the border), particularly the east coast north from Edinburgh, and with outliers on Orkney, Shetland and the Outer Hebrides. It is currently vying with *Pentatoma rufipes* to be Britain's most widespread shieldbug species.

Spathocera dalmanii **(Schilling, 1829) – Dalman's Leatherbug**

Description. Length 5.5–6.5 mm. This is a small, broad, flat, squat, brown squash bug. The short head is square, uneven and knobbly. The pronotum is rather long, strongly narrowed to the front, with the margins slightly concave, flattened into narrow ridges; the hind angles are slightly prominent. The scutellum is triangular, but because of a raised central line and dark patches it appears heart-shaped. The hemelytra are broad and flat, with the veins prominent; the membrane is smoky and wrinkled, with prominent veins. The abdomen is rather broadly rounded. Altogether it is a mostly dark-brown insect, sometimes mottled, but with the narrow edges of the pronotum strongly contrasting pale cream, and the two dark marks on the scutellum strong. The underside

Adult

Nymph

is brown, streaked paler. The short legs are brown. The relatively short antennae are brown, with the extreme tip of the third segment expanded and dark brown, and the fourth segment rounded, with a sharply pointed tip, also dark brown.

The nymph is rather short and squat, warty; dark brown except for the paler front of the head, legs and antennae.

Similar species. The pronotum is longer and narrower than that of *Bathysolen*, and it lacks the spines of *Arenocoris*. The body form is generally shorter and broader than *Ceraleptus* and *Coriomeris*. There are several other European members of this genus, distinguished by subtle differences in the shape of the pronotum and antennal segments; only *S. lobata*, with a bluntly swollen apex to the third antennal segment and a large blade-shaped fourth segment, occurs in northern France.

Ecology. Nymphs are recorded in July and August, with adults in almost any month of the year. It is generally found on Sheep's Sorrel (*Rumex acetosella*) in warm, dry, well-drained sandy places, usually heaths, breckland, dunes and coastal cliffs. Evans & Edmondson (2005) suggest it prefers recently burned places, or well-grazed areas around rabbit warrens.

Distribution and status. This is a scarce bug found mainly in the band of sandy soil extending from East Anglia, through London, Surrey, Hampshire and Dorset, but with outlier records from the dunes of the east Kent coast, and disused sand pits in Cambridgeshire and Oxfordshire.

As is pointed out by Hawkins (2003), descriptions and notes on this bug have been published through the years with six different spellings of the specific name. These permutations vary from *dalmani* to *dahlmannii*.

Syromastus rhombeus (Linnaeus, 1767) – Rhombic Leatherbug

Description. Length 9.0–11.5 mm. This insect is readily identifiable by its general appearance as a light, delicate, pale-brown squash bug with a distinctive angular abdomen. The head is long, with the central area produced into a short triangular point. The pronotum is broad, with the sides gently concave in front of slightly produced hind angles; each edge is produced into a narrow pale-edged carina. The scutellum is an equilateral triangle. The hemelytra are broad and flat, but neatly parallel-sided; the abdomen is very broad, expanded into a neat diamond shape; the membrane is smoky, and wrinkled. The entire

Adult

Nymph

upper side is a nearly uniform mid-brown colour, but with some vague mottling and with the margins of the pronotum strongly pale creamy; this coloration is sometimes narrowly extended onto the foremost edges of the hemelytra. The underside is a similar pale creamy brown. The legs are light brown. The antennae have the first segment brown, matching the upper body surface, with the second and third segments reddish, and the fourth segment dark above, sometimes nearly black.

The nymph is mid-brown, prickly, and although early instars look rather different with their large stout antennae, the later instars already show the distinct broad abdominal segmentation.

Similar species. The rhombic abdomen is highly distinctive, much broader and more angular than in other similar-sized, but narrow, coreid species like *Gonocerus*, *Coriomeris* and *Ceraleptus*. The pronotum is also narrower than in other large broad coreids like *Coreus* and *Enoplops*.

Ecology. Nymphs occur in July and August, with adults through other months of the year. *Syromastus* occurs in dry sandy or chalky places, usually recorded as feeding on spurreys (*Spergularia*) and sandworts (*Arenaria*), and although these are small low-growing plants, it can still sometimes be swept from general vegetation. It usually occurs in rough grassy places, including brownfield sites, roadside verges and sand pits.

Distribution and status. This is a rather scarce bug known mainly from south-east England – East Anglia, London, Thames estuary, Kent, Sussex, Hampshire – but apparently moving inland into counties like Berkshire and Oxfordshire, and with other slightly more coastal sites stretching through Lincolnshire, Dorset, Cornwall, and into south Wales.

FAMILY RHOPALIDAE

Brachycarenus tigrinus (Schilling, 1829)

Description. Length 6.0–7.0 mm. At first glance this is a very delicate pale bug, straw-coloured but mottled with small black spots. The head is short, broad and triangular, and the eyes are prominent. The pronotum is short, broad, slightly in-curved behind the front angles; the rear angles are raised, and rounded, but slightly prominent. The scutellum is triangular, extended into a slight lobe at the apex. The hemelytra are mostly transparent membranous, with only the veins coloured, and the very outermost and hindmost section of the corium. The membrane is long, extending slightly beyond the tip of the body, clear, wrinkled, showing the abdomen beneath. The background colour of the insect is a pale yellowish straw, with small black dots throughout, these forming irregular blotches on the head inside the eyes, along the front edge of the pronotum and across the centre of the pronotum curving back to the hind angles. The connexivum is yellow, and the upper side of the abdomen mostly black, visible through the membrane. The head, pronotum and scutellum are minutely, evenly and relatively densely punctured. The underside is yellow with small brownish spots. The legs are yellow to very pale brown, with a mixture of small and large black spots; they are slightly hairy. The antennae are yellow with the last segment a dirty grey.

Adult

The nymph is straw yellow with strong black spots on the abdominal segments and vague dirty stipple marks elsewhere.

Similar species. The stippled straw colour pattern is fairly distinctive. *Brachycarenus* is smaller, narrower and more elegant than *Stictopleurus*, generally more yellow than both *Rhopalus* and *Liorhyssus*, though teneral specimens of these species, freshly moulted into their adult stage, might appear pale like *Brachycarenus*, until they have coloured up.

Ecology. The life history of *Brachycarenus* in Britain is, as yet, poorly understood. It is recorded from a wide variety of plant species, but mostly on hot, dry, well-drained soils. In Britain it is most usually associated with sparsely vegetated brownfield sites.

Distribution and status. *Brachycarenus* is widespread through Europe, and by the time of Hawkins' Surrey atlas (2003) it was known to be spreading north (Aukema 2001) and was anticipated in the British Isles shortly. It was first recorded from London's Battersea Park (Jones 2004; my first heteropteran bug new to Britain), and is now spread widely in London and the Thames estuary, north Kent and Essex, particularly on brownfield sites. One was found crawling across Will George's desk at the RSPB's Lodge headquarters in Sandy, Bedfordshire, in 2018. The species will probably spread much further.

Corizus hyoscyami (Linnaeus, 1758) – Cinnamon Bug

Description. Length 8.5–10.5 mm. This is immediately a very striking and elegant black and red bug. The head is broad and triangular, the eyes are bulbous. The pronotum is short and broad, with the sides straight behind a large collar. The scutellum is large, triangular, slightly raised into a narrowed lobe at the apex. The hemelytra are broad and flat. The entire upper surface is bright red with strong black marks beside the eyes, a bold bar across the thoracic collar, two large kidney-shaped marks near the hind angles of the pronotum, a large blotch either side at the base of the scutellum, a streak down the clavus, and a large prominent circular or slightly four-lobed spot in the middle of the corium on each hemelytron. The membrane is slightly darkened, but the black abdomen is still visible through it. The underside is mostly red with some black spots. The legs and antennae are black. Many long hairs cover the head, pronotum, legs and antennae.

Adult

Nymph

The nymph is a pale to mid-brown, with the abdomen often greenish, speckled all over with dark brown or black spots; it is hairy all over and the abdominal segments are armed with large stout pale spines along the edges.
Similar species. There is nothing else like it, except perhaps the equally brightly red and black *Pyrrhocoris apterus*, which is broader and rounder, and also the several obviously much more shield-shaped *Eurydema* species. Several brightly coloured black, red and white ground bugs, family Lygaeidae, might appear superficially similar, but they obviously lack ocelli and the colour patterns are distinctive on close examination.
Ecology. Mating takes place in May and June and nymphs start to appear in July. Adults can be found almost all year long, but with peaks in June and September. *Corizus* is recorded from a wide range of plants, but especially restharrows (*Ononis*) and stork's-bills (*Erodium*), in a wide variety of habitats including coastal cliffs, dunes, brownfield sites, rough grassy places, parks and gardens.
Distribution and status. Throughout most of the nineteenth and twentieth centuries this attractive bug was considered a rare south-western coastal species usually associated with sandy soils, including areas of dune and seaside undercliffs. However, since the 1990s it has increased and spread, and it is now widespread and common across most of England to Northumberland, although it is still rather coastal in the West Country and Wales. It has just about got into Scotland, with records from Berwick-upon-Tweed and Dumfries. I knew it had become abundant when I found one sitting on the lavender pot-plant just outside my front door in 2010. For me its high point was when I found many hundreds, adults and nymphs, on various ornamental restharrows planted or which had colonised the verges and river banks of the Olympic Park in east London in 2014.

Liorhyssus hyalinus **(Fabricius, 1794)**

Description. Length 6.0–7.0 mm. This is a delicate pinkish bug, sometimes yellow or straw-coloured but usually with a reddish blush. The head is broad and triangular, with bulbous eyes. The pronotum is short and broad, trapezium-shaped; the side margins are straight, the front edge has a raised rounded collar. The scutellum is triangular, slightly extended into a small lobe at the end. The hemelytra are long, mostly transparent, and only the veins and cuneus are coloured. The membrane is very long, extending well beyond the apex of the abdomen. The head, pronotum, scutellum, cuneus and connexivum are reddish, with small, dense, even black punctures. The veins of the hemelytra are mostly pale straw or yellow, contrasting against the black of the abdomen, which is visible through the transparent parts of the wings. The abdomen is black with a series of small pale squarish spots, one on the centre of each segment. The legs are red with small black spots, especially on the femora. The antennae are red, with the basal segment often slightly darker, and the last segment very long and thin. The entire body is more or less covered with sparse, long, pale pubescence.

Adult

The nymph is a dirty reddish; the head, pronotum and developing wing buds are darker than the stippled abdomen.

Similar species. *Liorhyssus* is usually distinguishable from *Brachycarenus* and *Rhopalus* by the much longer membrane, which easily clears the end of the abdomen. Measuring the length of the membrane from the junction of the hemelytra, just behind the scutellum, about 3/5 covers the abdomen and 2/5 projects beyond, compared to about 4/5 and 1/5 in *Brachycarenus* and usually even less in *Rhopalus*. Pale-straw specimens lack the dark spotting found on the veins of *Brachycarenus*.

Ecology. Little is known, yet, of this insect's life history in the British Isles, but nymphs are likely to be found in July and August, adults almost all year round. This bug has been recorded from a wide variety of foodplants including St John's-wort (*Hypericum*), stork's-bill (*Erodium*) and various Asteraceae, usually on hot, dry, well-drained soils.

Distribution and status. Until the 1990s this was regarded as a very scarce migrant, with scattered records usually in coastal localities along the south of England, but it is now much more widely known and seems permanently, if erratically, established (Judd 2011). It is mainly recorded through East Anglia and south-east England, but seems to remain more coastal in the West Country, Wales and Ireland; currently Cumbria is the most northerly locality known here. This is a regular brownfield species in London and the Thames estuary, and it is not uncommon to find half a dozen of them flying around in the bottom of the sweep net on a hot July day. It is reputed to be one of the most widely distributed hemipterans in the world (Dolling 1991), being known from South Africa, Australia, and North and Central America. Butler (1923) wondered if the long wings had something to do with its ability to colonise so vast an area of the globe.

Rhopalus maculatus (Fieber, 1837) (formerly *Aeschyntelus maculatus*)

Description. Length 7.5–8.5 mm. This is a narrow, elegant reddish and black-speckled bug. The head is broad, triangular, and the eyes are bulbous. The pronotum is trapezium-shaped behind a rounded collar. The scutellum is triangular. The hemelytra are flat, narrow, with the fore-parts transparent and the rear parts coloured. Though mostly reddish orange in colour, the upper surface is stippled with small black marks, and covered all over with small, dense, evenly spread punctures. The rear portion of the hemelytra is a deeper red. The connexivum

Adult

is paler, yellowish, with small black marks. The abdomen is mostly pale yellow/pink in the centre, marked with black alongside the connexivum. The membrane is clear, transparent, showing the abdomen beneath. The underside is mostly pale dirty straw-yellowish with rows of small black spots. The legs are orange-red, marked with numerous small black spots; the last segment of the tarsus and claws are black. The antennae are red, the last segment vaguely darker. The body is more or less covered with sparse, long, pale hairs.

The nymph is a dirty yellowish red, stippled all over with black punctures.

Similar species. The coloured portion of the hemelytra is greater in this species than in others of the genus, with only three inner cells transparent, rather than the usual five. The bright red cuneus and mostly pale abdomen are also helpful identifiers. The membrane is much shorter than in *Liorhyssus*.

Ecology. Nymphs appear in July and August, with adults almost the rest of the year. Various foodplants are recorded, including cinquefoil (*Potentilla*), Marsh Thistle (*Cirsium palustre*) and willowherbs (*Epilobium*), and usually in marshy and boggy places.

Distribution and status. This is a scarce southern species, mostly confined to the wetter parts of lowland heaths in Surrey, Hampshire and Dorset, but with outliers in East Anglia, Sussex, Kent and Wales.

***Rhopalus parumpunctatus* Schilling, 1829**

Description. Length 6.5–7.5 mm. This is a narrow elegant black-speckled red bug. The head is broad, triangular, and the eyes are bulbous. The pronotum is trapezium-shaped, with sides straight; a deep groove runs across it near the front edge, but not quite reaching the side margins, and there are signs of a small

narrow longitudinal keel. The scutellum is triangular, slightly extended into a narrow lobe at the apex. The hemelytra have cells clear, transparent, with the veins contrasting. The colour throughout is reddish, sometimes paler straw, stippled all over with small black spots; larger blotches usually occur behind the eyes, at the rear edge and corners of the pronotum, and at the base of the scutellum. The underside is usually pale straw, but with a reddish blush. The body is covered with small, dense, even punctures and sparse hairs. The connexivum is red but is often marked with small black spots. The membrane is transparent, revealing the mainly black abdomen beneath. The legs are pale red to cream with black spots, these often coalescing into a dark streak behind each rear femur. The antennae are reddish, marked with tiny black spots.

The nymph is a dirty reddish yellow, marked with small black spots; the abdomen is edged with small dark spines.

Similar species. The general all-over reddish colour (no white as in *R. subrufus*) and dark spots on the veins of the corium (lacking in *R. rufus* and *R. lepidus*) are fairly distinctive, along with the mainly black abdomen. The membrane is much shorter than in *Liorhyssus.*

Ecology. Nymphs occur in June and July, with adults almost the entire rest of the year. This species is recorded from a wide variety of foodplants, but especially

Adult

mouse-ear (*Cerastium*), stork's-bills (*Erodium*), Corn Spurrey (*Spergula arvensis*) and Ling (*Calluna vulgaris*) in dry sandy or heathy places, chalk downland, brownfield sites, roadside verges and rough grasslands. In warm weather it is very active and readily takes to the wing.
Distribution and status. This is a scarce species of south-east England, with most records from East Anglia, Kent, Sussex, Surrey, Hampshire and Dorset, but with outliers in Cornwall and Wales, where it appears to be mainly coastal.

Rhopalus rufus Schilling, 1829

Description and similar species. Length 6.0–7.5 mm. Although very similar to *R. parumpunctatus* (indeed in the past it has been suggested they were together one species), it lacks the black spots along the veins of the corium. As in *R. parumpunctatus*, the connexivum is red or with some very small black tick marks, and the abdomen, visible through the transparent membrane, is slightly more red-streaked at the apex.

Adult

Ecology. Nymphs are recorded in July and August, and adults almost the rest of the year. This uncommon bug is recorded from Corn Spurrey (*Spergula arvensis*) and Sand Spurrey (*Spergularia rubra*) in sandy heathy places.
Distribution and status. This is a very rare species, mostly known from a few coastal East Anglian localities and the lowland heaths of Surrey, Hampshire and Dorset.

Rhopalus subrufus (Gmelin, 1790)

Description. Length 6.5–8.0 mm. This is the most distinctive species of the genus by virtue of its narrow red form with black and white chequered connexivum. The head is broad, triangular, and the eyes are bulbous. The pronotum is trapezium-shaped, with a deep transverse furrow near the front edge. The scutellum is triangular, but is slightly drawn out into a narrow spatulate lobe at the apex, which is bluntly abbreviated or slightly indented at the very tip. The hemelytra have the colouring confined to veins and cuneus. The general colour is

Adult

Nymph

red, but there are some dark blotches on the scutellum and cuneus, and the veins of the central area of the hemelytra are white stippled with black, contrasting with the reddish fore-body and cuneus. The connexivum is distinctively chequered black and white. The membrane is clear, showing the black-and-white-marked abdomen beneath. The underside is a dirty reddish orange. The entire body is shining, evenly and densely punctured, and with short pale hairs. The legs are reddish with black stipples, but tibiae and tarsi are pale cream with small black marks. The antennae are red, shining, the last segment covered with short pale pubescence.

The nymph is pinkish and green with dark stipples; the abdominal segments have small spines, dark in front, pale behind.

Similar species. This species is less red than others in the genus, with the white area of the hemelytra veins and large pale spots on the connexivum giving a tricoloured rather than bicoloured appearance. *R. distinctus*, a mainly southern European species (reputedly from Scandinavia), also has the connexivum fairly strongly marked, although it appears to be a duller, more overall brownish insect, with a strongly pale-marked central line (nearly a keel) running down the centre of the pronotum onto the scutellum.

Ecology. Adults start appearing from hibernation in May, but by July they have become scarce. Nymphs are recorded from July to August, and the new generation of adults from August, peaking in September. This species is reported from a very wide range of foodplants including St John's-worts (*Hypericum*), sage (*Salvia*), mint (*Mentha*), vetches (*Vicia*, *Lathyrus*) and Stinging Nettle (*Urtica dioica*) in rough grassy and flowery places like field edges, downs, verges, parks, gardens and brownfield sites.

Distribution and status. This is the commonest and most widespread member of the genus, and is frequent across England and Wales to Yorkshire and Arnside, south Cumbria.

Rhopalus lepidus Fieber, 1861

Description. Length 6.0–8.0 mm. This is a narrow, elegant, pinkish and greenish bug. The head is broad, triangular, with bulbous eyes. The pronotum is trapezium-shaped with straight sides and a deep groove across near the front margin. The scutellum is triangular, but drawn out into a small narrow lobe at the apex. The hemelytra are mostly transparent. The head, pronotum, scutellum and veins near the cuneus are mostly reddish. The connexivum is a contrasting green. The abdomen, visible through the transparent membrane, is black with greenish straw marks. The whole upper surface is strongly and evenly punctured all over. The underside is a dirty greenish straw. The legs are greenish, but with a pinkish blush, and they are stippled with tiny black spots. The antennae are yellowish green at the base, becoming pinkish on the apical segments.

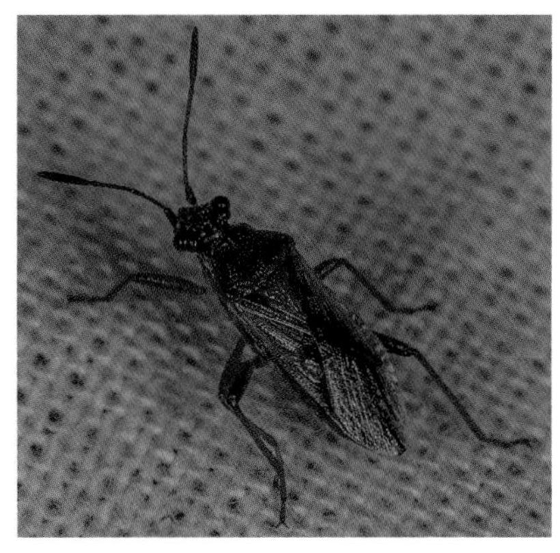

Adult

The nymph is a dirty greenish straw, with pink blush and black stipples.

Similar species. Like *R. rufus* this species also lacks the distinctive black spots on the veins of the corium found in most other British members of the genus. The green connexivum is quite distinctive, but it can fade to dull straw in dead specimens.

Ecology. Foodplants are uncertain; it is recorded by Moulet (1995) from a species of juniper, and is also reported from pinks (Caryophyllaceae) and crucifers, all in dry open places.

Distribution and status. This species is not recorded from the British mainland, but it occurs in the Channel Islands and Europe to France and Germany, so should be looked for.

***Stictopleurus abutilon* (Rossi, 1790)**

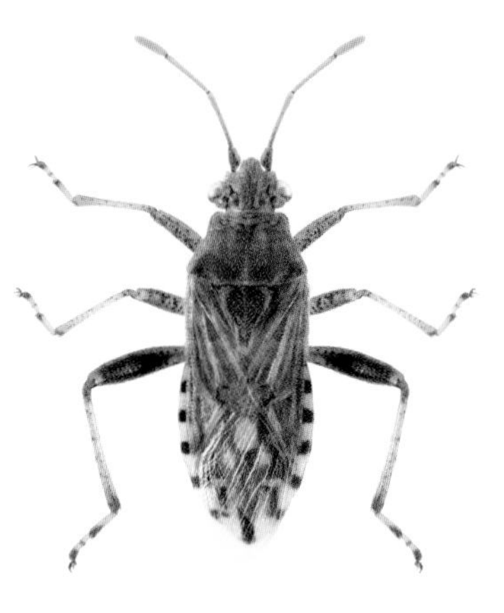

Description. Length 7.0–9.0 mm. This is a stout yellowish-brown rhopalid. The head is broad, nearly square, the eyes are bulbous, and the central lobe of the clypeus is longer than the side lobes. The pronotum is trapezium-shaped, with the sides slightly in-curved behind the front angles. A raised ridge crosses near the front of the pronotum, and behind this a narrow but deep groove, the ends of which circle round to enclose a small raised shining island tubercle on each side, and with a small short keel running backwards across it in the centre. The scutellum is triangular with a slightly rounded spatulate lobe expanded and raised at the apex. The hemelytra are mostly transparent, with only the outer edges infilled. Colour varies from dirty yellow through olive to vaguely orange-brown. The head, pronotum and scutellum are covered all over with dense heavy punctures, and with some vague dark shadow streaks. The connexivum is generally broad, pale and marked with large dark spots, although these vary greatly. The abdomen is dark-marked, nearly chequered, and is visible through the transparent membrane. The underside is more or less uniform dull olive/yellow/brown. The legs are pale, but stippled with tiny black spots coalescing into variable dark streaks on the rear femora; the last tarsal segment and the claws are dark. The antennae are pale, but stippled and often with the final segment darker.

Adult

The nymph is dull grey-brown mottled with vague marks and small black stipples. The raised dark nodes on the third and fourth visible tergites are different sizes, the anterior one larger than the posterior one.

Similar species. The small raised islands surrounded by the curled ends of the pronotal groove distinguish this species from the extremely similar *S. punctatonervosus*, which only has raised peninsulas. Other potential species in this genus (see below) really need close examination of the genital capsule.

Ecology. After the adults emerge from hibernation, nymphs start to appear in May. By late July and August they are nearly fully grown, and the new generation of adults peaks in September, though stragglers can occur well into late October. Although recorded from a wide variety of plants, mainly *Ammophila* grasses, I mostly find it on Mugwort (*Artemisia vulgaris*), ragworts (*Senecio*) and Yarrow (*Achillea millefolium*) in rough grassy flowery places including field edges, verges and brownfield sites.

Distribution and status. Although it was sporadically recorded from Britain during the nineteenth century, this species was thought to be extinct here until it started to turn up in south Essex and the London area in the 1990s. I got very excited when I found it on an embankment to the London Tube near Acton in 1999, but it had already been found in several London localities by then. It has now spread widely and occurs commonly (often very abundantly) across south-east England from Suffolk, Cambridgeshire and Oxfordshire through Kent, Sussex, Surrey and Hampshire into Dorset, with outliers in Gloucestershire and Yorkshire. It is probably set to continue its invasion north and west.

***Stictopleurus punctatonervosus* (Goeze, 1778)**

Description and similar species. Length 7.0–9.0 mm. This stout yellowish-brown rhopalid is very similar to *S. abutilon*, but the raised ridge and impressed groove across the front part of the pronotum are not so prominent and the groove is not curled around two small shining raised islands, but rather curves slightly then fades out. Perhaps it is generally slightly narrower and darker than *S. abutilon*, but coloration of both species is highly variable, including between the sexes of the same species.

The nymph is also very similar to *S. abutilon*, but the raised dark nodes on the third and fourth visible tergites are similar in size.

Adult

Ecology. The life history follows almost exactly the previous species, with nymphs June to August and new adults appearing in August through to October. Like *S. abutilon*, this species is recorded from a wide variety of plants, mainly Asteraceae, but seemingly especially Mugwort (*Artemisia vulgaris*), ragworts (*Senecio*) and Yarrow (*Achillea millefolium*) in rough grassy flowery places including field edges, verges and brownfield sites. The two species very often occur together.
Distribution and status. Like *S. abutilon*, this species was historically recorded from a few scattered localities in the nineteenth century, but was thought to be extinct here until it recolonised from 1997. It has now spread even further than *S. abutilon* and is widespread and common through most of England to Yorkshire and Bristol, just into Wales, and is set to move further.

***Stictopleurus crassicornis* (Linnaeus, 1758) and *S. pictus* (Fieber, 1861)**

Description and similar species. Length 7.0–8.5 mm. Both of these species are very similar to *S. abutilon* in having the pronotal groove curled around two small prominences, although this is a slightly finer line in *S. crassicornis*. In *S. crassicornis* the scutellum tip is also slightly constricted and more pointed at the tip, the central lobe of the clypeus and side lobes are similar lengths, and the front of the head is broadly stepped rather than being smoothly rounded. However, these are difficult and very comparative characters. *S. pictus*

Adult

can only be reliably separated by examination of the genital capsule.
Ecology. Like other *Stictopleurus* species, they are recorded from a wide variety of Asteraceae including mugworts (*Artemisia*), and Oxeye Daisy (*Leucanthemum vulgare*), along with many other plants.
Distribution and status. Neither species has yet been found in the British Isles, but *S. crassicornis* is widespread throughout Europe well into Scandinavia and *S. pictus* occurs in southern Europe extending along the Atlantic coast of France. Both are outside possibilities to look out for.

***Chorosoma schillingi* (Schummel, 1829) – Marram Bug**
Description. Length 13.0–16.0 mm. *Chorosoma* is a long, slim pale-coloured bug – not at all shield-shaped. Its body is very long and narrow, parallel-sided. The head is long, rectangular, parallel-sided, the central lobe of the clypeus longer than the side lobes; the eyes are bulbous, and the ocelli prominent. The pronotum is long and narrow, nearly parallel-sided; the centre line has a small narrow low keel. The scutellum is long triangular. The hemelytra are also long and narrow, and are transparent except for the veins. The membrane is long and narrow, reaching about halfway along the rest of the abdomen. The legs are very long and thin; the inside tips of the hind tibiae are darkened. The antennae are very long and thin; the first two segments are covered with a short dense pubescence.

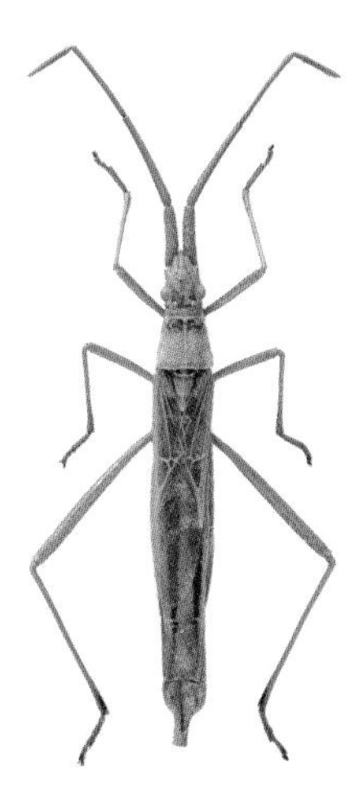

ABOVE: Nymph on net

Adult

The background colour for the entire insect is a pale straw-yellow. The abdomen has two parallel black marks along almost the entire length; these are more strongly marked in the male.

The nymph is long and narrow, of a mottled dirty yellow straw colour.

Similar species. This bug's peculiar narrow form makes it look rather like a mirid grass bug (*Stenodema* or *Leptopterna*), but these mirids lack the ocelli found in rhopalids. The very similar central and southern European *C. gracile* is smaller than *C. schillingi*: male 10.5–11.6 mm (versus 13.0–14.0 mm) and female 9.5–13.3 mm (versus 13.5–16.0 mm), and has shorter sparser hairs on the hind tibiae.

Ecology. The adults hibernate and mate in May and June. The eggs soon hatch and the nymphs occur in July into August, when the new adults start to appear. *Chorosoma* is found on various grasses, including Marram (*Ammophila arenaria*), in dry sandy and mostly coastal habitats, including dunes, coastal grazing meadows, sea walls, roadside verges and brownfield sites.

Distribution and status. This is a local, but not uncommon, insect; it is mostly coastal as far north as Yorkshire on the east coast and Cumbria on the west. It is well established in the Breckland, and there are increasing inland records, particularly in Yorkshire, East Anglia and along the River Thames right into west London.

Myrmus miriformis **(Fallén, 1807)**

Description. Length 6.5–9.0 mm. *Myrmus* is an elegant parallel-sided green rhopalid. The head is long, parallel-sided, nearly square, with the clypeus produced into a long triangle, the central lobe longer than the side lobes. The pronotum is nearly square, flat with side margins narrowly beaded. The scutellum is small, triangular, slightly raised at the apex into a small rounded spatulate lobe. The head, pronotum and scutellum are relatively densely punctured all over. The hemelytra can be long or short. In the commoner short-winged form (brachypter) the hemelytra are about twice as long as the scutellum, more or less transparent between the coloured veins, but with no membrane at the end. The less frequent long-winged form (macropter) has wings nearly completely covering the abdomen, and the membrane is long and transparent. The whole insect is mostly green, but with the hemelytra veins pink to bright red. It usually has a dark line down the centre of the abdomen; this is often suffused with a reddish blush. There are dark brownish smudges on the head, pronotum and scutellum. Males vary from bright green to nearly all-over dirty brown. Females are larger, with a broader, more oval abdomen, and are always green. The underside is green, varying to dirty smudged brown. The colours fade in dead specimens. The legs are green, varying to straw or brown. The antennae are greenish, reddish, brown or straw, often vaguely darkened on the last segment. The body is covered with long hairs, especially prominent on the legs and antennae.

Adult macropterous male

The nymph is greenish to brownish, stippled with black spots and covered with long hairs.

Similar species. *Myrmus* is not so long and narrow as *Chorosoma*, but it is still possible to mistake this slim species for a mirid grass bug. The presence of ocelli (in the adult stage) immediately distinguishes it as a rhopalid.

Ecology. Unusually for a 'shieldbug', *Myrmus* overwinters in the egg stage in the British Isles. Nymphs hatch from May and are mature by late June, peaking in August. *Myrmus* feeds on grasses in a wide variety of habitats such as field edges, rough grazing meadows, verges, downlands, parks and brownfield sites,

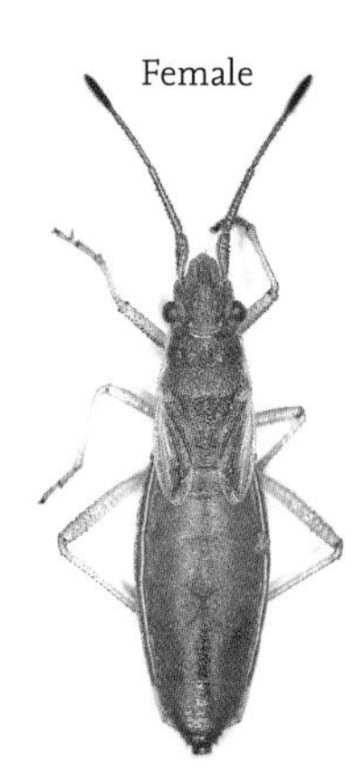
Female

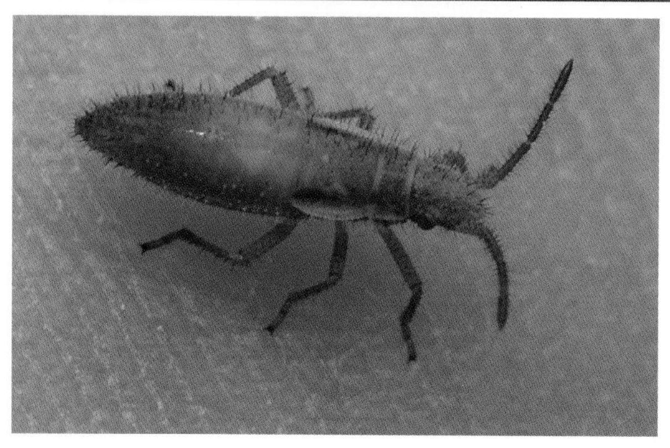

ABOVE: Female

Nymph

seeming to prefer well-drained sandy, chalky or acid grassland and avoiding more lush pastures.

Distribution and status. *Myrmus* is local, but widespread across most of England as far north as Yorkshire; it is rather coastal in the West Country, Wales and Cumbria, and there is, seemingly, just a single record from Ireland.

FAMILY STENOCEPHALIDAE

Dicranocephalus medius (Mulsant & Rey, 1870)

Description. Length 9.5–12.0 mm. This is a large dark, slim, handsome and elegant spurge bug. The head is rather long, with the side lobes of the clypeus extended into two small but sharp points ahead of the very short central lobe. The pronotum is trapezium-shaped, with sides straight ahead of hind angles which are hardly protruding; the scutellum is triangular, pointed. The hemelytra are broad and flat, nearly parallel-sided, the membrane is smoky and wrinkled, and the abdomen is gently rounded. The legs and antennae are long and slim.

Adult

The insect is almost entirely dark brown, except for a tiny white point at the scutellum tip, some large spots on the connexivum, broad rings on antennal segments 2 and 3, and with the legs, especially the femora, precisely bicoloured. The underside is a similar dark brown.

The nymph is slim, dark brown, but with clear broad pale rings already showing on the antennal segments, and the legs are mostly pale.

Similar species. This species is extremely similar to *D. agilis*, but is, perhaps, slightly smaller. The first antennal segment is shorter and slightly less hairy; the second antennal segment has a single broad central pale band, although this can sometimes be clouded in the middle to appear as two. The tibiae are also slightly less hairy. The rostrum is slightly longer, reaching to the hind coxae.

Ecology. Nymphs occur in July and August, with new adults appearing from August onwards. It is recorded feeding on Wood Spurge (*Euphorbia amygdaloides*) and possibly other similar ornamental garden species, usually in open woodland rides, grassy verges and parkland.

Distribution and status. This is a scarce bug of central and southern England, and although there are outlier localities in Yorkshire, Shropshire, Gloucestershire and Cambridgeshire, it is mostly recorded south of the River Thames.

Dicranocephalus agilis (Scopoli, 1763)

Description and similar species. Length 11.0–14.0 mm. This is a large dark, slim and elegant spurge bug almost identical to *D. medius*. The first segment of the antennae is longer, and the second segment has slightly longer hairs. The two pale bands on the second antennal segment are also usually more clearly and precisely defined. The hairs on the tibiae are also very slightly longer than those in *D. medius*. The rostrum is very slightly shorter, not reaching the hind coxae.

The nymph is extremely similar to that of *D. medius*.

Ecology. The life history is probably similar to that of the previous species. This is a strongly coastal species, usually found in sand dunes, on Portland Spurge (*Euphorbia portlandica*) and Sea Spurge (*E. paralias*).

Distribution and status. This very scarce bug is limited to the sandy coasts of Wales and western England from Anglesey to Hampshire, and there are old records from Liverpool and south-east Kent.

ABOVE: Adult

Nymph

***Dicranocephalus albipes* (Fabricius, 1781)**

Description and similar species. Length 10.0–13.5 mm. The adult is extremely similar to the two species listed above (particularly *D. agilis*), but it lacks the dark band across the middle of the second antennal segment and the membrane is smoother, less wrinkled, between the veins. The nymphs are also extremely similar.

Ecology. In Europe it is reported from various spurges, including *Euphorbia cyparissias*, the Cypress Spurge, which is a widespread naturalised species in the British Isles.

Distribution and status. Although reputed to have been taken from localities 'vaguely given as the coasts of Devon and Cornwall, and the New Forest' (Butler 1923), these are likely to have been misidentifications. It occurs sparingly in Normandy though, so is not completely beyond the realms of possibility.

CHAPTER 9

How to Study Shieldbugs

First find your bug. Being relatively large, shieldbugs are usually pretty easy to find without recourse to specialist techniques. Many sit brazenly on the leaves of their foodplants, relying on their bitter almond-and-diesel taste to deter would-be predators. Here they are variously calm and collected or busy and active. This often depends on whether they are already occupied in feeding, mating (Fig. 127) or egg-laying, or wandering about looking for a likely spot to bask in the sun, feed, or get ready to launch themselves into flight. Some can also be found sitting in flowers, sunning themselves on fences or tree trunks, climbing up and down grass stems, or sheltering under foliage close to the ground.

Since most shieldbugs are plant-feeders, finding the host plant is often the first stage in finding the bug. Appendix 1 lists all British (and nearly British) species

FIG 127. Occupied with each other, mating pairs of shieldbugs (here *Coreus marginatus*) can often be approached closely for study or photography. And if they try to run off, the fact that they are facing opposite directions results in stationary mutual antagonism.

with some of their preferred (or most often reported) foodplants, and for predatory species some of their regularly reported prey. Once a plant has been found, any bugs can be sought by visual searching, or dislodged into net or tray. There is a lot to be said for quietly and calmly examining the herbage, moving about slowly and carefully to see what is crawling across it (Fig. 128). Personally I find a great zen benefit in this – watching and admiring nature in all its diversity and complexity. This is sometimes the only way to fully appreciate insect behaviour – you might see a shieldbug courting or mating, egg-laying, feeding or engaged in a stand-off with predator or parasitoid. No amount of captive rearing or study in a glass container can match the genuine excitement of a real wildlife encounter out in the wild.

Shieldbugs are supremely photogenic (Fig. 129), and since many can be identified from a photograph, a camera is a very useful accessory. Technology continues to move on in astonishing leaps and bounds, so rather than try and prescribe any particular brands or equipment, which will quickly be out of date, the following is really just an attempt at user guidelines. A single-lens reflex (SLR) camera taking removable lenses can easily be adapted to close-up photography by the purchase of a macro lens. Unlike the telephoto used like a telescope to snap distant birds, this simply allows much closer focusing, often on an object just a few centimetres away, with concomitant enlargement of the image. Alternatively, cheap extension tubes sit between camera body and conventional lens to achieve the same end. Magnifications of ×1 and ×2 life size are achievable like this. Another option is a supplementary close-up lens, which screws into the end of the ordinary lens barrel and works as a hand lens does in front of your eye. Any combination of all three will work.

Some photographers use electronic flash to light the subject evenly, whilst others swear by ambient light to give a more natural picture. Various contraptions like ring-flash, double-flash, diffusers and reflectors are available to help avoid any stark shadows cast by flash guns. Whatever works for you and your budget. The more equipment you carry the less manoeuvrable is your setup. Compact cameras (sometimes rather inaccurately called snapshot) are a good alternative, offering high resolution with small size and easy portability; they're also much cheaper than SLRs. Many models have close-focus settings allowing you to get right up to the insect to get the macro shot (Fig. 130).

Shieldbugs are less flighty than bees and butterflies, but can still be alarmed at the sudden looming approach of a heavily laden photographer. They may fly off, or drop to the ground to escape the perceived threat. Move slowly, don't let your shadow fall over your quarry, and prepare to spend some time scrabbling on the ground on all fours as you struggle to get the right angle, and the bug into crisp focus.

FIG 128. *Adomerus biguttatus*, the Cow-wheat Shieldbug, sitting, aptly enough, on its foodplant Cow-wheat (*Melampyrum pratense*) in the Blean Woods of Kent.

FIG 129. Many shieldbugs make great photographic subjects, as demonstrated by this sublime picture of a slightly translucent *Palomena prasina* nymph by Maria Justamond.

FIG 130. Maria Justamond in typical entomologist photographer pose – grubbing about in the herbage on hands and knees to get that shot. Maria uses a compact camera which has a macro setting, and some of her excellent photos grace this book.

Phone cameras are also getting very good and they have more or less replaced the small compact cameras once in everyone's pocket; they are less adaptable than SLRs but clip-on macro lenses are available cheaply. Sometimes the auto-focus gets confused by the background if the bug is dangling on a stalk, but they are quick and convenient and the quality is improving with each new model. Several pictures in this book are photos I took with my phone. Obviously the key thing is – is it in focus? If not, try again until you succeed.

NO EXPENSIVE SPECIALIST TECHNICAL EQUIPMENT REALLY NECESSARY

Much of the equipment used by entomologists is simple, light, cheap, and can often be home-made, or cobbled together using familiar household objects. My choice of entomological equipment is the four-fold insect net, a light, compact and versatile tool that I mistreat horribly. Some entomologists swear by the industrially tough sweep net, with a reinforced metal ring opening and a hardy canvas bag, but this can be heavy and cumbersome if you are traipsing miles across the countryside. These nets are available from specialist suppliers, but for very many years I used one that was home-made from metal coat-hangers, offcuts of upholstery canvas, old net curtain and half a broom handle. I use my more delicate insect net to sweep through grass and herbage, and it works well enough if I am careful. The technique is analogous to trawling: as the net is dragged through the plants any insects perched on leaves, flowers or stems are dislodged and caught to be examined as they scurry around in the mess of petals,

FIG 131. *Peribalus strictus* in the net.

dead leaves, seeds and other assorted vegetable debris at the bottom of the bag (Fig. 131). I always carry a needle and thread to repair the inevitable tears and holes in the light net bag, and also sometimes in my trousers.

Trees, shrubs, gorse bushes and ivy thickets can be shaken and tapped with a stick to dislodge residents down into the net or onto a beating tray. Again, I mostly use my little net, but also sometimes carry a small pale folding umbrella for use under broad spreading tree branches. Specially constructed beating trays are available from commercial natural history suppliers and usually comprise a series of metal, wooden or bamboo struts that support a sail-like white cloth sheet. These are excellent for spotting the insects knocked down onto them, but they can be heavy and just another awkward piece of equipment to cart about. Dolling (1991) complained that they are 'more trouble than they are worth'.

Ground-dwelling shieldbugs can be found by fingertip grubbing around the bases of the foodplants, or in the general grass root thatch. Handfuls of loose leaf litter can be sieved or shaken over a plastic sheet. This is the only way I have ever found *Odontoscelis lineola*, on hands and knees, nose to the ground rootling about under stork's-bill on the sandy coast at Deal. Patience is often required, because many of these root- or soil-dwelling species are very slow-moving and cryptically coloured – virtually invisible against the soil. I adopt a middle-distance gaze when staring at the ground in these situations and achieve a peaceful reverie even if nothing is moving about at the time. *Podops inuncta* and *Thyreocoris scarabaeoides* are other ground-dwellers most often found by turning over logs or wooden planks resting on grass or herbage. On official environmental survey sites, these two species are also regulars under the squares of roofing felt left out by herpetologists to monitor reptiles.

Suction samplers have been enthusiastically embraced by many entomologists. These are two-stroke petrol-powered garden blower-vacs normally used to blow leaves from the lawn or path into a heap, then suck them up into a mulch-collecting bag. They can easily be adapted for entomology by affixing a small home-made canvas bag inside the intake tube and clamping it firmly in place using a large jubilee clip. The suction draws up tiny insects into the spout, where they are held in the bag by the pressure of the passing air; the bag also prevents the insects getting drawn up into the cutting and mulching machinery, and even the most delicate creatures remain unharmed. Turning off the machine allows the bag contents to be emptied onto a plastic tray or cloth sheet for sorting. Suction samplers are ideal for use on very short grass swards like sheep-grazed chalk hillsides, but also work well at extracting insects living further down into the herbage, and at root level, where they are inaccessible to the sweep net. My travel-light policy means I usually leave mine at home. But I might be persuaded to get it out more often – Graeme Lyons was using a suction sampler when he discovered the tiny and globose *Coptosoma scutellatum*, a shieldbug, indeed a family of shieldbugs, new to Britain in 2019 (Fig. 132).

Although useful in studying other insects (including other het bugs), the pooter (sometimes called an aspirator) is unlikely to be much utilised by the shieldbug enthusiast. Here are a few words of warning. Named after its inventor US entomologist William Poos (1891–1987), it comprises a small glass container with two flexible rubber tubes and comes in various larger or smaller designs (Leather 2015). Sucking on the long tube held between your teeth creates a

FIG 132. The tiny glossy globose *Coptosoma scutellatum*, recently found, new to Britain, in Sussex by Graeme Lyons and his trusty suction sampler.

vacuum and small delicate insects are siphoned up through the short inlet tube; they are held in the main body of the pooter, prevented from being gulped up into your mouth by a thin gauze barrier. Pooters are excellent for collecting many small insects, including leafhoppers and mirid bugs. But sucking up shieldbugs is apt to upset them enough for them to exude copious amounts of their defensive aromatic chemicals, which pass through the gauze and get inhaled – causing at best a bitter taste in the mouth, and at worst gagging and retching.

Many insects can be found using active or passive traps left for hours or days, and some of these may be applicable to shieldbug hunting. Carrion and dung traps work well for many insects, but since no British shieldbugs regularly feed on these substrates they need only be superficially mentioned here. Shieldbugs are occasionally found in the light traps that are routinely used by lepidopterists to attract moths. It has long been known that moths (and other insects) are attracted to bright lights, but nobody really knows why. Contrary to popular myth, moths do not migrate using the moon or stars to navigate (nor do shieldbugs), so they do not get disoriented by artificial lights. The latest idea is that the bright light overpowers the ability of the optic nerves of the moth to recover; effectively blinded, the insect heads towards the light source, which now appears as a dark patch in its visual field, potentially offering shelter in the alarming brightness all around. Running a powerful mercury vapour light is a fascinating activity and will bring in all manner of insects including moths, beetles, lacewings and a few het bugs, which either land on the white sheet background or hit vertical vanes and drop down through a funnel into the dark capture box beneath to be examined at leisure in the morning. A jumble of egg cartons or other rough material in the box allows them to find a place to sit tight, and they soon settle down to roost. Shieldbugs make up a small proportion of the by-catch, usually common species like *Pentatoma rufipes*, *Acanthosoma haemorrhoidale* or *Zicrona caerulea*, but unusual species can turn up. The Mediterranean *Mecidea lindbergi* was found in several moth traps along the south coast of England in December 2015, and in September 2021 *Pinthaeus sanguinipes* turned up in the moth trap of Antony Wren in Lowestoft, Suffolk, the first time it had been found in the British Isles. The Saucer Bug (*Ilyocoris cimicoides*), an aquatic species, is a regular in light traps, and because of its broad shield-like shape is sometimes mistaken for a shieldbug by non-specialists (see Fig. 14). Various moth traps are available, in several different designs, and to fit varying budgets, from commercial suppliers. I sometimes use a naked mercury vapour bulb on a small wooden mount, and place it on an old white cotton sheet on the ground in the garden.

Flight-interception traps, including the tent-like Malaise trap, will occasionally produce shieldbugs. These work by catching or stopping flying insects and funnelling or corralling them into a collection chamber or pot, but since flying shieldbugs are easy to find anyway, these are best used as part of general entomological surveys across all orders rather than just the Pentatomomorpha.

Pitfall trapping is an easy technique involving the setting of small plastic cups into the ground so that the lip is level with the surface. The idea is that ground-dwelling insects walk across and fall down into them, then cannot get out. All passive traps have the disadvantage that you need to be able to go back and examine the catch a day or two later; this applies especially to pitfalls, where carnivorous ground beetles which also fall down in large numbers soon start demolishing all the other insects. To prevent this you can lace the bottom of the trap with a few centimetres of water and a drop or two of detergent (washing-up liquid) to kill everything. A preservative also needs to be added if the traps are not going to be examined for several days. Various options are available including ethylene glycol (antifreeze), though this is poisonous to foxes, badgers and dogs attracted to the scent of decay in the trap; alternatively concentrated salt solution or white vinegar will pickle the trap contents. In controlled scientific studies pitfall trapping means that analyses of catch weight and numbers across and between sites are statistically more robust, but this technique comes with the ethical dilemma associated with the destruction of the unwanted by-catch. I've never had much success pitfall trapping for shieldbugs, although it works very well for lygaeid ground bugs (and those ground beetles). Having said that, I was sent a series of pitfall samples collected on the living roofs of some of the tall buildings in London's Canary Wharf office complex, and amongst these were specimens of *Syromastus rhombeus* and *Bathysolen nubilus*, both scarce species normally associated with brownfield sites in the area.

Although it is slightly left-field for anyone studying shieldbugs, I can't really pass by an opportunity to mention the only device I know of specifically made to catch het bugs. Designed by Sheila Brooke (2003), it is for catching shore bugs (Saldidae), which are active, flighty little insects of shore- stream- and pond-side mud. Trying to catch them by net inevitably means wet mud everywhere, but they are too quick to simply have a small glass tube plopped over them. Comprising a small plastic kitchen funnel with the large opening slightly narrowed by a rim-like internal flange, this device can be plonked over the saldid, which now runs around on the inside of the flange. Turning the whole caboodle upside down allows you to drop the bug down the narrow funnel shaft into a waiting glass tube below.

WHAT IS THIS BUG?

Shieldbug in hand, or in the net, or in the photograph, there now comes the process of trying to identify the species. Some shieldbugs are utterly distinctive and not likely to be mistaken for any other. We only have one bright metallic-blue species (*Zicrona caerulea*), and only one black- and red-striped (*Graphosoma italicum*) in the British Isles. *Aelia acuminata* has a highly distinctive shape (Fig. 133), and *Adomerus biguttatus* a highly distinctive pattern. But many others require a closer look to confirm their identities. Most shieldbugs are resilient little insects and are quite capable of surviving gentle fingertip handling. Here, between finger and thumb, they can be examined under the hand lens. I always carry a selection of small glass tubes about my person, and popping a shieldbug into one confines it enough to have a first close look with the lens. Plastic tubes tend to get scuffed and tarnished and are harder to squint through. A hand lens is perhaps the most important piece of field equipment for the serious hemipterist.

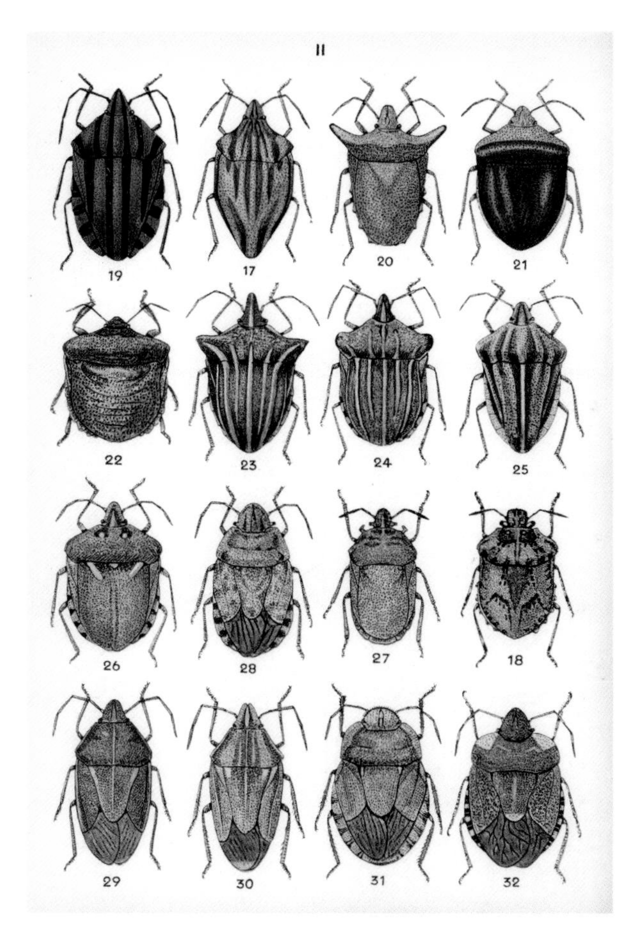

FIG 133. Colour plate from Villiers (1951), one of the first picture books on bugs I ever picked up. Distinctive British species illustrated here include *Graphosoma italicum* (19), *Podops inuncta* (27), *Aelia acuminata* (30), *Menaccarus arenicola* (31) and *Dyroderes umbraculatus* (32).

Do not attempt to use the large Sherlock Holmes type magnifier on a stem-handle, which is really only good for reading the small print. Entomological hand lenses are easily available from specialist natural history, geological, medical or jewellery suppliers, hobby shops and optical stores. A few pounds gets a folding glass lens (clearer and brighter than plastic) in a stout plastic or metal surround that can be secured by a lanyard around the neck or to the lapel button-hole. Entomologists (and botanists) can often be identified in the field because they have a hand lens slung round their neck rather than the birdwatcher's binoculars. Be careful, though: my father used to have lens, biro and notebook all attached to his jacket by pieces of twine, and after a day falling out of hedges backwards he often looked as if he was held together by pieces of string. Low magnification of about ×10 is more than enough for most purposes. This is usually perfectly sufficient to see the hairs on the body of *Dolycoris baccarum*, or check the pronotal groove shape on a specimen of *Stictopleurus*.

There comes a point, however, where a live wriggling specimen under a hand lens defeats the viewer and a decision has to be made whether to escalate the study and examine a dead specimen under the microscope. This is the central paradox of entomology – why kill the things you love? In truth you do not need to kill insects to study entomology, but if you do, then your contribution will be more profound and longer lasting. The arguments for taking a few sample specimens are manifold and well reasoned. The removal of a small number of individuals will not damage any given population; insects are small and secretive and the majority are completely overlooked and never seen anyway. They also have prodigious fecundity, and even at the lower end of shieldbug egg-laying ability, populations numbering thousands can result from a very few generations. Insects suffer far greater mortality from each other, and from an endless variety of predators, parasites, diseases and unfortunate accidents. And given that so many insects are destroyed thoughtlessly by humans anyway, it is hypocritical to chastise entomologists for killing a few for scientific research.

Although I may have to update it sometime soon, the statistic I frequently bandy about is that more insects were killed in the construction of the 2012 London Olympic Park than by all the entomologists in the world who have ever lived. On the back of the proverbial envelope, the calculations go as follows. The Olympic Park in East London is 2.5 km^2, that's 2.5×10^6 m^2, so with a nominal insect density of 1,000/m^2 we would get 2.5×10^9 (two and a half billion) insects destroyed as the area was razed by the intense bulldozer activity necessary to prepare the ground for building and landscaping. This number is roughly 40 times the size of the Natural History Museum's insect holdings (60 million specimens). The museum has one of the largest accumulated collections anywhere in the world, so 40 of them ought to be enough to equate to worldwide entomological activity during, say, the last 300 years. Similar million/billion/trillion estimates are given for insects splattered on cars each year. And don't get me started on the countless gazillions of insects blasted into oblivion because their presence is deemed unwelcome in the crop field. Even popular and friendly shieldbugs are wantonly swatted, sprayed and otherwise destroyed just because some of them decide to feed on a few broad beans up at the allotment.

Another myth to get out of the way here is that entomologists are just like stamp collectors – all busy trying to get the complete set, or fill pretty glass-topped display cases for the sake of stoking private vanity. Today the reference collection is a necessary workaday tool to aid identification and maintain taxonomic conformity. The baseline currency of modern biological conservation is the species. No biological significance was ever attributed to a site simply because of what it looked like, or because it seemed to have quite a number of nice-looking plants and animals. Biodiversity is measured in the abundance and variety of the individual species, and these species need to be accurately and consistently identified. With roughly 25,000 insect species in Britain, the vast majority of them under 10 mm long, many with different forms in male and female, different colour morphs, and different larval or nymphal stages, accurate identification can be a major challenge. No picture-book identification guide can be adequate. Specialist keys and complex scientific papers must be consulted, and the specimens must be observed under the microscope to examine the sometimes awkward characters. Field and laboratory identification skills need to be learned and honed, and I suggest they usually need to be honed down the microscope. A reference collection is often the only way to provide the certainty of identification necessary (Fig. 134). The net-wielding entomologist is not the enemy of conservation here, but the herald.

It is often the scarcer species that have a greater impact on management of a site for conservation than the common or garden species, but because they are scarce, they are more likely to be missed or mistaken simply because an entomologist may never have come across them before. Some large and common insect species can be identified easily in the field, but this is a tiny minority – the majority of small fry defy even the expert until a specimen can be examined under the microscope and compared to other specimens laboriously collected and curated over many years. True, most British and Irish shieldbugs fall into the first category, but when *Geotomus petiti* was discovered at Dungeness, it took dissection of the male genitalia from collected specimens to confirm that it was, truly, different from *G. punctulatus*. If the currently mainland European *Coriomeris affinis* or *C. scabricornis* are ever found in the British Isles it will only be because

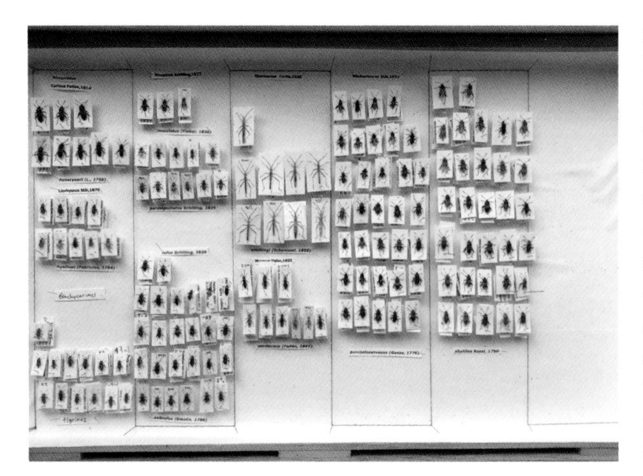

FIG 134. Entomology books often show beautifully curated drawers of immaculately set insect specimens neatly arrayed in precise and regimented decorative order. By way of contrast I offer up here my rather scrappy but perfectly usable reference collection of the Rhopalidae.

someone has collected a specimen and spotted that it looked different to the row of C. *denticulatus* in the reference collection. And if heteropteran interest is kindled enough to move attention on from shieldbugs and towards their relatives the ground bugs (Lygaeidae) and leaf bugs (Miridae), it will soon become clear that a reference collection is an absolute necessity.

The ethical decision to collect, kill and mount a specimen for microscopic examination is one that each of us has to make on a day-to-day basis, according to the unique set of circumstances that present themselves at that moment – it is a complex algorithm based on personal knowledge (and private doubt), how uncertain an identification might be in the field, and whether its presence in the reference collection will better aid future identification decisions. Once that decision is made, though, the good news is that entomology is an easy and cheap science to follow.

Shieldbugs, indeed all insects, can be killed easily and quickly (normally in 5–10 minutes) using a few drops of ethyl acetate on a tissue placed into a small glass bottle with them (not a plastic bottle, as the solvent will react with it). This chemical can be bought from specialist suppliers, but is also the major constituent of acetone-free nail varnish remover, easily found on the supermarket shelf. The specimens next need to be mounted for the reference collection (Fig. 135), and the two traditional techniques are direct pinning or carding. Entomological pins can be bought cheaply from commercial suppliers. They are finer than haberdashery pins and usually made of stainless steel – standard sewing pins are thick and ungainly and often use alloys that will eventually react with the body chemicals of the insect, causing them to corrode, discolour or break over time. For consistency, the pin should pass through the top-right quadrant of the bug's scutellum. A

FIG 135. Three ways to mount a shieldbug, in this case *Coreus marginatus*. Left to right: direct pinning through the insect and data labels (the simplest solution), insect pinned onto Plastazote mounting strip (which dampens vibrations when the insect is being manipulated under the microscope), gluing the insect to a card (protective, but hiding some underside characters). There is no hard and fast rule, but most entomologists follow their own style or adapt it according to the body size and fragility of each specimen. My style is to be erratic, so I use all forms of mounting whenever I please.

pinned specimen means the underside can easily be examined, but legs and antennae are exposed and they can be damaged by knocks and vibrations. To guard against this the bug can be pinned into a mounting strip which itself takes a bigger pin at the other end for handling. Traditionally mounting strips were cut from dry brackets of *Polyporus*, the birch polypore fungus, which is clean, white, tough, resilient, long-lasting, flexible and can be sliced precisely. It acts like a suspension/damping device to reduce jarring on the brittle insect when it is being manipulated. Nowadays strips of Plastazote, a tough expanded polythene foam, are readily available at low cost from specialist suppliers.

Gluing the entire insect onto card offers a more protective solution, but some underside characters may be obscured. The standard entomological textbooks often recommend four-sheet white Bristol board for the card, and for the glue they give recipes for brewing your own gum tragacanth and various other obscure products from substances bought from the local apothecary, but modern and cheap alternatives are easily available. Any thick white card around 250–300 g/m^2 is sufficient – watercolourist's board or similar from an art shop or stationer, or even chopped-up Christmas cards, but use the matt rather than the glossy side. Entomological gum is another product available cheaply from specialist suppliers, but thick wallpaper paste, slightly watered-down PVA glue or other craft gums will suffice at a pinch. Lay the bug on its back and brush out legs and antennae with a small soft dry watercolour brush, then lightly glue up a segment of card and place the bug on it belly-side down. Next align the legs and antennae neatly in the moist glue. I use a small headless pin mounted in the tip of a matchstick, but any long pin or needle might work. Those old textbooks helpfully suggest a horse bristle or a cat's whisker. Splaying the appendages out neatly will allow any obscure characters to be better seen down the microscope. And should the need arise for comparison between specimens, this is easier if the insects are all neatly arrayed to look like the diagrams in the identification guides. When the gum has dried, the excess card can be carefully clipped away to leave the bug neatly displayed in the middle of a small rectangle. This card can now be pierced at the base with a supporting pin.

Whether it is supporting a mounting strip or a card, or directly piercing the insect's body, the pin allows the specimen to be firmly anchored into the storage case or tray, and also allows it to be manipulated carefully by hand under the microscope. The pin should also pass through any data labels. As I complained in Chapter 6, a specimen without data is scientifically useless. Every specimen should have at least one data label, and that label should give at the very minimum the locality and date the insect was found. I'm pleased that I knew to label the first het bug I ever collected (specimen number 14 in my hand-written catalogue, water scorpion *Nepa cinerea*): 'Rodmill, River Ouse, 1 June 1968, in dyke'. I can live with the misspelling of Rodmell – I was only nine years old. Nowadays I would hope a label to give: name of the locality, geographical region, Ordnance Survey grid reference or coordinates, vice-county (see box opposite), date found, name of collector, how the specimen was found or collected, any foodplant or habitat, and any other pertinent information. When the bug is identified, a second data label should record that name, and also the name of the person determining it. If it is re-identified in the future the old labels should remain, but new labels should be added. No museum curator ever complained that there was too much information, or too many data labels attached to a specimen.

Adult insects are excellent subjects in terms of long-term preservation. Simply by virtue of being dried out, an insect's tough chitin carapace retains its shape, structure, surface sculpture, hairs, bristles and much of its colour. There is no need to work on taxidermy skills, no rotting innards need be dissected out, and no fluids like alcohol or formalin are needed for embalming. With luck an adult insect specimen will survive for centuries, and this gives it lasting value in the personal reference collection and later in the museum to which it is bequeathed on your death. Specimens can be examined and re-examined again and again over the years as interpretations of species change, new species are described, or genera are revised. Having said this, pinned het bug nymphs do not preserve that well, and have a tendency to shrivel to 'an unsightly mess' (Butler 1923) because much of the body remains soft and elastic to accommodate gluttonous larval feeding, and the fully sclerotised hard and stiff integument

VICE-COUNTIES – QUAINTLY ARCHAIC YET HIGHLY PRACTICAL

The well-known counties of Britain and Ireland, established under the feudal system of the Middle Ages, are useful for some administrative purposes, but their hugely varying geographical sizes make them awkward when trying to compare and contrast biological diversity. To counter this, the vice-county (VC) system of biological recording was devised by botanist H. C. Watson in 1852. This was based on the political county boundaries of the day, but subdivided or adjusted to give geographical zones all of very approximately equal area. Thus Sussex is divided into East (VC14) where I grew up, and West (VC13), which I sometimes visited, whilst tiny Rutland is tacked onto Leicestershire to give a similar-sized VC55.

There have been significant boundary changes since – not least being the abolition of Middlesex, and the absorption of the metropolitan segments of Surrey, Kent and Essex when they were swallowed up by Greater London in 1965. But by using those original 1852 vice-counties as recording units, a unity and geographic continuity is maintained. The *Shieldbugs of Surrey* (Hawkins 2003) records the 46 species of these families known from vice-county 17, which still includes the south-west quadrant of what is now Greater London, right up to the Thames near Greenwich. This modern Surrey list can genuinely be compared to the 41 species in the Victoria County History volume for Surrey published a hundred years earlier (Saunders 1902), and the 45 species in the county distribution summary produced by Massee (1955). By my reckoning the Surrey total is now 59 species.

Aligning the Watsonian vice-county to modern geography can sometimes be awkward. Even a few years after Watson adopted them, parliamentary boundary movements altered the 1852 county borders, so anyone wanting to know details of historical localities often had to examine old maps. The Ray Society published a vice-county map (Dandy 1969), but it was small-scale and difficult to appreciate the precise boundaries, which often followed erratic field margins and meandering streams, or crossed blank open hillsides. A series of 1-inch Ordnance Survey maps were hand-annotated by Dandy in 1947; these were available to be consulted at the Natural History Museum, but this was far from ideal. The county boundaries have recently been digitised, and organisations such as the Botanical Society of Britain and Ireland now offer interactive map and search facilities on their websites.

The boundary between VC17 (Surrey) and VC16 (West Kent) used to pass through the back 2 metres of my garden when I lived in Nunhead (Jones 1997), so it was sometimes quite a challenge correctly labelling specimens I found there. And on 15 April 2003 I was delighted to see a specimen of the Box Bug (*Gonocerus acuteangulatus*) take off from my hand in the gardens of the Horniman Museum, Forest Hill (VC17, Surrey), travel across the vice-county boundary which runs along the crest of the hill there, and fly a few metres into VC16, West Kent (Jones 2003), to become just the second record for the vice-county.

really only fully forms in the final adult stage. This is where photography really comes into its own; otherwise ethanol and formalin are required to pickle them.

Sadly my first collected shieldbug seems not to have survived – according to my original manuscript catalogue it was specimen number 130, *Acanthosoma dentatum*. My dad probably helped me to name it using his copy of Saunders (1892); it's a common birch-feeder now called *Elasmostethus interstinctus*. I found it on our family holiday to Prospect Farm Caravan Park (Caravan number 90), Swanage, Dorset, on 17 August 1969. It is no longer in my collection; nor are 180 *Piezodorus lituratus*, Denton, East Sussex, 31 August 1969, or 626 *Syromastus* (*Coreus*) *marginatus*, St Lawrence, Isle of Wight, 20 June 1971. I suspect they all fell to the ravages of the Museum Beetle (*Anthrenus verbasci*). This notorious pest feeds on animal fibres in carpets (it's also called Carpet Beetle), and as its name suggests it is an important nuisance in museums, feeding on the preserved remains of stuffed animals and birds, and pinned insect specimens. It is against this critter that well-constructed cork- and paper-lined drawers with close-fitting framed glass lids are made to house insect collections. This level of workmanship does not come cheap, and before I could afford my first proper insect cabinet I used a series of four small unglazed drawers in a flimsy desk-top cabinet more likely to have been made for stationery or knick-knacks. And I paid the price. Elegant entomological cabinets are serious pieces of furniture and are priced well into three and often into four figures; a cursory internet search in early 2021 suggests you'll need to pay between £60 and £250 per drawer depending on the size and quality. Stand-alone wooden store boxes are a good alternative; although new boxes sell for in the region of £40–50 each, second-hand ones can often be picked up cheaply, and local museums sometimes sell off excess stock that they have received from ill-matched donations and legacies. There is also a wide variety of plastic lock-top Tupperware-style containers available from stationery, craft and hobby shops, and even some supermarkets. These need to be lined with a tight-fitting floor of Plastazote, but are good cheap and cheerful alternatives. Because they are completely airtight all specimens placed inside must be absolutely dry, otherwise the trapped moisture will encourage mould, which damages and destroys specimens just as effectively as *Anthrenus* beetles.

Whatever your style of insect storage, low-end picnic boxes or high-end mahogany cabinets, you still have to be vigilant though. Every museum in the world has a tragic tale or two about examining drawers or boxes long left unopened, only to find a forest of naked pins standing amidst piles of dust after a colony of the beetles has spent decades destroying the irreplaceable specimens. An annual inspection regime to look for the tell-tale dusty *Anthrenus* excrement should catch any infestation early. These drawers or boxes can then be treated with chemicals (traditionally paradichlorobenzene or naphthalene), or placed into a chest freezer for a week – but before you do that, clear away any completely or partly destroyed specimens, as well as live beetles and larvae.

The point of collecting, setting, curating and storing insect specimens is to examine them at leisure under the microscope. This needs to be a stereomicroscope with magnification around ×10 to ×30. Although professional pieces will set you back thousands of pounds, a simple budget desktop stereoscope can be picked up for around the £100–200 mark, and this will be perfectly adequate for the beginner entomologist. I still use my cheap first purchase occasionally. Many basic models have a fixed magnification, something like ×10. This doesn't sound like much, no more than most pocket lenses, but a stereoscope gives a clarity, three-dimensionality and stability beyond any hand-held device and allows much better visualisation of the sometimes obscure and awkward features that need to be examined to achieve satisfactory identification. Magnification can be altered by buying extra eye-pieces, though this usually darkens the image because the glass of the higher-power lenses is smaller, thus letting less light through. Slightly more sophisticated microscope designs have a swivel-rotating turret, allowing you to switch easily between, say, ×10 and ×30. Seamless zoom optics and larger brighter lens designs come with added premiums. Built-in lights also come at extra cost, but a small desk-top lamp is often sufficient to light the specimen perfectly adequately. Shop around before you buy. Many microscope suppliers attend trade fairs for entomologists, geologists and stamp-collectors, so it is often possible to try out something and quiz the sellers about their ranges, or pick up second-hand or demonstration models at bargain prices. A microscope will last a lifetime. I still have the old black and chrome Watson 'Greenough' model stereomicroscope made in the 1930s that my father bought second-hand in the early 1970s. I get it out for demonstrations occasionally. The internal lenses and prisms have acquired a slight flecking of dust but it still works well enough to identify an insect. And it's an elegant prop for any entomology photo-shoot.

CAPTIVE REARING

After the Victorian surge of publishing about finding and describing the adult insects, the twentieth century saw numerous entomological books enthusiastically urging readers to study the early life stages, since there was still so much work to be done identifying and describing larvae and nymphs. Even Butler (1923) in his deeply researched monograph on the early stages and life cycles of bugs states 'scarcely a single life history can be regarded as completely worked out, and in some cases the early stages are quite unknown'.

The exhortation to study these early stages still stands true for so many groups – shieldbugs included. Actually, shieldbugs especially, because shieldbug nymphs are relatively easy to rear through to adulthood.

Unlike the standard schoolroom rearing projects involving butterfly and moth caterpillars, shieldbug nymphs have the advantage of hemimetaboly in that they are already adult-like in their feeding and locomotion. They have no need to pass through the vulnerable chrysalis or pupa stage which any Lepidoptera text will tell you is the time that your charges are most likely to succumb to mould and fungal diseases. Unlike when rearing, say, moth caterpillars, there is no need to have a layer of soil in which the nymphs can pupate. Instead the final-instar shieldbug nymphs will simply grasp onto the plant substrate and moult into adulthood, just as they did between nymphal stages. You still need to keep their foodplants fresh and mould-free, though. The usual rearing setup involves some sort of container – large jam jar, cardboard or plastic box, vivarium, aquarium – with muslin (or fine netting) over the top to allow fresh air in and moisture out. Throughout my career I've used old takeaway food containers, Kilner jars, a cracked fish tank, yoghurt pots, and for the most ambitious projects a giant cardboard box with large razored holes in the sides resealed with old net curtain glued in place. Keep it somewhere bright, but out of direct sunlight, ideally in a shed or unheated glasshouse so as to mimic outdoor conditions. Regularly check the wilt status of the foodplant and replenish it as necessary.

Record your findings. Photograph or describe the instars. How big are the individual nymphs? When are the moults, and how long were times between them (Fig. 136)? How and when are the nymphs active on the plants? Note foodplants, feeding sites and behaviours, and any interactions between the

FIG 136. An eight-day-old nymph of *Gonocerus acuteangulatus* moults to the third instar. Details meticulously recorded during captive rearing by Yvonne Couch.

FIG 137. Fresh from its final moult, a teneral *Tritomegas sexmaculatus* is confusingly white for several hours before it darkens to its characteristic black-and-white pied form. Images like this, posted on Twitter by Yvonne Couch, are highly informative to a non-specialist audience who may not recognise such instances, or realise the nature of shieldbug moulting.

nymphs or with other invertebrates. When they achieve adulthood, watch for courtship and mating behaviours, then observe egg-laying. Record any parasitoids that emerge. Record any abnormalities in leg or antenna structure. Record the proportions of different colour morphs and how long colour patterns take to appear on the initially pale newly emergent bugs (Fig. 137). Record how the bugs walk or fly; how they stand still or roost. In fact record anything and everything. This is how science advances: not through sudden seismic revelations, but by small incremental steps in observation and understanding.

PASSING ON THE KNOWLEDGE FOR THE FUTURE

Finally, report your findings. There is an active Facebook group dedicated to terrestrial Hemiptera, and another on Flickr, where people post observations and sightings, or just put up pictures to ask for help with identification. Other platforms like Twitter and Instagram are also useful, either to say 'look what I've found', or simply to ask 'anyone know what this is?' Mention of these early-twenty-first-century social media phenomena will surely date this book firmly, but whatever fads come next, be assured that entomologists will be able to make good use of them.

As reported back in Chapter 6, entomologists have been excitedly publishing their new discoveries in the entomological press ever since the entomological press started up. Sometimes this is a species new to Britain (for example *Dyroderes*

FIG 138. Since its first discovery in west London in 2013, the distinctive *Dyroderes umbraculatus* has spread, and is easily identified from photographs. This one from Hedge End near Southampton, May 2021, by Ian Harding.

umbraculatus, Nau *et al.* 2014; Fig. 138), a record outside a known geographic range (*Rhacognathus punctatus* in heathland-free Oxfordshire, Ryan 2016), or on a new foodplant (*Canthophorus impressus* feeding on Marjoram, Binding & Binding 2015), the unexpected appearance of a very rare species (*Dicranocephalus medius*, Denton 2016), sometimes a complete local list (the het bugs of London, by Groves 1964–86). Several entomological journals regularly publish short notes and longer reports on bugs, and *The Heteropterist's Newsletter* (1983–1989), *Het News* (2003–2016) and *The Hemipterist* (2014–present) are dedicated solely to the Hemiptera.

The language of these reports can often seem a little staid, imbued with a gravitas seemingly beyond their content, and this can dissuade some people from sending in their observations. All I can say is, please don't be put off. One of the very first articles I sent to a scientific journal was a report that I had found the flatbug *Aradus atterimus* in West Sussex in April 1977 (Jones 1978); it was previously known only from a few specimens found in Kent. My 'paper' had a grand total of 103 words and made eight lines of small type at the bottom of the page. But I was so pleased. These short notes are interesting and important. They preserve observations; they disseminate knowledge; they break up what might otherwise be the rather intense blocks of text of the longer articles. Just as those early hemipterists published their fascinating snippets of discovery and knowledge in the burgeoning entomological journals of the nineteenth century, so today's journals need these shorts too. As well as formally recording observations, they also make lively and entertaining reading. And they connect people. When

C. G. Hall sent in a list of the 'Hemiptera–Heteroptera at Dover and its vicinity' to the *Entomologist's Monthly Magazine* in 1890, he reported some very interesting finds, including *Eurygaster niger* (= *austriaca*), *Coreus* (= *Enoplops*) *scapha* 'abundant', and *Pseudophloeus* (= *Arenocoris*) *fallenii*, the 'remarkable black variety', but ended with a plaintive statement: 'There must be a great many more species to be recorded from this district ... I have the disadvantage of working alone.' With the connectivity of the twenty-first century it is much easier to make contact with like-minded individuals – and I speak from personal experience when I say there is a friendly hemipterist crowd out there ready and willing to help with identification, and to welcome new and even well-worn observations on shieldbugs.

This modern connectivity is not just for the social interaction of exchanging a bit of bug gossip. It feeds into the modern obsession with big data. There was a time when anybody wanting to draw together the national records for a species (or indeed for a whole genus, family or order) had to trawl through private and museum collections transcribing data labels, and then trawl through entomological journals looking for published records. There is still a bit of useful work to be done combing through web-based announcements on Facebook and Twitter, but there are more direct ways of getting observations onto the databases (see Appendix 2). Top amongst these for the casual observer are iSpot, where photos can be uploaded for identification by experts, and iRecord, if a fairly confident identification is already made. Experts vet both of these schemes and 'clean' the records to ensure high levels of confidence and accuracy.

FIG 139. *Camptopus lateralis* photographed by Maria Justamond, in Kent, in December 2019. It was found inside a car that had recently journeyed from France. All the details were attached at the Flickr photo-sharing site she uses, making this a valuable record of the insect's occurrence in the British Isles.

The idea is that these records can then be fed into monitoring schemes to measure the geographic spread of individual species, and whether they are increasing or decreasing. There is a shieldbug recording scheme up and running, which has already produced provisional distribution maps (Bantock 2018). Hawkins' work in Surrey (2003) showed how increasing ranges of the Box Bug (*Gonocerus acuteangulatus*) and Juniper Bug (*Cyphostethus tristriatus*) could be measured and mapped, and the chapter on bugs by Kirby *et al.* (2001) comments on the changes (both positive and negative) in the status of various British species. With the current threat of climate change affecting wildlife in ways as yet unknown, and mainstream-media insectageddon headlines, insects are guaranteed to be at the forefront of wider research into nature conservation, environmental resilience, rewilding of the wider countryside, and the greening of cities. Shieldbugs are one of the flagship groups that can help offer insights into these changes.

Shieldbugs are already being referenced in climate change studies. Musolin & Fujisaki (2006) summarise several studies on the expanding ranges of various shieldbugs including the 'southern' green shieldbug *Nezara viridula* moving further north in Europe and Japan, and the South African scutellerid *Calidea dregii* moving south, following an assumed opposite geographic direction modulated by climate changes in the southern hemisphere. They point out that about 60% of 'new' species reported in the faunas of Britain, the Netherlands and Austria between 1978 and 2003 are definite new colonist arrivals expanding their geographic ranges (i.e. not previously overlooked species), and the implication is that these expansions are all moving northwards from warmer southern climes to warming northerly ones. There is more to this than curiously watching these spreads from the sidelines – with changes in distribution come changes in behaviour, and these may yet impact our understanding and assessment of shieldbugs. Powell (2020) reports how the common and widespread (and previously 'harmless') forest shieldbug *Pentatoma rufipes* is emerging as a pest of apples, cherries, strawberries and raspberries in northern Europe, and that this may be a result of climate change. The bug overwinters as a 3 mm second-instar nymph, sheltering in the wrinkled bark, branch junctions or twig nodes of its host trees (usually reported as oak, beech, hazel and others), and milder winters and earlier springs appear to be allowing it better survival rates, and earlier maturity, with concomitant population increases. Likewise Carvajal *et al.* (2018) predict that the African shieldbug *Bagrada hilaris* is likely to become a serious pest of cabbages and other brassicas in Mediterranean Europe and North and South America with global warming. These are exactly the sorts of observations and reports that field entomologists can feed into by recording what they find, and what they see (Figs. 140, 141).

FIG 140. The supposedly warm-region colour form *torquata* of the normally all-green *Nezara viridula*, found at the Tower of London after the record-breakingly hot summer of 2022. If hot weather during the nymphal stages is a factor in producing this morph, monitoring its future occurrence in the British Isles may illustrate a genuine climatic effect on the insect.

FIG 141. Melanic specimen of *Pyrrhocoris apterus* photographed at Dungeness in October 2022 by Simon Warry. This is thought to be the first time that this unusual colour form has been found in the British Isles. It was posted on Facebook, a convenient and easily accessible means of spreading the word.

Exchanging records, observations and notes is the first step in getting to know other entomologists. Despite the authoritative style of biological monographs and the clear photos and diagrams in identification guides, there is nothing to beat a friendly discussion between like-minded aficionados, whether at a field meeting to look at live insects in the countryside, or with a microscope and a tray of specimens in the study with tea and biscuits. Expertise is exchanged and bolstered by these interactions. Facebook, Twitter and Instagram are easy to search for active entomologists. In 2022 the Field Studies Council ran the first of what will hopefully be a regular series of bug identification workshops. National and local natural history societies often run field events and indoor lecture meetings. It is about one such lecture meeting that I now have to make a personal confession. In the 1980s I organised the twice-monthly lectures of the British Entomological and Natural History Society, then held in the rooms of the Alpine Club in London's Mayfair district. When I approached Hemiptera expert Peter Kirby he readily agreed to give a talk about bugs to the society on 22 October 1987 – 'The art of bugging'. This was not, however, his first suggestion. Being a relative youngster in the society, I fretted that some of the older and more conservative members (retired army colonels, senior civil servants, men of

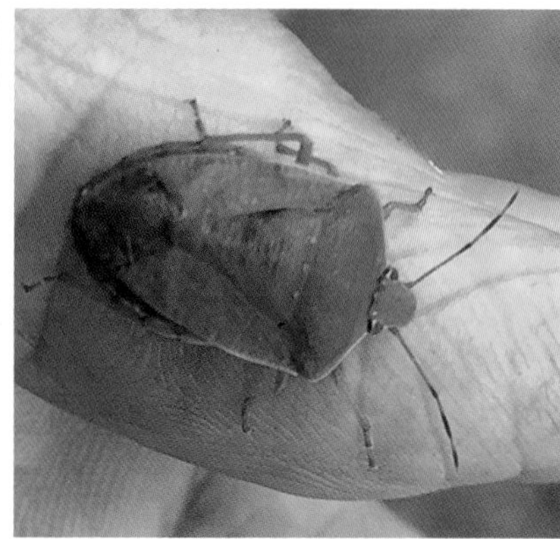

FIG 142. A shieldbug in the hand is worth any number of shieldbugs in a textbook and makes the perfect object for sharing one's Heteroptera enthusiasm with any non-entomologists who happen to be about. *Nezara viridula*, the Southern Green Shieldbug, turns brown for winter, but reverts to bright green in spring.

the cloth) would balk at his original title of 'How to be a real bugger'. Sorry Peter, with hindsight it's obvious we should have run with that one.

And, of course, part of the responsibility of the entomologist passing on that expertise is not just keeping the knowledge inside the clique – secret, arcane, academic – it is spreading it out to all. This doesn't mean you have to actively seek people out and ram it down their throats. Anyone who has ever walked out with an insect net will have attracted curious glances or questions. And this is the ideal opportunity for a bit of entomological outreach (Fig. 142). Dispel the notion that entomologists are merely eccentric collectors, greedily raiding the countryside for prizes to display in their cabinets. Simply by explaining how a knowledge of insects (shieldbugs or whatever) feeds into wildlife conservation, land management, environmental monitoring and ecological understanding, you can help to bolster a much wider appreciation of these tiny, poorly understood, often overlooked and underestimated creatures.

Together, even single records of single species build up into a bigger picture. And it's an important picture. Ecology is the scientific study of the interconnectedness of every organism on this planet – that has ever lived. We can't study it all. We can't take it all in. But we can look closely at individual facets and use these studies as a window to view the world. Even if it is a small window, we can at least see part of the picture and use it to inform our understanding of life on Earth. Simply by admiring a shieldbug walking clumsily across a leaf or the open palm of your hand you have joined this endeavour. Welcome.

APPENDIX 1

Selected British and Irish Shieldbug Foodplants and Prey

Most shieldbugs are plant-feeders, and while some have broad tastes, others occur on just a single plant species. Rather than clutter the pages with a huge table listing all the many known foodplants for all British shieldbugs, the following is an abbreviated list of the more usual host plants. As ever, these associations are not definitive, merely guides, and there is always the chance that a local population of a bug can be found feeding on an unusual foodplant. Likewise predators famously eat whatever they can get, so this list includes just a few dietary regulars.

Acanthosomatidae

Acanthosoma haemorrhoidale	Hawthorn (*Crataegus* spp.), but also many other trees
Cyphostethus tristriatus	Nominally Juniper (*Juniperus communis*), but mostly garden cypress trees (*Cupressus* spp.)
Elasmostethus interstinctus	Birch (*Betula* spp.), but also Hazel (*Corylus avellana*) and Aspen (*Populus tremula*)
Elasmucha ferrugata	Bilberry (*Vaccinium myrtillus*), Cowberry (*V. vitis-idaea*), Fly Honeysuckle (*Lonicera xylosteum*) and hawthorn (*Crataegus* spp.)
Elasmucha grisea	Birch (*Betula* spp.), also Alder (*Alnus glutinosa*)
Elasmucha fieberi	Birch (*Betula* spp.), Alder (*Alnus glutinosa*) and Hazel (*Corylus* spp.)

Cydnidae

Aethus flavicornis	Unknown, but found under low-growing plants in sandy places

Geotomus punctulatus	Lady's Bedstraw (*Galium verum*)
Geotomus petiti	Unknown, but found under low-growing plants in sandy places
Adomerus biguttatus	Cow-wheat (*Melampyrum pratense*)
Canthophorus aterrimus	Spurges (*Euphorbia* spp.)
Canthophorus impressus	Bastard Toadflax (*Thesium humifusum*)
Legnotus limbosus	Bedstraws (*Galium* spp.)
Legnotus picipes	Bedstraws (*Galium* spp.)
Ochetostethus nanus	Unknown, but found under low-growing plants in dry sandy places
Sehirus luctuosus	Forget-me-nots (*Myositis* spp.)
Sehirus morio	Alkanet (*Anchusa officinalis*), Hound's-tongue (*Cynoglossum officinale*)
Tritomegas bicolor	White Dead-nettle (*Lamium album*), also woundworts (*Stachys* spp.) and Black Horehound (*Ballota nigra*)
Tritomegas sexmaculatus	Black Horehound (*Ballota nigra*)

Thyreocoridae

Thyreocoris scarabaeoides	Violets (*Viola* spp.) (?)

Pentatomidae

Mecidea lindbergi	Grasses
Arma custos	Lepidoptera and sawfly caterpillars, beetle larvae especially Alder Leaf Beetle (*Agelastica alni*)
Jalla dumosa	Lepidoptera caterpillars and beetle larvae
Picromerus bidens	Lepidoptera and sawfly caterpillars, beetle larvae etc.
Pinthaeus sanguinipes	Arboreal moth caterpillars
Rhacognathus punctatus	Larvae (and adults) of Heather Leaf Beetle (*Lochmaea suturalis*)
Troilus luridus	Moth and sawfly caterpillars
Zicrona caerulea	Larvae (and adults) of *Altica* flea beetles
Aelia acuminata	Grasses
Aelia klugii	Grasses
Aelia rostrata	Grasses
Neottiglossa pusilla	Grasses
Neottiglossa leporina	Grasses
Halyomorpha halys	Many plant species, sometime pest of apple, pear, peach and soft fruits

Carpocoris purpureipennis	Asteraceae/Compositae
Carpocoris mediterraneus	Asteraceae/Compositae, Umbelliferae and others
Chlorochroa juniperina	Juniper (*Juniperus communis*)
Dolycoris baccarum	Many different plant species
Holcogaster fibulata	Conifers including garden cypresses (*Cupressus* spp.) and pines (*Pinus* spp.)
Peribalus strictus	Wide variety of plant species
Palomena prasina	Wide variety of plants, sometime pest of cultivated broad beans
Palomena viridissima	Wide variety of plants
Eysarcoris aeneus	Slender St John's-wort (*Hypericum pulchrum*)
Eysarcoris ventralis	Grasses
Eysarcoris venustissimus	Woundworts (*Stachys* spp.), also White Dead-nettle (*Lamium album*) and Black Horehound (*Ballota nigra*)
Stagonomus bipunctatus	Various Lamiaceae
Nezara viridula	Wide variety of plants, sometime pest of broad beans
Pentatoma rufipes	Wide range of broadleaved trees
Rhaphigaster nebulosa	Various broadleaved trees
Piezodorus lituratus	Gorse (*Ulex* spp.), also broom (*Cytisus* spp.) and Dyer's Greenweed (*Genista tinctoria*)
Dyroderes umbraculatus	Bedstraws (*Galium* spp.)
Menaccarus arenicola	Marram Grass (*Ammophila arenaria*)
Sciocoris cursitans	Mouse-ear Hawkweed (*Pilosella officinarum*)
Sciocoris homalonotus	Unknown, but is ground-dwelling under low plants
Sciocoris sideritidis	Unknown, but is ground-dwelling under low plants
Eurydema oleracea	Cruciferae/Brassicaceae
Eurydema dominulus	Lady's Smock (*Cardamine pratensis*)
Eurydema ornata	Sea Radish (*Raphanus raphanistrum*)
Eurydema herbacea	Sea Rocket (*Cakile maritima*)
Eurydema ventralis	Various Cruciferae
Graphosoma italicum	Wild Carrot (*Daucus carota*) and other umbellifers
Podops inuncta	Uncertain, but is ground-dwelling in root thatch and under low plants

Plataspidae

Coptosoma scutellatum	Dyer's Greenweed (*Genista tinctoria*)

Scutelleridae

Eurygaster testudinaria	Grasses, sedges and rushes
Eurygaster maura	Grasses, sedges and rushes
Eurygaster austriaca	Grasses, sedges and rushes
Odontoscelis fuliginosa	Stork's-bills (*Erodium* spp.)
Odontoscelis lineola	Stork's-bills (*Erodium* spp.)
Odontotarsus purpureolineatus	Knapweeds (*Centaurea* spp.) and thistles (*Cirsium* spp. and *Carduus* spp.)

Pyrrhocoridae

Pyrrhocoris apterus	Mallows (*Malva* spp.), particularly Tree Mallow (*Malva arborea*); also lime (linden) (*Tilia* spp.)

Alydidae

Alydus calcaratus	Fabaceae, including restharrow (*Ononis* spp.) and medicks (*Medicago* spp.)
Camptopus lateralis	Bird's-foot trefoils (*Lotus* spp.), restharrow (*Ononis* spp.), Gorse (*Ulex europaeus*)
Micrelytra fossularum	Grasses

Coreidae

Arenocoris fallenii	Stork's-bills (*Erodium* spp.)
Arenocoris waltlii	Stork's-bills (*Erodium* spp.)
Bathysolen nubilus	Black Medick (*Medicago lupulina*)
Bothrostethus annulipes	Sainfoin (*Onobrychis* spp.), brooms (*Cytisus* and *Genista* spp.)
Ceraleptus lividus	Clovers (*Trifolium* spp.)
Coreus marginatus	Docks (*Rumex* spp.)
Haploprocta sulcicornis	Sheep's Sorrel (*Rumex acetosella*)
Coriomeris denticulatus	Medicks (*Medicago* spp.)
Coriomeris scabricornis	Various low Fabaceae
Coriomeris affinis	Uncertain, but rough flowery places
Enoplops scapha	Mayweeds (*Tripleurospermum* spp.)
Gonocerus acuteangulatus	Box (*Buxus sempervirens*), hawthorn (*Crataegus* spp.), honeysuckle (*Lonicera* spp.) and others
Gonocerus juniperi	Junipers and cypresses
Leptoglossus occidentalis	Cones of various conifer trees

Spathocera dalmanii	Sheep's Sorrel (*Rumex acetosella*)
Syromastus rhombeus	Spurreys (*Spergularia* spp.) and sandworts (*Arenaria* spp.)

Rhopalidae

Brachycarenus tigrinus	Unknown in Britain, but in rough flowery places
Corizus hyoscyami	Restharrow (*Ononis* spp.), stork's-bills (*Erodium* spp.)
Liorhyssus hyalinus	St John's-worts (*Hypericum* spp.), stork's-bills (*Erodium* spp.)
Rhopalus maculatus	Cinquefoil (*Potentilla* spp.), Marsh Thistle (*Cirsium palustre*), willowherbs (*Epilobium* spp.)
Rhopalus parumpunctatus	Mouse-ear (*Cerastium* spp.), stork's-bills (*Erodium* spp.)
Rhopalus rufus	Corn Spurrey (*Spergula arvensis*) and Sand Spurrey (*Spergularia rubra*)
Rhopalus subrufus	St John's-worts (*Hypericum* spp.), sage (*Salvia* spp.), mint (*Mentha* spp.) etc.
Rhopalus lepidus	Uncertain, perhaps juniper pinks (Caryophyllaceae)
Stictopleurus abutilon	Asteraceae/Compositae
Stictopleurus punctatonervosus	Asteraceae/Compositae
Stictopleurus crassicornis	Asteraceae
Stictopleurus pictus	Asteraceae
Chorosoma schillingi	Marram (*Ammophila* spp.) and other grasses
Myrmus miriformis	Grasses and sedges

Stenocephalidae

Dicranocephalus medius	Wood Spurge (*Euphorbia amygdaloides*)
Dicranocephalus agilis	Portland Spurge (*Euphorbia portlandica*)
Dicranocephalus albipes	Cypress Sprurge (*Euphorbia cyparissias*)

APPENDIX 2

Shieldbug-related Websites and Apps

British Bugs

An online photo identification guide to British Hemiptera. Also checklists, back issues of *Het News* and links to other useful Hemiptera websites, including information on shieldbugs recording scheme and atlas. ***britishbugs.org.uk***

iNaturalist UK

UK version of international citizen scientist/social media app and website where individual records and observations can be uploaded, verified and discussed by other naturalists. Covers all groups of animals, plants and fungi. ***uk.inaturalist.org***

iRecord

Citizen science/social media app and website to upload individual or multiple records. Observations will be verified by experts before data is input into the national databases. Covers all groups of animals, plants and fungi. ***irecord.org.uk***

iSpot

An identification help website for uploading queries. Photos are regularly checked and named by experts, leading to identification by consensus. Covers all groups of animals, plants and fungi. ***ispotnature.org***

National Biodiversity Network

Searchable database showing provisional distributions of British species, and up-to-date nomenclature and classification. Covers all groups of animals, plants and fungi. ***nbn.org.uk***

Glossary

Every scientific discipline has its own particular terminology, often a kind of shorthand by which experts communicate quickly and easily – it's called jargon. No science can flourish if it hides behind or gets bogged down in its own jargon, but equally it would be patronising to talk down to an audience on the assumption that they cannot understand long or unusual words. If a word has a precise meaning and is useful or necessary to convey an accurate interpretation or difficult concept then it should be used, but used sparingly and carefully. It is easy to lapse into obscure technical terms when you get into full flight, and I thank the several editors who have forensically pointed out where I have started to become needlessly convoluted. Most specialist terms are explained in the text, on the first use; this glossary is a reminder.

Aedeagus Male insect sexual organ, usually a unit made of one or more hardened plates which together form a roughly cylindrical structure used to penetrate the female during mating. Difficult species pairs often need to be dissected to examine shapes of the sexual organs, which are often precisely species-specific.
Antenna (plural **antennae**) Two long sensory appendages, four- or five-segmented in shieldbugs, used for chemical detection ('smelling').
Antennifer (sometimes antenniferous tubule) Mound on the front, top and side of the head on or under which the antenna is mounted.
Aposematic Warningly marked with strong patterns or colours which act as a visual deterrent from attack – usually indicating a danger from toxins, venom or bite.
Apterous Lacking wings; mostly used of wingless adults of species where wings would be the norm. There are few truly apterous British shieldbugs, but the Firebug, *Pyrrhocoris apterus*, takes its name from its short-winged, and therefore non-flying, form. See also Brachypterous.

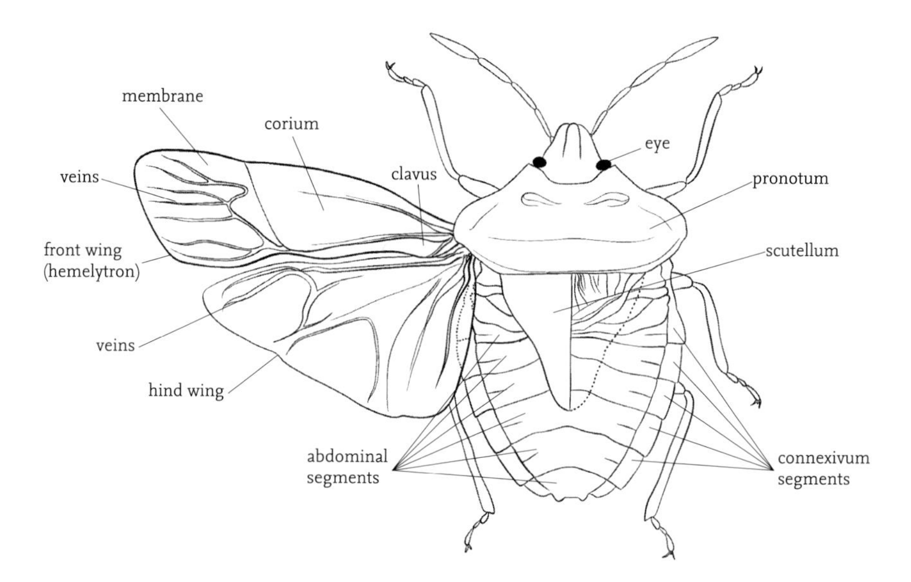

Generalised shieldbug structure: from above.

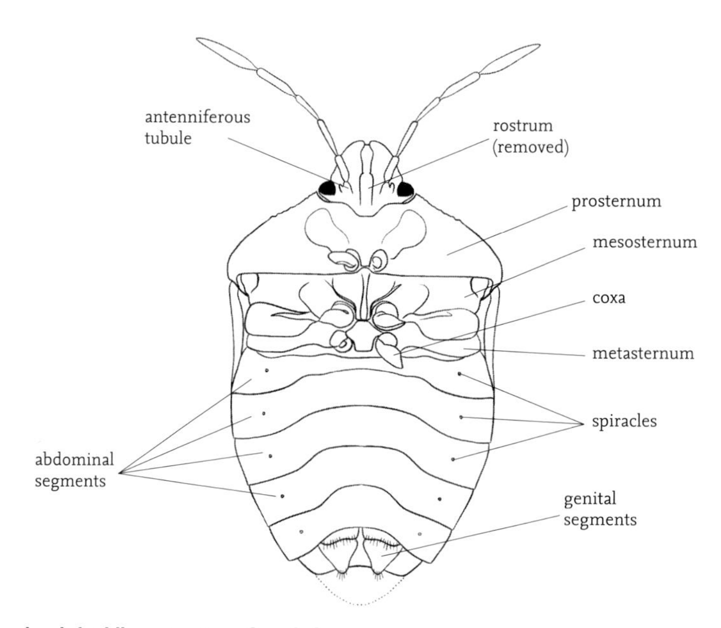

Generalised shieldbug structure: from below.

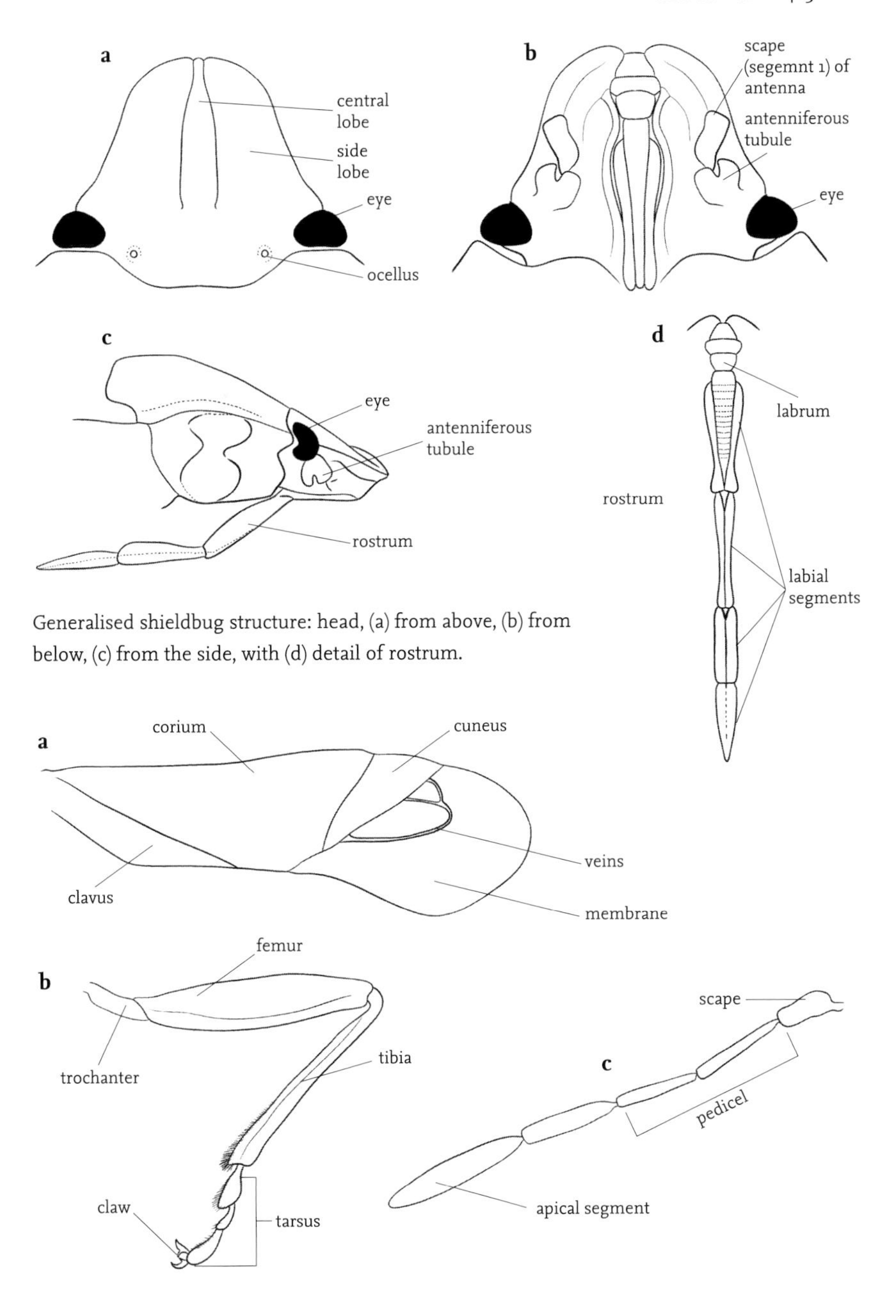

Generalised shieldbug structure: head, (a) from above, (b) from below, (c) from the side, with (d) detail of rostrum.

Generalised bug structure: (a) hemelytra (forewing) of a leaf bug (Miridae) – shieldbug families lack the fold marking off the cuneus as a distinct portion; (b) leg; (c) typical five-segmented antenna of a pentatomid bug.

Bivoltine Having two generations per year. See also Multivoltine, Univoltine.
Brachypterous An adult form (a 'brachypter') having short, usually non-functional wings, meaning the insect cannot fly. In het bugs this usually means the membrane of the hemelytron is shortened or missing, and hind wings may be reduced or absent. Amongst British species *Myrmus miriformis, Micrelytra fossularum* and *Pyrrhocoris apterus* normally show brachypterous forms.
Buccula (plural **bucculae**) Small raised flange-like ridges running underneath the head, and acting as a resting slot for the rostrum when not extended in feeding mode.
Central lobe The central prominence of the front of the upper part of the head, bounded either side by the side lobes. The shape and proportions of all these lobes are often used in identification keys.
Chitin The main component of insect integument, made from many parallel polysaccharide fibres laid down in multiple overlapping layers. Its structure gives the insect body its hard but slightly flexible strength.
Claval commisure Meeting point of the clavus of each wing from tip of scutellum to membrane.
Clavus The hind edge of the hardened front wing of a bug, usually long and parallel-sided, and resting hard up against the scutellum when the wings are furled.
Clypeus The upper front plate of the head of an insect, immediately behind and above the mouthparts. In shieldbugs it is divided into three parts – a central lobe and two side lobes.
Connexivum Outer edge of the abdomen of a bug, usually visible from above outside the folded front wings (hemelytra) when the insect is at rest. The segmentation of the abdomen is reflected in the segmentation of the connexivum.
Corium Large triangular leading portion of the hardened part of the forewing of a het bug, excluding the clavus and membrane. It is sometimes divided into the outer exocorium and inner endocorium.
Coxa (plural **coxae**) The first segment of the leg, where it articulates against the thoracic segment, often short and cylindrical or globose.
Cuneus Small triangular segment of the corium, where it meets the membrane. This is an important wing part in many het bugs, but is not clearly differentiated in shieldbugs.
Diapause Physiologically enforced stoppage during an insect's development, often during hibernation, beyond which point growth will not proceed unless triggered by an environmental event such as a prolonged cold period. This helps synchronise maturity, with the appearance of new plant growth, the appearance of particular prey, or the appearance of potential mates.

Dorsal abdominal glands (DAGs) Scent glands opening onto the back of the abdominal segments in shieldbug nymphs. In most groups these cease to be functional when wings develop and cover them in the adult stage.
Druckknopf Flap-like knob on the side of the second thoracic segment, onto which a tufted ridge of minute hairs near the base of the hemelytron clip hold when the wings are folded.
Elytron (plural **elytra**) Hardened wing case found in beetles. The analogous hard part of a shieldbug wing is the hemelytron – half a wing case.
Embolium Narrow section at the leading edge of the hardened bug hemelytron.
Evaporatorium Area of matt-textured cuticle around the pore of the metathoracic scent gland of shieldbugs. Once thought to aid evaporation of the volatile chemicals, it is now thought to act as a holding area to concentrate the secretion in one area and keep the chemicals from spreading all over the body.
Femur (plural **femora**) The third and largest segment of the insect leg, analogous to the thigh.
Frena Groove-and-ridge closing mechanism by which a strengthened strut running along the back edge of the clavus is held in place against the edge of the scutellum by a slot fitted with minute downward pointing scale-like hairs that lock tight when the bug is not flying.
Haemocoel The main internal body cavity of the insect, in which organs are bathed in a sea of liquid containing nutrients, hormones etc.
Haemolymph The insect equivalent of blood, containing nutrients, waste products and hormones, but not part of the gas-exchange mechanism.
Hemelytron (plural **hemelytra**, sometimes **hemielytron/hemielytra**) Partially hardened front wing of a bug, which acts as a protective sheath over the membranous hind wings folded underneath when the insect is at rest. I often use the term to mean just the hardened, sclerotised, coloured, basal section comprising corium, clavus, embolium and cuneus combined.
Hemimetaboly/Hemimetabolous Growth Development process by which wingless nymphs grow by stages, but always resemble the adults in their body segmentation, leg and antennal structure, eyes and mouthparts. They do not pass through the major change of the chrysalis or pupa stage. Shieldbugs are hemimetabolous.
Holometaboly/Holometabolous Growth Development process by which larva, grub, maggot or caterpillar looks and behaves very differently to the winged adult, and which passes through the major restructuring of the pupa or chrysalis stage.
Instar One of several discrete growth stages in the immature development of an insect. Shieldbug nymphs pass through five larval instars before finally emerging as a winged adult.

Jugum (plural **juga**) One of the pair of side lobes of the clypeus at the front of the head of a shieldbug.
Labium The lower lip of an insect. In shieldbugs this forms a stout but flexible three- or four-segmented sheath protecting the piercing stylet mouthparts, which can be hinged backwards when the bug is feeding.
Labrum The upper lip of an insect. In shieldbugs this forms a short reinforcing hem under the front of the rostrum mouthparts.
Lorum (plural **lora**) Another name for the side lobes (juga) of the clypeus at the front of the head.
Macropterous Having long, fully developed wings. This is the normal body form of most British and Irish shieldbugs.
Mandibular stylets Two long, narrow, sharp, piercing mouthparts in bugs, used to puncture plant or prey. A narrow cylindrical gap between them takes saliva down and a large cylindrical gap allows part-digested food to be sucked up.
Maxillary stylets Two long mouthparts that form a lubricated sheath around the piercing mandibular stylets.
Membrane The end of the front wing of a bug, which remains membranous, not stiffened or hardened, but which may still be clouded, darkened, spotted or streaked.
Mesoscutellum Another name for the scutellum, since it arises from the meso (second) segment of the thorax.
Metathoracic glands Stink or scent glands in the third segment of the thorax of a shieldbug, opening onto each side of the bug.
Micropterous Having extremely short wings. Micropters occur in various bug species, but in shieldbugs like *Micrelytra fossularum* and *Myrmus miriformis* even those with the shortest wings are still brachypters.
Micropylar process Point on an egg where a micropyle has developed into a narrow point, tube, or node, seemingly concerned with gas exchange.
Micropyle Narrow opening in a bug egg through which sperm pass.
Multivoltine Having multiple generations per year. See also Bivoltine, Univoltine.
Nymph Immature bug, often a miniature, but wingless, version of the adult.
Ocellus (plural **ocelli**) Two simple eyes, near the back of the head, found in most adult shieldbugs, but not *Pyrrhocoris apterus*. They are thought to aid the bug in flight by giving a measure of general light levels to help maintain the body level.
Operculum Flip-top lid on an insect egg with narrow rim above a narrowing of the eggshell giving a line of weakness that allows the emerging nymph to push it off. Technically shieldbugs have a pseudo-operculum, since the rim is not weakened underneath.

Ovariole Egg-producing tube in the ovary. Shieldbugs usually have seven ovarioles in each of their two ovaries so often lay fourteen eggs in a batch, or multiples of seven or fourteen.
Palp (sometimes also **palpus**, pleural **palpi**) Small single- or multi-segmented appendages attached to the mouthparts of many insects, used in sensing and food-manipulation. Shieldbugs, indeed all Hemiptera, have lost these appendages through evolution, in the development of a sucking rostrum form.
Paraclypeal lobe. Central lobe of the clypeus at the front of the shieldbug head.
Paramere One of a pair of hardened plates either side of the male sexual organ, used in clasping the female genital tract during mating.
Paurometaboly Form of hemimetabolous growth with nymphs of gradually increasing size, eventually turning into an adult, as found in the Hemiptera, rather than the sudden change from aquatic naiad to airborne adult found in dragonflies.
Pendergrast's organs Specialised secretory areas of the cuticle under either side of abdominal segments 5, 6 or 7 found in Acanthosomatidae, and thought to contain symbiont bacteria which are smeared onto the eggs by the mother's back legs during laying.
Peritreme Ear- or spout-shaped orifice of the metathoracic scent gland.
Pleuron (plural **pleura**) any of the side plates of the three body segments that make up the thorax.
Prolarva Embryonic larva found inside the shieldbug egg. The skin is moulted and left inside the egg as the first-instar nymph emerges.
Pronotum Large reinforced plate covering most of the exposed upper thorax of a shieldbug between head and scutellum.
Pseudo-operculum Flip-top lid of a shieldbug egg. Technically not quite an operculum, which has a ring of weakness around it. The shieldbug embryo is required to use a hard and sharp egg-opener on its head to rupture and open the egg.
Pulvillus (plural **pulvilli**) Small flat pad structure arising between the shieldbug's tarsal claws, thought to aid climbing on smooth flat surfaces.
Pygophore Genital capsule of a male shieldbug, usually abdominal segment 9.
Remigium The front part of an insect wing. This is usually the main portion containing the supportive struts of the wing veins. These are obscured in shieldbug front wings because of the hardening and colouring of the hemelytra, but here the remigium is more or less equivalent to the corium.
Resilin Tough but elastically flexible protein found in insect wings and wing mechanisms.
Rostrum Segmented snout-like feeding mouthparts found in all Hemiptera. In shieldbugs it is usually only visible in side view, unless the insect is feeding.

Sclerotised Of insect cuticle that has become hardened and coloured. When a shieldbug moults it takes a few hours before sclerotisation is complete, and while this occurs the bug is often soft and pale.
Scutellum Large triangular plate behind the pronotum, against which the wings of a shieldbug are anchored when the insect is not flying. Often shield-shaped and frequently cited as the reason shieldbugs are so called.
Sensillum (plural **sensilla**) Microscopic spine, cone, knob or slit structures on the antennal segments which contain chemosensitive receptors – giving an insect its sense of smell.
Spermatheca Sperm-storage organ found in the abdomen of female insects.
Spiracle Opening of a trachea breathing tube on the side of thorax or abdominal segment, usually a round or kidney-shaped hole.
Stadium (plural **stadia**) Alternative name for instar.
Sternite Plate on the underside of each abdominal segment.
Stridulitrum File of minute pegs, grooves or ridges against which a hard node (the plectrum) is rubbed, to create sound.
Stylet See Mandibular stylets, Maxillary stylets.
Tarsomere Each of the segments that make up the tarsus.
Tarsus (plural **tarsi**) The foot portion of the leg. In shieldbugs the tarsus is made up of two or three segments depending on which family it belongs to.
Teneral Soft, and pale in colour after freshly emerging from moulted skin, and before hardening sclerotisation and deposition of colour.
Tergite Plate on the upper surface of each abdominal segment.
Tibia (plural **tibiae**) Fourth portion of the leg of an insect, analogous to the shin. Sometimes longer than the femur, but usually slimmer.
Trachea (plural **tracheae**) Insect breathing tubes which enter on each side of most thoracic and abdominal segments and which branch finer and finer to facilitate gas exchange to the internal organs and tissues.
Trichobothrium (plural **Trichobothria**) Long sensory bristles that detect air movement. In shieldbugs these are mainly on the underside of the abdomen, and their arrangements are constant in the various different bug families.
Trochanter Small second portion of an insect leg, anchoring the femur to the coxa. Sometimes called the fulcrum in older books.
Tylus Alternative name for the central lobe of the clypeus.
Univoltine Having just a single generation per year. See also Bivoltine, Multivoltine.
Vannus Hind portion of an insect wing, usually lacking or with reduced veins.
Wing bud (or **wing pad**) Small lobes attached to the pronotum of a shieldbug nymph, in which the embryonic wings are developing. Usually only present, or noticeable, in fourth or fifth instar.

References

Ahmad, A., Parveen, S., Brozec, J. & Dey, D. (2016). Antennal sensilla of phytophagous and predatory pentatomids (Hemiptera: Pentatomidae): a comparative study of four genera. *Zoologischer Anzeiger – A Journal of Comparative Zoology* 261, 48–55.

Aldrich, J. R. (1988). Chemical ecology of the Heteroptera. *Annual Review of Entomology* 33, 211–238.

Aldrich, J. R. & Blum, M. S. (1978). Aposematic aggregation of a bug (Hemiptera: Coreidae): the defensive display and formation of aggregations. *Biotropica* 10, 58–61.

Aldrich, J. R., Oliver, W. R., Lusby, J. P., Kochansky, J. P. & Lockwood, J. A. (1987). Pheromone strains of the cosmopolitan pest *Nezara viridula* (Hemiptera, Pentatomidae). *Journal of Experimental Zoology* 244, 171–175.

Alexander, D. E. (2015). *On the Wing: Insects, Pterosaurs, Birds, Bats and the Evolution of Animal Flight.* Oxford University Press, Oxford.

Alexander, K. N. A. (2008). *The Land and Freshwater Bugs (Hemiptera) of Cornwall and the Isles of Scilly.* CISFBR & ERCCIS, Cornwall.

Amyot, C. J.-B. & Serville, A. (1843). *Histoire naturelle des insectes. Hémiptères.* Librairie Encyclopédique de Roret, Paris.

Arnold, J. W. (1964). Blood circulation in insect wings. *Memoirs of the Entomological Society of Canada* 96, 5–60.

Ashwell, D. & Denton, J. (1998). Firebugs *Pyrrhocoris apterus* (L.) (Hemiptera: Pyrrhocoridae) breeding in Surrey. *British Journal of Entomology and Natural History* 10, 219.

Aukema, B. (1988). *Orsillus depressus* nieuw voor Nederland en België (Heteroptera: Lygaeidae). *Entomologische Berichten* 48, 181–183.

Aukema, B. (2001). Recent changes in the Dutch Heteroptera fauna (Insecta: Hemiptera). In: *Changes in Ranges: Invertebrates on the Move* (ed. M. Reemer, P. J. van Helsdingen & R. M. J. C. Kleukers). Proceedings of the 13th International Colloquium of the European Invertebrate Survey, Leiden, 2–5 September 2001. European Invertebrate Survey, Leiden, pp. 39–52.

Aukema, B. & Rieger, C. (eds) (2006). *Catalogue of the Heteroptera of the Palaearctic region. Volume 5: Pentatomomorpha* 2. Netherlands Entomological Society, Amsterdam.

Aukema, B., Duffels, H., Günther, H., Rieger, C. & Strass, G. (2013). New data on the Heteroptera fauna of La Palma, Canary Islands (Insects: Hemiptera). *Acta Musei Moraviae, Scientiae Biologicae* 98, 459–493.

Baker, H. (1744). *The Microscope Made Easy.* Dodsley, London.

Bantock, T. (2016a). A review of the Hemiptera of Great Britain: the shieldbugs and allied families. Coreoidea, Pentatomoidea and Pyrrhocoroidea. Species Status Number 26. Natural England Commissioned Report number 190.

Bantock, T. (2016b). [Species new to Britain] *Mecidea lindbergi* Wagner, 1954 (Pentatomidae). *Het News* 2, 3–4.

Bantock, T. (2017). *Tritomegas sexmaculatus* (Cydnidae) arrives in Britain. *Het News* 17/18, 4.

Bantock, T. (2018). Provisional atlas of shieldbugs and allies. www.britishbugs.org.uk/Provisional_atlas_of_shieldbugs_and_allies_2018.pdf (accessed December 2022).

Bantock, T. (2019). Official and sectional reports for 2018 – Hemiptera. *London Naturalist* 98, 26.

Bantock, T. & Kenward, H. (2017). *Sciocoris homalonotus* (Heteroptera: Pentatomidae) – a species new to Britain. *British Journal of Entomology and Natural History* 30, 27–29.

Bantock, T. M., Notton, D. & Barclay, M. V. L. (2011). *Rhaphigaster nebulosa* (Pentatomidae: Pentatomini) arrives in Britain. *Het News* 17/18, 5.

Barclay, M. V. L. & Nau, B. S. (2001). A recent record of *Carpocoris purpureipennis* (De Geer) (Hem., Pentatomidae) from the West of England. *Entomologist's Monthly Magazine* 137, 72.

Barnard, J. (2016). Pied shieldbug (*Tritomegas bicolor*) and Rambur's pied shieldbug (*Tritomegas sexmaculatus*) in Britain – is there a conflict of interests? *Het News* 23, 9.

Barnard, J. (in preparation). A provisional atlas of the shieldbugs and allies of Kent.

Bates, H. W. (1877). On *Ceratorrhina quadrimaculata* (Fabr.), and descriptions of two allied species. *Transactions of the Entomological Society* 1877, 201–203.

Baugnée, J. Y. (1999). Quelques Hétéroptères peu communs observés récemment en Lorraine belge (province de Luxembourg). *Bulletin et Annales de la Société Royale Belge d'Entomologie* 135, 61.

Beavis, I. C. (1988). *Insects and Other Invertebrates in Classical Antiquity*. Exeter University Press, Exeter.

Bedwell, E. C. (1909). *Odontoscelis dorsalis*, Fabricius, a British insect. *Entomologist's Monthly Magazine* 45, 253–254.

Berenbaum, M. R. & Miliczky, E. (1984). Mantids and milkweed bugs: efficacy of aposematic coloration against invertebrate predators. *American Midland Naturalist* 111, 64–68.

Betts, C. R. (1986). The comparative morphology of the wings and axillae of selected Heteroptera. *Journal of Zoology* 1, 255–282.

Binding, A. & Binding, A. (2015). *Canthophorus impressus* feeding on marjoram. *Het News* 22, 4.

Boardman, P. (2014). *A Provisional Atlas of the Shieldbugs and Allies of Shropshire*. Field Studies Council, Telford.

Bouldrey, S. M. & Grimnes, K. (1995). An allometric study of the boxelder bug, *Boisea trivittata* (Heteroptera: Rhopalidae). *The Great Lakes Entomologist* 28, 207–212.

Brooke, S. E. (2003). Gadget corner: the saldid catcher. *Het News* 2, 3.

Brookes, R. (1763). *The Natural History of Insects, with Their Properties and Uses in Medicine*. Vol. 4. Newbery, London.

Butler, E. A. (1911). *Stenocephalus medius*, M. et R.: an addition to the list of British Hemiptera. *Entomologist's Monthly Magazine* 47, 134–135.

Butler, E. A. (1923). *A Biology of the British Hemiptera–Heteroptera*. Witherby, London.

Carroll, S. P. (1987). Contrasts in reproductive ecology between temperate and tropical populations of *Jadera haematoloma*, a mate-guarding hemipteran (Rhopalidae). *Annals of the Entomological Society of America* 81, 54–63.

Carvajal, M. A., Alaniz, A. J., Núñez-Hidalgo, I. & González-Césped, C. (2018). Spatial global assessment of the pest *Bagrada hilaris* (Burmeister) (Heteroptera: Pentatomidae); current and future scenarios. *Pest Management Science* 75, 809–820.

Champion, G. C. (1870). Captures of Hemiptera–Heteroptera during 1869. *Entomologist's Monthly Magazine* 6, 232.

Champion, G. C. (1871). Note on the occurrence in Britain of *Corizus abutilon* Rossi, a species of Hemiptera–Heteroptera new to our lists. *Entomologist's Monthly Magazine* 7, 208.

China, W. E. (1927). *Eurygaster testudinaria* (Geoffroy), an addition to the list of British Heteroptera, with notes on the nomenclature of *E. maura* (L.), *E. borealis* Péneau and *E. meridionalis* Péneau. *Entomologist's Monthly Magazine* 63, 251–254.

China, W. E. (1941). Systematic notes on the British species of *Corizus* auct. (Hem., Coreidae). *Entomologist's Monthly Magazine* 77, 273–278.

Cirino, L. A. & Miller, C. W. (2017). Seasonal effects on the population, morphology and reproductive behaviour of *Narnia femorata* (Hemiptera: Coreidae). *Insects* 8, 13.

Cobb, M. (2000). Reading and writing *The book of nature*: Jan Swammerdam (1637–1680). *Endeavour* 24, 122–128.

Coe, R. L. (1953). Diptera Syrphidae. *Handbooks for the Identification of British Insects*, Vol. 10, part 1. Royal Entomological Society, London.

Cokl, A., Zunic, A. & Virant-Doberlet, M. (2011). Predatory bug *Picromerus bidens* communicates at different frequency levels. *Central European Journal of Biology* 6, 431–439.

Conradi, A. F. (1904). Variations in the protective value of the odoriferous secretions of some Heteroptera. *Science* 19, 393–394.

Constant, J. (2007). Note on coprophily and necrophily in the Hemiptera Heteroptera.

Bulletin de l'Institut Royal des Sciences Naturelles de Belgique 77, 107–112.

Costa, J. T. (2006). *The Other Insect Societies.* Belknap Press, Cambridge, MA.

Coulianos, C.-C. (1976). *Sciocoris homalonotus* Fieb. in Sweden, a shield bug (Hem.-Het., Pentatomidae) new to northern Europe. *Entomologisk Tidskrift* 97, 3–4.

Cumming, R. T. & Le Tirant, S. (2021). Drawing the Excalibur bug from the stone: adding credibility to the double-edged sword hypothesis of coreid evolution (Hemiptera: Coreidae). *ZooKeys* 1043, 117–131.

Curtis, J. (1823–40). *British Entomology: being illustrations and descriptions of the genera of insects found in Great Britain and Ireland: containing coloured figures from nature of the most rare and beautiful species, and in many instances of the plants upon which they are found.* 8 volumes. Printed for the author, London.

Curzon, E. R. (1871). *Stenocephalus agilis* in South Wales. *Entomologist's Monthly Magazine* 7, 157.

Czaja, J. (2012). The wing-to-wing coupling mechanism of Scutelleridae (Hemiptera: Heteroptera). *Zootaxa* 3198, 54–62.

Dahanukar, A., Hallem, E. A. & Carlson, J. R. (2005). Insect chemoreception. *Current Opinion in Neurobiology* 15, 423–430.

Dandy, J. E. (1969). *Watsonian Vice-counties of Great Britain.* Ray Society, London.

De Geer, C. (1773). *Mémoires pour servir a l'histoire des insectes.* Pierre Hesselberg, Stockholm, Vol. 3, pp. 168–171.

de Souza-Firmino, T. S., Alevi, K. C. C. & Itoyama, M. M. (2020). Chromosomal divergence and evolutionary inferences in Pentatomomorpha infraorder (Hemiptera, Heteroptera) based on chromosomal location of ribosomal genes. *PLOS ONE* 15, e0228631.

Dennis, D. S., Lavigne, R. J. & Dennis, J. G. (2010). Hemiptera (Heteroptera/Homoptera) as prey of robber flies (Diptera: Asilidae) with unpublished records. *Journal of the Entomological Research Society* 12, 27–47.

Denton, J. (2016). *Dicranocephalus medius* (Mulsant & Rey) (Hemiptera: Stenocephalidae) in North Hampshire (VC12). *British Journal of Entomology and Natural History* 29, 117.

Denton, J. (2022). Is *Podops inunctus* (Fabricius) (Hemiptera: Pentatomidae) myrmecophilous? *The Hemipterist* 9, 286.

Derjanschi, V. & Péricart, J. (2005). *Hémiptères Pentatomoidea Euro-Méditerranéens. Volume 1. Généralités systématique: première partie.* Faune de France 90. Fédération française des Sociétés de Sciences Naturelles, Paris. Second edition 2016.

Devillers, C. & Lupoli, R. (2016). *Elasmucha ferrugata* (Fabricius, 1787): observations de son développement sur Cassis (*Ribes nigrum* L.) (Hemiptera: Acanthosomatidae). *L'Entomologiste* 72, 329–332.

Distant, W. L. (1877). Descriptions of two new species of Hemiptera–Heteroptera from West Africa. *Entomologist's Monthly Magazine* 14, 62–63.

Distant, W. L. (1879). On some African species of the lepidopterous genus *Papilio. Proceedings of the Zoological Society* 1879, 647–649.

Distant, W. L. (1883). First report on the Rhynchota collected in Japan by Mr. George Lewis. *Transactions of the Entomological Society* 1883, 413–443, plates XIX–XX.

Distant, W. L. (1892). *A Naturalist in the Transvaal.* R. H. Porter, London.

Dolling, B. (2008). *Crocistethus waltlianus* – news from the supermarkets. *Het News* 11, 11.

Dolling, W. R. (1991). *The Hemiptera.* Oxford University Press, Oxford.

Donisthorpe, H. St J. K. (1927). *The Guests of British Ants: Their Habits and Life-Histories.*Routledge, London.

Donovan, E. (1792–1817). *The Natural History of British Insects: explaining them in their several states, with the periods of their transformations, their food, oeconomy, etc, together with the history of such minute insects as require investigation by the microscope: the whole illustrated by coloured figures, designed and executed from living specimens.* 16 volumes. Donovan & Rivington, London.

Douglas, J. W. (1869). Note on *Asiraca clavicornis,* Fa. *Entomologist's Monthly Magazine* 6, 162–163.

Douglas, J. W. & Scott, J. (1865). *The British Hemiptera. Vol. 1. Hemiptera – Heteroptera.* Ray Society, London.

Drickamer, L. C. & McPherson, J. E. (1992). Comparative aspects of mating behavior patterns in six species of stink bugs (Heteroptera: Pentatomidae). *The Great Lakes Entomologist* 25, 287–295.

Druce, H. (1880). Description of new species of Heterocera from West Africa. *Entomologist's Monthly Magazine* 16, 268–269.

Du, B.-J., Chen, R., Tao, W.-T. *et al.* (2020). A Cretaceous bug with exaggerated antennae might be a double-edged sword of evolution. *iScience* 24, 101932.

Dudley, R. (2000). *The Biomechanics of Insect Flight: Form, Function, Evolution.* Princeton University Press, Princeton, NJ.

Eberhard, W. G. (1975). The ecology and behaviour of a subsocial pentatomid bug and two scelionid wasps: strategy and counterstrategy in a host and its parasites. *Smithsonian Contributions to Zoology* 205, 1–39.

Edwards, J. (1896) *The Hemiptera – Homoptera of the British Islands.* L. Reeve and Co., London.

Eger, J. E., Brailovsky, H. & Henry, T. J. (2015). Heteroptera attracted to butterfly traps baited with fish or shrimp carrion. *Florida Entomologist* 98, 1030–1035.

Emberts, Z., St Mary, C. M. & Miller, C. W. (2016). Coreidae (Insecta: Hemiptera) limb loss and autotomy. *Annals of the Entomological Society of America* 109, 678–683.

Endo, N., Sasaki, R. & Muto, S. (2010). Pheromonal cross-attraction in true bugs (Heteroptera): attraction of *Piezodorus hybneri* (Pentatomidae) to its pheromone versus the pheromone of *Riptortus pedestris* (Alydidae). *Environmental Entomology* 39, 1973–1979.

Esquivel, J. F. (2019). Stink bug rostrum length vs stylet penetration potential. *Entomologia Experimentalis et Applicata* 167, 323–329.

Esquivel, J. F., Droleskey, R. E., Ward, L. A. & Harvey, R. B. (2018). Morphometrics of the southern green stink bug [*Nezara viridula* (L.) (Hemiptera: Pentatomidae)] stylet bundle. *Neotropical Entomology* 48, 78–86.

Eubanks, M. D., Styrsky, J. D. & Denno, R. F. (2003). The evolution of omnivory in heteropteran insects. *Ecology* 84, 2549–2556.

Evans, M. & Edmondson, R. (2005). *A Photographic Guide to the Shieldbugs and Squashbugs of the British Isles.* WGUK, Wakefield.

Exnerová, A., Štys, P., Krištín, A., Volf, O. & Pudil, M. (2003). Birds as predators of true bugs (Heteroptera) in different habitats. *Biologia (Bratislava)* 58, 253–264.

Fabre, J.-H. (1901). Les pentatomes. *Revue des Questions Scientifiques* 1, 158–176.

Fabricius, J. C. (1794). *Rhyngota. Entomologgica systematica ememattica emendate et aucta: secundun classes, ordines genera species, adjectis synonimis, locis, observationibus, descriptionibus.* Gottlob, Copenhagen.

Fabricius, J. C. (1803). *Systema Rhyngotorum.* Apud Carolum Reichard, Brunswick.

Falk, S. (1992). *A Review of the Scarce and Threatened Flies of Great Britain (Part 1).* JNCC, Peterborough.

Fan, S., Chen, C., Zhao, Q., Wei, J. & Zhang, H. (2020). Identifying potentially climatic suitability areas for *Arma custos* (Hemiptera: Pentatomidae) in China under climate change. *Insects* 11, 674.

Faúndez, E. I. & Carvajal, M. A. (2016). El género *Planois* Signoret, 1864 (Heteroptera: Ancanthosomatidae) en la Patagonia Chilena. *Anales Instituto Patagonia (Chile)* 44, 55–59.

Fischer, C. (2006). The biological context and evolution of Pendergrast's organs of Acanthosomatidae (Heteroptera: Pentatomoidea). *Denesia* 19, 1041–1054.

Fisher, H. L. & Watson, J. (2015). A fossil insect egg on an early Cretaceous conifer shoot from the Wealden of Germany. *Cretaceous Research* 53, 38–47.

Frantsevich, L. (1995). Optimal leg design in a hexapod walker. *Journal of Theoretical Biology* 175, 561–566.

Frantsevich, L., Mokrushov, P., Shumakova, I. & Gorb, S. (1996). Insect rope-walkers: kinematics of walking on thin rods in a bug, *Graphosoma italicum* (Heteroptera, Pentatomidae). *Journal of Zoology* 238, 713–724.

Genevcius, B. C. & Schwertner, F. (2017). Strong functional integration among multiple parts of complex male and female genitalia of stink bugs. *Biological Journal of the Linnean Society* 22, 1–13.

Gorb, S. N. & Perez Godwyn, P. J. (2004). Frictional properties of contacting surfaces in the hemelytra–hindwing locking mechanism in the bug *Coreus marginatus* (Heteroptera, Coreidae). *Journal of Comparative Physiology A* 190, 575–580.

Grimaldi, D. & Engel, M. S. (2005). *Evolution of the Insects.* Cambridge University Press, Cambridge.

Grosso-Silva, J. M. (2004). Contribuicao para a catalogacao e cartograpfia da fauna de Acanthosomatidae e Nabidae (Insecta, Hemiptera) de Portugal continental. *Bolletin de la Sociedad Entomologica Aragonesa* 34, 131–138.

Groves, E. W. (1964–89). Hemiptera–Heteroptera of the London area. *The London Naturalist* 43, 34–66; 44, 82–110; 45, 60–88; 46, 82–104; 47, 50–80; 48, 86–120; 50, 87–94; 52, 31–59; 54, 21–34; 55, 6–15; 56, 32–43; 61, 72–87; 62, 69–86; 63, 97–120; 64, 63–94; 65, 119–152.

Guilbert, E. (2003). Habitat use and maternal care of *Phloea subquadrata* (Hemiptera: Phloidae) in the Brasilian Atlantic forest (Espirito

Santo). *European Journal of Entomology* 100, 61–63.

Hall, C. G. (1890). Hemiptera–Heteroptera at Dover and its vicinity. *Entomologist's Monthly Magazine* 26, 81–82.

Hanelová, J. & Vilímová, J. (2013). Behaviour of the central European Acanthosomatidae (Hemiptera: Heteroptera: Pentatomoidea) during oviposition and parental care. *Acta Musei Moraviae, Scientiae Biologicae* 98, 433–457.

Haug, C. & Haug, J. T. (2017). The presumed oldest flying insect: more likely a myriapod? *PeerJ* 5: e3402.

Hawkins, R. D. (1989). *Orsillus depressus* Dallas (Hem., Lygaeidae) an arboreal groundbug new to Britain. *Entomologist's Monthly Magazine* 125, 241–242.

Hawkins, R. D. (2003). *Shieldbugs of Surrey*. Surrey Wildlife Trust, Pirbright.

Heckmann, R., Strass, G. & Rietschel, S. (2015). Die Heteropteranfauna Kretas. *Carolinea* 23, 83–130.

Heller, J. L. (1943). The etymology of *Aphis*. *Classical Weekly* 1 November 1943, 53–55.

Hellins, J. (1870). A fragment of a life-history of *Acanthosoma grisea*. *Entomologist's Monthly Magazine* 7, 53–55.

Hemala, V. & Hanzlik, V. (2015). First record of *Elasmucha ferrugata* (F.) (Hemiptera: Heteroptera: Acanthosomatidae) from Montenegro. *Ecologica Montenegrina* 2, 147–149.

Hischen, F., Buchberger, G., Plamadeala, C. *et al.* (2018). The external scent efferent system of selected European true bugs (Heteroptera): a biomimetic inspiration for passive unidirectional fluid transport. *Journal of the Royal Society Interface* 15, 20170975.

Hoebeke, R. & Carter, E. (2003). *Halyomorpha halys* (Stål) (Heteroptera: Pentatomidae): a polyphagous plant pest from Asia newly detected in North America. *Proceedings of the Entomological Society of Washington* 105, 225–237.

Hoefnagel, J. (1630). *Diversae insectarum volatilium icones ad vivum accuratissime depictae*. Visscher, Amsterdam.

Hoke, S. (1926). Preliminary paper on the wing-venation of the Hemiptera (Heteroptera). *Annals of the Entomological Society of America* 19, 13–29, plates I–V.

Honek, A., Martinkova, Z. & Pekár, S. (2020). How climate change affects occurrence of a second generation in the univoltine *Pyrrhocoris apterus* (Heteroptera: Pyrrhocoridae). *Ecological Entomology* 45, 1172–1179.

Hutchinson, G. E. (1974). Marginalia. Aposematic insects and the master of the Brussels initials. *American Scientist* 62, 161–171.

Iley, R. G. (2011). Southwood's Heteroptera collection. *British Journal of Entomology and Natural History* 24, 33–37.

Javahery, M. (1994). Development of eggs in some true bugs (Hemiptera–Heteroptera). Part 1. Pentatomoidea. *The Canadian Entomologist* 126, 401–433.

Johansen, A. I., Exnerová, A., Hotová Svádova, K. *et al.* (2010). Adaptive change in protective coloration in adult striated shieldbugs *Graphosoma lineatum* (Heteroptera: Pentatomidae): test of detectability of two colour forms by avian predators. *Ecological Entomology* 35, 602–610.

Johnson, K. P., Dietrich, C. H., Friedrich, F. *et al.* (2018). Phylogenomics and the evolution of hemipteroid insects. *Proceedings of the National Academy of Sciences of the United States of America* 115, 12775–12780.

Jones, R. A. (1978). *Aradus atterimus* Fieber (Hem., Aradidae) in Sussex. *Entomologist's Monthly Magazine* (1977) 113, 120.

Jones, R. A. (1991). [Exhibit at British Entomological and Natural History Society, 14 November 1990, not a beetle, but *Irochrotos maculiventris* (Germ.)!] *British Journal of Entomology and Natural History* 4, 56.

Jones, R. A. (1997). Life on the edge – a caution on the precise demarcation of Watsonian vice-county boundaries in the London Area. *London Naturalist* 76, 79–81.

Jones, R. A. (2000). The juniper mirid *Dichrooscytus gustavi* Josifov (Hem. : Miridae) found on Cypress. *Entomologist's Record and Journal of Variation* 112, 133–134.

Jones, R. A. (2003). *Gonocerus acuteangulatus* (Goeze) (Hemiptera: Coreidae) new to Kent. *British Journal of Entomology and Natural History* 16, 102.

Jones, R. A. (2004). *Brachycarenus tigrinus* (Schilling) (Hemiptera: Rhopalidae) new to Britain. *British Journal of Entomology and Natural History* 17, 137–141.

Jones, R. & Ure-Jones, C. (2021). *A Natural History of Insects in 100 Limericks*. Pelagic Publishing, Exeter.

Jordan, K. H. C. (1958). Die Biologie von *Elasmucha grisea* L. (Heteroptera: Acanthsomidae). *Beiträge zur Entomologie* 8, 385–397.

Judd, S. (2006). Status and distribution of the shieldbug *Odontoscelis fuliginosa* (L.)

and seedbug *Pionosomus varius* (Wolff) (Hemiptera: Heteroptera) associated with bare and partially-vegetated dunes on the Castlemartin Peninsula. *British Journal of Entomology and Natural History* 19, 97–103.

Judd, S. (2011). The scent-less plant bug *Liorhyssus hyalinus* (Hemiptera: Rhopalidae) – regular migrant or established British species? *British Journal of Entomology and Natural History* 24, 227–234.

Kirby, P. (1992). *A Review of the Scarce and Threatened Hemiptera of Great Britain.* JNCC, Peterborough.

Kirby, P., Stuart, A. J. A. & Wilson, M. R. (2001). True bugs, leaf- and plant-hoppers, and their allies. In: *The Changing Wildlife of Great Britain and Ireland* (ed. D. L. Hawksworth). Taylor & Francis, London, pp. 262–299.

Kirby, W. & Spence, W. (1815–26). *Introduction to Entomology, or, elements of the natural history of insects: comprising an account of noxious and useful insects, of their metamorphoses, food, stratagems, habitations etc.* 4 volumes. Longman, Brown, Green and Longman, London.

Kirkaldy, G. W. (1904a). Upon maternal solicitude in Rhynchota and other nonsocial insects. *Annual Report of the Smithsonian Institute* 1904, 577–585.

Kirkaldy, G. W. (1904b). Bibliographical and nomenclatorial notes on the Hemiptera. No. 3. *The Entomologist* 37, 279–283.

Kirkaldy, G. W. (1907). Biological notes on the Hemiptera of the Hawaiian Isles, No. 1. *Proceedings of the Hawaiian Entomological Society* 1, 135–161.

Kirkaldy, G. W. (1909). *Catalogue of the Hemiptera (Heteroptera) with biological and anatomical references, list of food plants and parasites etc. vol 1. Cimicidae* [=Pentatomidae]. F. L. Dames, Berlin.

Kment, P. & Vilímová, J. (2010). Thoracic scent efferent system of Pentatomoidea (Hemiptera: Heteroptera): a review of terminology. *Zootaxa* 2706, 1–77.

Konvicka, M., Cizek, O., Filipova, L. *et al.* (2005). For whom the bells toll: demography of the last population of the butterfly *Euphydryas maturna* in the Czech Republic. *Biologia* 60, 551–557.

Krenn, H. W. & Pass, G. (1994). Morphological diversity and phylogenetic analysis of wing circulatory organs in insects, part 1: non-Holometabola. *Zoology* 98, 7–22.

Kühl, G. & Rust, J. (2009). *Devonohexapodus bocksbergensis* is a synonym of *Wingertshellicus backesi* (Euarthropoda) – no evidence for marine hexapods living in the Devonian Hunsrück Sea. *Organisms Diversity and Evolution* 9, 215–231.

Kühn, A. C. (1775). Anecdoten zur Insecten Geschichte. Von einer mit den Bettwanzen anzustellenden Jagd. *Der Naturforscher* 6, 80–82.

Larivière, M.-C. & Larochelle, A. (1989). *Picromerus bidens* (Heteroptera: Pentatomidae) in North America, with a world review of distribution and bionomics. *Entomological News* 100, 133–146.

Latreille, P. A. (1810). *Considérations générales sur l'ordre naturel des animaux.* Schoell, Paris.

Leather, S. R. (2015). An entomological classic – the pooter or insect aspirator. *British Journal of Entomology and Natural History* 28, 52–54.

Lesieur, V., Lombaert, E., Guillemaud, T. *et al.* (2018). The rapid spread of *Leptoglossus occidentalis* in Europe: a bridgehead invasion. *Journal of Pest Science* 92, 189–200.

Leston, D. (1954). [Exhibit of female of *Birketsmithia anomala* at meeting of 3 March 1954]. *Proceedings of the Royal Entomological Society of London (C)* 19, 7.

Leston, D. (1957). The stridulatory mechanisms in terrestrial species of Hemiptera Heteroptera. *Proceedings of the Zoological Society of London* 137, 89–106.

Li, H., Shao, R., Song, N., Song, F. *et al.* (2014). Higher-level phylogeny of paraneopteran insects inferred from mitochondrial genome sequences. *Scientific Reports* 5, 8527.

Linnaeus, C. (1758). *Systema naturae per regna tria naturae, secundum classes, ordines, genera, species, cum characteribus, differentiis, synonymis, locis*, 10th edition. [1st edition 1735.] Alurentii Salvii, Stockholm.

Linnavuori, R. E. (2008). Studies on the Acanthosomatidae, Scutelleridae and Pentatomidae (Heteroptera) of Gilan and adjacent provinces in northern Iran. *Acta Entomologica Musei Nationalis Prague* 48, 1–21.

Lis, J. A. & Heyna, J. (2001). The metathoracic wing stridulation of the Cydnidae (Hemiptera: Heteroptera). *Polski Pismo Entomologizne* 70, 221–245.

Lis, J. A. & Webb, M. (2007). Redescription of the burrower bug *Adrisa sepulchralis* (Erichson, 1842) (Insecta: Hemiptera: Cydnidae), based on the only known male (recently introduced to the UK from Australia), and the lectotype from Tasmania. *Entomologist's Monthly Magazine* 143, 59–65.

Lis, J. A. & Whitehead, P. F. (2019). Another alien bug in Europe: the first case of transcontinental introduction of the Asiatic burrower bug *Macroscytus subaeneus* (Dallas, 1851) (Hemiptera: Heteroptera: Cydnidae) to the UK through maritime transport. *Zootaxa* 4555, 588–594.

Lister, M. (1671). Concerning an insect feeding upon henbain, the horrid smell of which is in that creature so qualified thereby, as to become in some measure aromatical, together with the colour yielded by the eggs of the same. *Philosophical Transactions giving some Accompt of the Present Undertakings, Studies, and Labours of the Ingenious in Many Considerable Parts of the World* 6, 2176–2177.

Liu, Y., Li, H., Song, F., Zhao, Y., Wilson, J. J. & Cai, W. (2019). Higher-level phylogeny and evolutionary history of Pentatomomorpha (Hemiptera: Heteroptera) inferred from mitochondrial genome sequences. *Systematic Entomology* 44, 810–819.

Lodos, N. & Önder, F. (1979). Contribution to the study on the Turkish Pentatomoidea (Heteroptera). IV. Family: Acanthosomatidae Stal 1864. *Türkiye Bitki Koruma Dergisi* 3, 139–160.

Lomholdt, O. (1984). *The Sphecidae (Hymeoptera) of Fennoscandia and Denmark.* Fauna Entomologica Scandinavica vol. 4, 2nd edition. Brill, Leiden.

Lo Verde, G. & Carapezza, A. (2018). An exceptional outbreak of *Macroscytus brunneus* (Fabricius, 1803) (Hemiptera: Heteroptera: Cydnidae) on Linosa Island (Pelagian Islands), Sicily. *Heteroptera Poloniae – Acta Faunistica* 12, 29–32.

Lucini, T. & Panizzi, A. R. (2018). Electropenetrography (EPG): a breakthrough tool unveiling stink bug (Pentatomidae) feeding on plants. *Neotropical Entomology* 47, 6–18.

Lupoli, R. & Dusoulier, F. (2015). *Les punaises Pentatomoidea de France.* Editions Ancyrosoma, Fontenay-sous-Bois.

Malumphy, C., Botting, J., Bantock, T. & Reid, S. (2008). Influx of *Leptoglossus occidentalis* Heidemann (Coreidae) in England. *Het News* 12, 7–9.

Mappes, J. & Kaitala, A. (1994). Experiments with *Elasmucha grisea* L. (Heterotomera: Acanthosomatidae): does a female parent bug lay as many eggs as she can defend? *Behavioural Ecology* 5, 314–317.

Mappes, J., Kaitala, A. & Alatalo, R. V. (1995). Joint brood guarding in parent bugs – an experiment on defence against predation. *Behavioural and Ecological Sociobiology* 36, 343–347.

Massee, A. M. (1937). *The Pests of Fruit and Hops.* Crosby Lockwood, London.

Massee, A. M. (1955). The county distribution of the British Heteroptera, second edition. *Entomologist's Monthly Magazine* 91, 7–27.

Massee, A. M. (1963). The Hemiptera–Heteroptera of Kent. 2. *Proceedings of the South London Entomological and Natural History Society* 1962, 123–183.

McLean, D. L. & Kinsey, M. G. (1964). A technique for electronically recording aphid feeding and salivation. *Nature* 202, 1358–1359.

McPherson, J. E., Packauskas, R. J., Taylor, S. J. & O'Brien, M. F. (1990). Eastern range extension of *Leptoglossus occidentalis* with a key to *Leptoglossus* species of America north of Mexico (Heteroptera: Coreidae). *The Great Lakes Entomologist* 23, 99–104.

Merian, M. S. (1705). *Metamorphosis insectarium surinamensis ofte verandering der surinaamsche insecten.* Gerarde Valck, Amsterdam.

Miller, N. C. E. (1956). *The Biology of the Heteroptera.* Leonard Hill, London.

Modéer, A. (1764). Några måkvärdigheter hos insectet Cimex ovatus pallidegriseus, abdominis lateribus albo nigroque variis albis, basi scutelli nigricante. *Kongliga Svenska Vetenskaps Akademiens Handlingar* 25, 41–57.

Montaga, M., Strada, L., Dioli, P. & Tintori, A. (2018). The Middle Triassic Lagetstätte of Monte San Giorgio reveals the oldest lace bugs (Hemiptera: Tingidae): *Archetingis ladinica* gen. n. sp. n. *Rivista Italiana di Paleontologia e Stratigrafia* 124, 35–44.

Monteith, G. B. (1982). Dry season aggregations of insects in Australian monsoon forests. *Memoirs of the Queensland Museum* 20, 533–543.

Moraes, M. C. B., Pareja, M., Laumann, R. A. & Borges, M. (2008). The chemical volatiles (semiochemicals) produced by neotropical stink bugs (Hemiptera: Pentatomidae). *Neotropical Entomology* 37, 489–505.

Mouffet, T. (1658). *The Theater of Insects: or, lesser living creatures, as bees, flies, caterpillars, spidrs* [sic], *worms, &c. a most elaborate work.* E. Cotes, London.

Moulet, P. (1995). *Hémiptères Coreoidea (Coreidae, Rhopalidae, Alydidae), Pyrrhocoridae, Stenoceophalidae Euro-Méditerranéens.* Faune de France. Fédération française des Sociétés de Sciences Naturelles, Paris.

Musolin, D. L. (2011). Life-history responses to simulated climate warming of *Nezara viridula*. *Het News* 17/18, 10–13.

Musolin, D. L. & Fujisaki, K. (2006). Changes in ranges: trends in distribution of true bugs (Heteroptera) under conditions of the current climate warming. *Russian Entomological Journal* 15, 175–179.

Musolin, D. L. & Saulich, A. K. (1996). Photoperiod control of seasonal development in bugs (Heteroptera). *Entomological Review* 76, 849–864.

Nakamura, K. (1990). Maternal care and survival in a Sumatran bug, *Physomerus grossipes* (Hemiptera, Coreidae). In: *Natural History of Social Wasps and Bees in Equatorial Sumatra* (ed. S. F. Sakagami, R. Phgushi & D. W. Roubik). Hokkaido University Press, Sapporo, pp. 233–243.

Nau, B. S. (1996). Shield-bugs in Bedfordshire. *Bedfordshire Naturalist* 50, 63–70.

Nau, B. S. (2004). *Shieldbugs of the British Isles.* Field Studies Council, Shrewsbury.

Nau, B. S., Merrifield, R. K. & Merrifield, R. M. (2014). *Dyroderes umbraculatus* (Heteroptera: Pentatomidae: Sciocorini) new to Britain, and related possible migrants. *British Journal of Entomology and Natural History* 27, 13–19.

Nielsen, A. I. & Hamilton, G. C. (2009). Life history of the invasive species *Halyomorpha halys* (Hemiptera: Pentatomidae) in northeastern United States. *Annals of the Entomological Society of America* 102, 608–616.

Nielsen, O. F. & Skipper, L. (2015). *Danmarks bredtæger, randtæger or ildtæger.* Apollo Books, Ollerup.

Noge, K., Prudic, K. L. & Becerra, J. X. (2012). Defensive roles of (E)-2-alkenals and related compounds in Heteroptera. *Journal of Chemical Ecology* 38, 1050–1056.

Ogur, E. & Tuncer, C. (2019). Chemical analysis of the metathoracic scent gland of *Eurygaster maura* (L.) (Heteroptera: Scutelleridae). *Journal of Agricultural Science and Technology* 21, 1473–1484.

Oliveira, P. S. (1985). On the mimetic association between nymphs of *Hyalymenus* spp. (Hemiptera: Alydidae) and ants. *Zoological Journal of the Linnean Society* 83, 371–384.

Ota, D. & Cokl, A. (1991). Mate location in the southern green stink bug *Nezara viridula* (Heteroptera: Pentatomidae), mediated through substrate-borne signals on ivy. *Journal of Insect Behaviour* 4, 441–447.

Packauskas, R. J. & Schaefer, C. W. (1998). Revision of the Cyrtocoridae (Hemiptera: Pentatomoidea). *Annals of the Entomological Society of America* 91, 363–386.

Papeschi, A. G. & Bressa, M. J. (2006). Evolutionary cytogenetics in Heteroptera. *Journal of Biological Research* 5, 3–21.

Pass, G. (2018). Beyond aerodynamics: the critical roles of circulatory and tracheal systems in maintaining wing functionality. *Arthropod Structure and Development* 47, 391–407.

Penney, D. & Jepson, J. E. (2014). *Fossil Insects: an Introduction to Palaeoentomology.* Siri Scientific Press, Manchester.

Péricart, J. (1998). *Hémiptères Lygaeidae Euro-Méditerranéens.* 3 volumes. Fédération Française des Sociétés de Science Naturelles, Paris.

Péricart, J. (2010). *Hémiptères Pentatomoidea Euro-Méditerranéens. Volume 3. Systématique: troisième partie, sous-familles Podopinae et Asopinae.* Faune de France 93. Fédération française des Sociétés de Sciences Naturelles, Paris.

Perry, I. (2006). *Opesia grandis* (Egger, 1860) (Diptera, Tachinidae) new to Britain. *Dipterists Digest* 13, 93–95.

Pinzari, M., Ciaferoni, F., Fabiana, A. & Dioli, P. (2019). Predation by nymphs of *Picromerus bidens* (Heteroptera Pentatomidae Asopinae) on caterpillars of *Euphydryas aurinia provincialis* (Lepidoptera Nymphalidae) in Italy. *Journal of Zoology* 102, 89–94.

Poe, E. A. (1843) The gold-bug. *Dollar Newspaper (Philadelphia).*

Poinar, G. P. Jr & Chambers, K. L. (2016). Mimosoideae (Fabaceae) diversity and associates in mid-Tertiary Dominican amber. *Journal of the Botanical Research Institute of Texas* 10, 121–136.

Poinar, G. Jr & Thomas, D. B. (2011). A stink bug, *Edessa protera* sp. n. (Pentatomidae: Edessinae) in Mexican amber. *Historical Biology* 2011, 1–5.

Powell, G. (2020). The biology and control of an emerging shield bug pest, *Pentatoma rufipes* (L.) (Hemiptera: Pentatomidae). *Agricultural and Forest Entomology* 22, 298–308.

Powell, G., Barclay, M. V. L., Couch, Y. & Evans, K. A. (2021). Current invasion status and potential for UK establishment of the brown marmorated stink bug, *Halyomorpha halys* (Hemiptera: Pentatomidae). *British Journal of Entomology and Natural History* 34, 9–21.

Ramsay, A. J. (2013). Coprophagous feeding behaviour of two species of nymphal pentatomid. *British Journal of Entomology and Natural History* 26, 145–147.

Ramsay, A. J. (2014). The history and status of the hawthorn shieldbug *Acanthosoma haemorrhoidale* (Hemiptera: Acanthosomatidae) in Scotland, 1946–2008. *British Journal of Entomology and Natural History* 27, 81–91.

Ramsay, A. J. (2016). Nocturnal mating behaviour in *Pentatoma rufipes* (L.) (Hemiptera: Pentatomidae). *British Journal of Entomology and Natural History* 29, 40.

Ramsay, A. J. (2019). A review of Scottish records of *Adomerus biguttatus* (L.) (Hemiptera: Cydnidae) and comments on distribution in Scotland. *British Journal of Entomology and Natural History* 32, 321–325.

Ratzlaff, C. & Scudder, G. G. E. (2018). First records of the juniper shield bug, *Cyphostethus tristriatus* (Fabricius, 1787) (Hemiptera: Acanthosomatidae), in North America. *The Pan-Pacific Entomologist* 94, 67–74.

Ray, J. (1710). *Historia insectorum.* Churchill, London.

Reichling, L. (1988). Punaises des genévriers trouvées sur faux cyprès (Heteropt.). *L'Entomologiste* 44, 46.

Reinhardt, K. (2018). *Bedbug.* Reaktion Books, London.

Ribes, J. & Pagola-Carte, S. (2013). *Hémiptères Pentatomoidea Euro-Méditerranéens. Volume 2. Systématique: deuxième partie, sous-famille Pentatominae (suite et fin).* Faune de France 96. Fédération française des Sociétés de Sciences Naturelles, Paris.

Rider, D. A. (2000). Stirotarsinae, a new subfamily for *Stirotarsus abnormis* Bergroth (Heteroptera: Pentatomidae). *Annals of the Entomological Society of America* 93, 802–806.

Rintala, T. & Rinne, V. (2011). *Suomen luteet.* Tibiale, Helsinki.

Rogers, D. R. (1642). *Matrimoniall honour, or, the mutuall crowne and comfort of godly, loyal, and chaste marriage wherein the right way to preserve the honour of marriage unstained, is at large described, urged and applied: with resolution of sundry materiall questions concerning this argument.* Th. Harper for Philip Nevel, London.

Ryan, R. P. (2016). The heather shieldbug, *Rhacognathus punctatus* (L.) (Hemiptera: Pentatomidae) new to Oxfordshire (VC23). *British Journal of Entomology and Natural History* 29, 43–44.

Ryan, R. P. (2019). The most widespread members of the Hemiptera–Heteroptera in the British Isles. *British Journal of Entomology and Natural History* 32, 239–240.

Salcedo, M. K. & Socha, J. J. (2020). Circulation in insect wings. *Integrative and Comparative Biology* 60, 1208–1220.

Salerno, G., Rebora, M., Gorb, E. & Gorb, S. (2018). Attachment ability of the polyphagous bug *Nezara viridula* (Heteroptera: Pentatomidae) to different host plant surfaces. *Scientific Reports* 8, 10975.

Salisbury, A., Barclay, M. V. L., Reid, S. & Halstead, A. (2009). The current status of the southern green shield bug, *Nezara viridula* (Hemiptera: Pentatomidae), an introduced pest species recently established in south-east England. *British Journal of Entomology and Natural History* 22, 189–194.

Salvetti, M. & Dioli, P. (2015). Prima segnalazione in Lombardia di *Elasmucha ferrugata* (Fabricius, 1787) (Hemiptera: Heteroptera: Acanthosomatidae) e note sulla distribuzione della specie in Italia e in Europa. *Il Naturalista Valtellinese* 20, 113–118.

Saulich, A. K. & Musolin, D. L. (2012). Diapause in the seasonal cycle of stink bugs (Heteroptera, Pentatomidae) from the temperate zone. *Entomological Review* 92, 1–26.

Saunders, E. (1875). British Hemiptera – an additional species. *Entomologist's Monthly Magazine* 12, 154.

Saunders, E. (1875–76). Synopsis of British Hemiptera–Heteroptera. *Transactions of the Entomological Society of London* 1875, 117–159, 245–309; 1876, 613–655.

Saunders, E. (1892). *The Hemiptera Heteroptera of the British Islands. A Descriptive Account of the Families, Genera, and Species Indigenous to Great Britain and Ireland, with Notes as to Localities, Habitats, etc.* London: L. Reeve.

Saunders, E. (1900). *Peribalus vernalis*, Wolff, in Slindon Woods, Sussex. *Entomologist's Monthly Magazine* 36, 132.

Saunders, E. (1902). *Heteroptera. A History of the County of Surrey. 3 Zoology.* Constable, London.

Schlee, M. A. (1986). Avian predation on Heteroptera: experiments on the European blackbird *Turdus m. merula* L. *Ethology* 73, 1–18.

Schuh, R. T. & Weirauch, C. (2020). *True Bugs of the World (Hemiptera: Heteroptera): Classification and Natural History*, 2nd edition. Siri Scientific Press, Manchester.

Schulz, K. (2018). When twenty-six thousand stinkbugs invade your home. *New Yorker*, March 12.

Scott, J. (1862). On Hemiptera, commonly called bugs. *Entomologist's Annual* 1862, 150–157.

Scott, J. (1873). Notes on the capture of *Pentatoma juniperina*. *Entomologist's Monthly Magazine* 9, 292.

Scott, J. K. & Yeo, P. B. (1996). Assessment of potential biological control insects associated with *Emex spinosa*. *Plant Protection Quarterly* 11, 165–167.

Shardlow, M. E. A. & Taylor, R. (2004). Is the southern green shieldbug, *Nezara viridula* (L.) (Hemiptera: Pentatomidae), another species colonising Britain due to climate change? *British Journal of Entomology and Natural History* 17, 143–146.

Sharp, D. (1909). *The Cambridge Natural History. Insects, Part 2*. Macmillan, London.

Sharp, W. E. (1900). *Elasmostethus ferrugatus*, Fab., in Wales. *Entomologist's Monthly Magazine* 36, 131.

Shcherbakov, D. E. (2010). The earliest true bugs and aphids from the Middle Triassic of France (Hemiptera). *Russian Entomological Journal* 19, 179–182.

Shirt, D. B. (ed.) (1987). *British Red Data Books: 2. Insects*. Nature Conservancy Council, Peterborough.

Silva, C. C. A., de Capdeville, G., Moraes, M. C. B. *et al.* (2010). Morphology, distribution and abundance of antennal sensilla in three stink bug species (Hemiptera: Pentatomidae). *Micron* 41, 289–300.

Slade, D., Collins, A. R. & Nau, B. S. (2005). *Eurydema ornatum* (L.) (Hem. : Pentatomidae) established on the Dorset coast and a key to European *Eurydema* species. *Entomologist's Record and Journal of Variation* 117, 221–227.

Southall, J. (1730). *A Treatise of Buggs. Shewing when and how they were first brought into England. How they are brought into and infect houses. Their nature, several foods, times and manner of spawning and propagating in this climate. etc.* J. Roberts, London.

Southampton Natural History Society (2007). Shieldbugs of Southampton. https://sotonnhs.net/wp-content/uploads/Documents/Survey-Shieldbugs.pdf (accessed December 2022).

Southwood, T. R. E. (1956). The structure of the eggs of the terrestrial Heteroptera and its relationship to the classification of the group. *Transactions of the Royal Entomological Society of London* 108, 163–221.

Southwood, T. R. E. (1963). *Chamaecyparis nootkatensis* (D. Don) Spach. A new host plant for *Cyphostethus tristriatus* (F.) (Hem., Acanthosomatidae). *Entomologist's Monthly Magazine* 98 (1962), 250.

Southwood, T. R. E. & Leston, D. (1959). *Land and Water Bugs of the British Isles*. Frederick Warne & Co., London.

Staddon, B. W. (1979). The scent glands of Heteroptera. *Advances in Insect Physiology* 14, 351–418.

Stainton, H. T. (1861). A list of British Hemiptera. *Entomologist's Annual* 1861, 47–51.

Stehlik, J. L. (1998). The heteropteran fauna of introduced Cupressaceae in the southern part of Moravia (Czech Republic). *Acta Musei Moraviae, Scientiae Biologicae* 82 (1997), 127–155.

Stephens, J. F. (1829). *A Systematic Catalogue of British insects: being an attempt to arrange all the hitherto discovered indigenous insects in accordance with the natural affinities. Containing also the references to every English writer on entomology, and to the principal foreign authors, with all the published British genera to the present time*. Baldwin and Cradock, London.

Stevens, N. M. (1905). *Studies in Spermatogenesis, with Especial Reference to the Accessory Chromosome*. Carnegie Institution, Washington, DC.

Strekopytov, S. (2021). Corrosive sublimate and its introduction as an insecticide for preserving natural history specimens in the eighteenth century. *Archives of Natural History* 48, 22–41.

Strid, T. & Forshage, M. (2012). *Bärfisar i Sverige – en fälthandbok*. Malmö: Entomologiska föreningen i Stockholm.

Stubbs, A. E. & Falk, S. J. (1983). *British Hoverflies: an Illustrated Identification Guide*. British Entomological and Natural History Society, Reading. Second edition, 2002.

Stubbs, A. E., Drake, M. & Wilson, D. (2001). *British Soldierflies and Their Allies: an Illustrated Identification Guide*. British Entomological and Natural History Society, Reading.

Swammerdam, J. (1758). *The Book of Nature, or, the History of Insects*. Seyffert, London.

Sweet, M. H. (1979). On the original feeding habits of the Hemiptera (Insecta). *Annals of the Entomological Society of America* 72, 575–579.

Szwedo, J. (2017). The unity, diversity and conformity of bugs (Hemiptera) through time. *Earth and Environmental Science: Transactions of the Royal Society of Edinburgh* 107, 109–128.

Takai, R., Yamaguchi, T. & Kurihara, T. (1975). Mass occurrence of *Aethus indicus* (Hem. Cydnidae, Heter.) as a house frequenting pest in the Amami Islands (Japan). *Japanese Journal of Sanitary Zoology* 26, 61–63.

Tietz, D. & Zrzavy, J. (1996). Dorsoventral pattern formation: morphogenesis of longitudinal coloration in *Graphosoma lineatum* (Heteroptera: Pentatomidae). *European Journal of Entomology* 93, 15–22.

Todd, J. W. (1989). Ecology and behaviour of *Nezara viridula*. *Annual Review of Entomology* 34, 273–292.

Tolsgaard, S. (2001). Status over danske bredtaeger, randtaeger og ildtaeger (Heteroptera: Pentatomoidea, Coreoidea & Pyrrhocoridea). *Entomologiske Meddelelser* 69, 3–46.

Tolsgaard, S. & Jensen, J. K. (2010). New records of true bugs (Heteroptera) on the Faroe Islands. *Entomologiske Meddelelser* 78, 21–28.

Vesely, P., Veselá, S., Fuchs, R. & Zrzavy, J. (2006). Are gregarious red-black shieldbugs *Graphosoma lineatum* (Hemiptera: Pentatomidae) really aposematic? An experimental approach. *Evolutionary Ecology Research* 8, 881–890.

Villiers, A. (1951). *Atlas des hémiptères de France. I. Hétéroptères Gymnocérates.* Boubée, Paris.

Virant-Doberlet, M. & Cokl, A. (2004). Vibrational communication in insects. *Neotropical Entomology* 33, 121–134.

Walker, D. & Hollamby, G. (2020). The discovery and sightings of the shieldbug *Geotomus petiti* (Hemiptera: Cydnidae) at Dungeness Kent in 2019. *British Journal of Entomology and Natural History* 33, 133–136.

Wang, Y., Du, S., Yao, Y. & Ren, D. (2019). A new genus and species of burrower bugs (Heteroptera: Cydnidae) from the mid-Cretaceous Burmese amber. *Zootaxa* 4585, 351–359.

Waterhouse, C. O. (1882–90). *Aid to the Identification of Insects*, Vol. 2. Janson, London.

Wedman, S., Kment, P., Campos, L. A. & Hörnschemeyer, T. (2021). Bizarre morphology in extinct Eocene bugs (Hemiptera: Pentatomidae). *Royal Society Open Science* 8, 211466.

Weirauch, C. & Schuh, R. T. (2011). Systematics and evolution of Heteroptera: 25 years of progress. *Annual Review of Entomology* 56, 487–510.

Weiss, H. B. (1931). John Southall's 'Treatise of buggs'. *Journal of the New York Entomological Society* 39, 253–260.

Werner, D. J. (2002). Die Verbreitung der Bauchkielwanze *Cyphostethus tristriatus* (Heteroptera: Acanthosomatidae) an Zypressengewächsen (Cupressaceae) in Deutschland. *Heteropteron* 14, 7–25.

Westwood, J. O. (1839–40). *An Introduction to the Modern Classification of Insects: Founded on their Natural Habits and Corresponding Organisation of the Different Families*, Vols 1–2. Longman, Orme, Brown, Green and Longmans, London.

White, F. B. (1877). New and rare Hemiptera observed during the years 1874, 1875, 1876. *The Entomologist* 10, 9–15.

White, F. B. (1883). Report on the pelagic Hemiptera procured during the voyage of H.M.S. *Challenger*, in the years 1873–1876. *Zoology* 7, 1–85. HMSO, London.

Wilson, D. M. (1966). Insect walking. *Annual Review of Entomology* 11, 103–122.

Wilson, E. B. (1905). The chromosomes in relation to the determination of sex in insects. *Science* 22, 500–502.

Wootton, R. J. & Betts, C. R. (1986). Homology and function in the wings of Heteroptera. *Systematic Entomology* 11, 389–400.

Wyatt, A. K. (1926). Obituary [E. A. Butler, and others]. *Entomological News* 37, 126.

Yoshizawa, K. & Lienhard, C. (2016). Bridging the gap between chewing and sucking in the hemipteroid insects: new insights from Cretaceous amber. *Zootaxa* 4079, 229–245.

Zhang, J., Zhang, X., Liu, C., Meng, L. & Zhou, Y. (2014). Fine structure and distribution of antennal sensilla of stink bug *Arma chinensis* (Heteroptera: Pentatomidae). *Entomological Fennica* 25, 186–198.

Zorovic, M. (2011). Temporal processing of vibratory communication signals at the level of ascending interneurons in *Nezara viridula* (Hemiptera: Pentatomidae). *PLOS ONE* 6, 1–8.

Picture Credits

All photographs by the author with the exception of those listed below. References are to page numbers. Multiple images on a page are referenced with numbers in parentheses, left to right and top to bottom.

Baldin, Sybil: 181(3). **Bantock, Tristan:** 11, 62(2), 63(2), 115(1), 119(2), 119(3), 252(1), 252(2), 320(2), 331(2), 331(3), 350(2). **British Library:** 147, 68(1), 68(2), 68(3), 140, 302(1), 302(2), 339(2), 347(2), 347(3), 388, 389. **Couch, Yvonne:** 51(2). **Covey, Steve:** 243. **Du *et al.*:** 130. **Featherstone, Alan Watson:** 99(3). **George, Will:** 239. **Harding, Ian:** 390. **Heijerman:** 246(1). **Justamond Maria:** 21(2), 22, 51(1), 56, 62(1), 65, 67(2), 72, 78, 85, 86, 369(2), 373(2), 391. **Lyons, Graeme:** 313(1), 376(1), 376(2). **Metal, Penny:** 2(1), 2(2), 2(3), 2(4), 2(5), 2(6), 63(1), 112, 120, 175, 245(1), 329(2). **Metropolitan Museum:** 179(2). **Oliver, Simon:** 241(1). **Rijksmuseum:** 150(1), 150(2), 179(1). **Robson, Simon:** 148, 238, 270(1), 270(2), 280, 287(1), 299(2), 307(2), 330(1), 330(2), 352(2), 357(1), 359. **Russel, Jony:** 59(2). **Shutterstock:** Abhishek Raviya 30(1); Alen thien 14; alslutsky 188(1), 188(3), 189(4), 190(3), 192(10), 192(2), 192(5), 192(6), 192(8), 192(9), 193(2), 193(3), 194(2), 194(4), 194(5), 199(2), 200(6), 200(7), 201(3), 201(4), 201(7), 202(2), 202(4), 203(1), 203(2), 203(3), 203(5), 203(7), 205(1), 205(2), 205(5), 206(4), 206(8), 207(2), 208(3), 208(5), 209(2), 209(3), 209(4), 209(6), 210(2), 210(3), 210(4), 210(5), 211(1), 211(2), 215(1), 215(3), 215(5), 217(6), 218(1), 218(4), 219(2), 219(3), 220(3), 220(4), 221(2), 221(3), 221(6), 221(7), 222(3), 222(7), 222(8), 223(5), 224(4), 226, 250(1), 250(3), 251, 256, 259(1), 260, 263(1), 264, 269, 282(1), 282(2), 284, 285(1), 285(3), 286, 289(1), 290(1), 290(2), 291(2), 297(2), 299(1), 300(1), 300(2), 306, 318(2), 320(1), 321(1), 324, 325(1), 325(2), 326, 327(1), 327(2), 332, 333(2), 338(1), 338(2), 342(1), 347(1), 348, 349(2), 351, 354, 357(2), 361, 362(2); Amy Lutz 8(3); Andrea Izzotti 111; Anton Kozyrev 191(1), 191(3), 191(4), 191(5), 192(1), 199(1), 200(5), 203(6), 205(6), 206(3), 206(5), 207(1), 208(1), 208(2), 209(9), 211(6), 212(2), 213(3), 214(1), 214(2), 215(2), 215(4), 221(5), 222(6), 248, 261(2), 268(1), 272, 294(1), 296(1), 303(1), 308, 322, 336(1), 341(1), 360(1), 368, 370; asturfauna 358(1); Bildagentur Zoonar GmbH 108; Branislav Ivcic 117; ColleenSlater Photography 84; D. Kucharski and K. Kucharska 18; Dan Olsen 297(1); Danny Radius 181(2); Dark Egg 274(3); Davide Bonora 271, 273(1); DeRebus 318(1); Dirk Daniel Mann 369(1); Eric Isselee 262(2), 336(2); ErikaFoto 255(1); Ernie Cooper 8(5); Ferenc Speder 265(2); goran_safarek 310(3); Grigorii Pisotsckii 344(1); gstalker 191(6), 192(7), 200(4), 201(8), 202(5), 203(4), 206(7), 209(8), 211(4), 219(4), 221(4), 254, 273(2), 292, 335(1), 344(3); gubernat 323(2); Guillermo Guerao Serra 345(3), 349(1); HartmutMorgenthal 366(2); Henri Koskinen 227(3); Henrik Larsson 34(1), 115(2), 233(2), 262(1); Hhelene 232; Hiske Boon 283(2); Holger Kirk 352(1); Hwall 13(1), 30(2), 267(1), 277(2), 279(2), 283(1), 283(3), 288(2), 311(2), 339(1), 342(2), 342(3), 358(2), 362(1); Ihor Hvozdetskyi 34(2) ,231(2), 310(1); imageBROKER.com 227(2), 315(1); Ireneusz Waledzik 28; Jgade 8(2); Joao Clerigo 8(4); JorgeOrtiz_1976 310(2); Keith Hider 249(1), 249(2); LABETAA Andre 307(1), 360(2); Macronatura.es 211(5), 212(1), 267(3), 274(1), 321(2); Marek Velechovsky 317(1); Milan Vachal 323(1); Mironmax Studio 291(3), 91; MrKawa 345(1); Muddy knees 283(4); NattapanKansena 188(2); Olga Shum 21(3), 303(2); Ox Karol 255(2); Paco Moreno 328; pingphoto17 15(2); Protasov AN 195(6), 316; Raw360 235(2); Reinier

Blok 188(4), 196(6), 200(8), 201(5), 202(1), 209(7), 211(3), 231(1), 257(2), 277(1); Rose Ludwig 263(2); Skyprayer2005 15(3); Squarelens 49; sruilk 211(8), 212(3), 270(3); Texturis 227(1); thatmacroguy 229(1), 247, 296(3); Thijs de Graaf 235(1), 258, 281(3), 296(2); Tomasz Klejdysz 265(1), 317(2); Vinicius R. Souza 204(6), 253; Wirestock Creators 13(2), 189(2), 194(3), 195(2), 196(1), 201(6), 202(3), 203(8), 206(6), 209(10), 213(2), 228, 234, 246(2), 250(2), 261(3), 274(2), 277(3), 281(1), 281(2), 288(1), 291(1), 293(1), 294(2), 295, 298, 304(1), 314, 315(2), 335(2), 341(2). **Slade, David:** 143. **Ure-Jones, Lillian:** 8(6), 48. **van der Noll, Leon:** 133. **Vincent, Sue:** 174. **Warry, Simon:** 83, 142, 241(3), 373(1), 393. **Wikimedia:** Accipiter (CC-BY-SA-3.0) (https://commons.wikimedia.org/wiki/File:Chorosoma_schillingii_23_080830.jpg) 364(1); AfroBrazilian (CC-BY-SA-4.0) (https://commons.wikimedia.org/wiki/File:Aradus_depressus_01.JPG) 12; AfroBrazilian (CC-BY-SA-4.0) (https://commons.wikimedia.org/wiki/File:Coriomeris_scabricornis_10.JPG) 340; AfroBrazilian (CC-BY-SA-4.0) (https://commons.wikimedia.org/wiki/File:Dicranocephalus_medius_02.JPG 367; AfroBrazilian (CC-BY-SA-4.0) (https://commons.wikimedia.org/wiki/File:Elasmucha_fieberi_01.JPG) 237(1); AfroBrazilian (CC-BY-SA-4.0) (https://commons.wikimedia.org/wiki/File:Gastrodes_abietum_03.JPG) 20; AfroBrazilian (CC-BY-SA-4.0) (https://commons.wikimedia.org/wiki/File:Jalla_dumosa_01.JPG) 257(1); AfroBrazilian (CC-BY-SA-4.0) (https://commons.wikimedia.org/wiki/File:Myrmus_miriformis_02.JPG) 365(2); AfroBrazilian (CC-BY-SA-4.0) (https://commons.wikimedia.org/wiki/File:Neottiglossa_pusilla_02.JPG) 268(2); AfroBrazilian (CC-BY-SA-4.0) (https://commons.wikimedia.org/wiki/File:Odontoscelis_fuliginosa_02.JPG) 319; AfroBrazilian (CC-BY-SA-4.0) (https://commons.wikimedia.org/wiki/File:Rhopalus_maculatus_01.JPG) 355; AfroBrazilian (CC-BY-SA-4.0) (https://commons.wikimedia.org/wiki/File:Rhopalus_parumpunctatus_01.JPG) 356(2); AfroBrazilian (CC-BY-SA-4.0) (https://commons.wikimedia.org/wiki/File:Sciocoris_cursitans_03.JPG) 300(3); Alciphron-Enka (CC-BY-SA-4.0) (https://commons.wikimedia.org/wiki/File:Eurydema_ventralis_kopula_Enka.jpg) 309(1); Aleksandar Đukić (CC-BY-SA-4.0) (https://commons.wikimedia.org/wiki/File:Chorosoma_schillingii_-_nimfa_2.jpg) 364(2); Anatoly Mikhaltsov (CC-BY-SA-4.0) (https://commons.wikimedia.org/wiki/File:Нимфа_клопа_Microvelia_reticulata.jpg) 189(10); B. Schoenmakers (CC-BY-3.0) (https://commons.wikimedia.org/wiki/File:Eysarcoris_venustissimus_(Pentatomidae)_-_(imago),_Elst_(Gld),_the_Netherlands.jpg) 208(4), 209(1); B. Schoenmakers (CC-BY-3.0) (https://commons.wikimedia.org/wiki/File:Eysarcoris_venustissimus_(Pentatomidae)_-_(imago),_Elst_(Gld),_the_Netherlands.jpg) 287(2); B. Schoenmakers (CC-BY-3.0) (https://commons.wikimedia.org/wiki/File:Holcogaster_fibulata_(Pentatomidae)_-_(imago),_Molenhoek,_the_Netherlands_-_4.jpg) 211(7), 213(1); B. Schoenmakers (CC-BY-3.0) (https://commons.wikimedia.org/wiki/File:Holcogaster_fibulata_(Pentatomidae)_-_(imago),_Molenhoek,_the_Netherlands_-_4.jpg) 279(1); B. Schoenmakers (CC-BY-3.0) (https://commons.wikimedia.org/wiki/File:Myrmus_miriformis_(Rhopalidae)_-_(female_imago),_Molenhoek,_the_Netherlands.jpg) 214(6); B. Schoenmakers (CC-BY-3.0) (https://commons.wikimedia.org/wiki/File:Myrmus_miriformis_(Rhopalidae)_-_(female_imago),_Molenhoek,_the_Netherlands.jpg) 366(1); B. Schoenmakers (CC-BY-3.0) (https://commons.wikimedia.org/wiki/File:Myrmus_miriformis_(Rhopalidae)_-_(male_imago_brachypteer),_Molenhoek,_the_Netherlands.jpg) 214(7); B. Schoenmakers (CC-BY-3.0) (https://commons.wikimedia.org/wiki/File:Myrmus_miriformis_(Rhopalidae)_-_(male_imago_brachypteer),_Molenhoek,_the_Netherlands.jpg) 365(1); Cathy Dzerefos (CC-BY-SA-3.0) (https://commons.wikimedia.org/wiki/File:Encosternum_delegorguei_harvesting_2.jpg) 16; Darius Baužys (CC-BY-SA-3.0) (https://commons.wikimedia.org/wiki/File:Eysarcoris_aeneus_LT.jpg) 285(2); David Short (CC-BY-2.0) (https://commons.wikimedia.org/wiki/File:Shieldbug_family_(BG)_(7313310026).jpg) 67(1); Dick Belgers (CC-BY-3.0) (https://commons.wikimedia.org/wiki/File:Arenocoris_fallenii_(Coreidae)_(Fallén's_Leatherbug),_Meijendel,_the_Netherlands.jpg) 222(1), 223(3); Dick Belgers (CC-BY-3.0) (https://commons.wikimedia.org/wiki/File:Arenocoris_fallenii_(Coreidae)_(Fallén's_Leatherbug),_Meijendel,_the_Netherlands.jpg) 329(1); Didier Descouens (CC-BY-SA-4.0) (https://commons.wikimedia.org/wiki/File:Liorhyssus_hyalinus_MHNT_Léguevin.jpg) 353(2); Donald Hobern (CC BY 2.0) (https://commons.wikimedia.org/wiki/File:Trissolcus_sp._(14409074733).jpg) 100; gailhampshire (CC-BY-2.0) (https://commons.wikimedia.org/wiki/File:Juniper_Shieldbug_

(Cyphostethus_tristriatus)._Acanthosomatidae._Heteroptera_-_Flickr_-_gailhampshire_(1).jpg) 40; Gilles San Martin (CC-BY-SA-2.0) (https://commons.wikimedia.org/wiki/File:Baby_bugs_(51071430338).jpg) 99(1); Gilles San Martin (CC-BY-SA-3.0) (https://commons.wikimedia.org/wiki/File:Rhacognathus_punctatus_1285M.JPG) 261(1); Gilles San Martin (CC-BY-SA-4.0) (https://commons.wikimedia.org/wiki/File:Stagonomus_bipunctatus_ssp._pusillus.jpg) 289(2); Göran Liljeberg (CC BY 3.0) https://commons.wikimedia.org/wiki/File:Legnotus_picipes.jpg 245(2); Göran Liljeberg (CC-BY-3.0) (https://commons.wikimedia.org/wiki/File:Adomerus_biguttatus1.jpg) 199(3), 199(4), 200(1); Göran Liljeberg (CC-BY-3.0) (https://commons.wikimedia.org/wiki/File:Adomerus_biguttatus1.jpg) 241(2); Göran Liljeberg (CC-BY-3.0) (https://commons.wikimedia.org/wiki/File:Aelia_klugii.jpg) 266; Göran Liljeberg (CC-BY-3.0) (https://commons.wikimedia.org/wiki/File:Aelia_rostrata.jpg) 206(2); Göran Liljeberg (CC-BY-3.0) (https://commons.wikimedia.org/wiki/File:Aelia_rostrata.jpg) 267(2); Göran Liljeberg (CC-BY-3.0) (https://commons.wikimedia.org/wiki/File:Chlorochroa_juniperina1.jpg) 209(5), 210(1); Göran Liljeberg (CC-BY-3.0) (https://commons.wikimedia.org/wiki/File:Chlorochroa_juniperina1.jpg) 275; Göran Liljeberg (CC-BY-3.0) (https://commons.wikimedia.org/wiki/File:Elasmucha_ferrugata.jpg) 196(7); Göran Liljeberg (CC-BY-3.0) (https://commons.wikimedia.org/wiki/File:Elasmucha_ferrugata.jpg) 233(1); Göran Liljeberg (CC-BY-3.0) (https://commons.wikimedia.org/wiki/File:Elasmucha_fieberi.jpg) 236; Göran Liljeberg (CC-BY-3.0) (https://commons.wikimedia.org/wiki/File:Graphosoma_lineatum1.jpg) 309(2); Göran Liljeberg (CC-BY-3.0) (https://commons.wikimedia.org/wiki/File:Podops_inunctus.jpg) 189(1), 200(3); Göran Liljeberg (CC-BY-3.0) (https://commons.wikimedia.org/wiki/File:Podops_inunctus.jpg) 311(1); Göran Liljeberg (CC-BY-3.0) (https://commons.wikimedia.org/wiki/File:Sciocoris_homalonotus.jpg) 301(1); Grzegorz W. Tężycki (CC-BY-SA-4.0) (https://commons.wikimedia.org/wiki/File:Pyrrhocoris-apterus-19LJJMUZ.jpg) 74; Hectonichus (CC-BY-SA-3.0) (https://commons.wikimedia.org/wiki/File:Coreidae_-_Bothrostethus_annulipes..JPG) 333(1); Hectonichus (CC-BY-SA-3.0) (https://commons.wikimedia.org/wiki/File:Pentatomidae_-_Sciocoris_(Aposciocoris)_homalonotus.-1.JPG) 301(2); Hectonichus (CC-BY-SA-3.0) (https://commons.wikimedia.org/wiki/File:Rhopalidae_-_Liorhyssus_hyalinus.JPG) 139; Hectonichus (CC-BY-SA-3.0) (https://commons.wikimedia.org/wiki/File:Rhopalidae_-_Stictopleurus_crassicornis.jpg) 363(1); Hectonichus (CC-BY-SA-4.0) (https://commons.wikimedia.org/wiki/File:Plataspididae_-_Coptosoma_scutellatum_(mating).JPG) 313(2); Heiss E. (CC-BY-SA-4.0) (https://commons.wikimedia.org/wiki/File:Aradus_macrosomus_holotype_baltic_amber_Fig_1.jpg) 129; Iwan Van Hoogmoed (CC BY 3.0) (https://commons.wikimedia.org/wiki/File:Brachycarenus_tigrinus.jpg) 217(7); Iwan Van Hoogmoed (CC-BY-3.0) (https://commons.wikimedia.org/wiki/File:Aneurus_laevis.jpg) 190(7); Iwan Van Hoogmoed (CC-BY-3.0) (https://commons.wikimedia.org/wiki/File:Aradus_depressus.jpg) 190(8); Iwan Van Hoogmoed (CC-BY-3.0) (https://commons.wikimedia.org/wiki/File:Bathysolen_nubilus.jpg) 222(2), 223(4), 224(3); Iwan Van Hoogmoed (CC-BY-3.0) (https://commons.wikimedia.org/wiki/File:Bathysolen_nubilus.jpg) 331(1); Iwan Van Hoogmoed (CC-BY-3.0) (https://commons.wikimedia.org/wiki/File:Brachycarenus_tigrinus.jpg) 350(1); Iwan Van Hoogmoed (CC-BY-3.0) (https://commons.wikimedia.org/wiki/File:Chorosoma_schillingii1.jpg) 214(5); Iwan Van Hoogmoed (CC-BY-3.0) (https://commons.wikimedia.org/wiki/File:Chorosoma_schillingii1.jpg) 363(2); Iwan Van Hoogmoed (CC-BY-3.0) (https://commons.wikimedia.org/wiki/File:Liorhyssus_hyalinus.jpg) 217(4); Iwan Van Hoogmoed (CC-BY-3.0) (https://commons.wikimedia.org/wiki/File:Liorhyssus_hyalinus.jpg) 353(1); jacilluch (CC-BY-SA-2.0) (https://commons.wikimedia.org/wiki/File:Aelia_-_Chinche_de_los_cereales_(13994791381).jpg) 173; James St. John (CC-BY-2.0) (https://commons.wikimedia.org/wiki/File:Fossil_giant_water_bug_(Crato_Formation,_Lower_Cretaceous;_Chapada_do_Araripe,_Brazil)_1_(47576999752).jpg) 127; janet graham (CC-BY-2.0) (https://commons.wikimedia.org/wiki/File:Saldula_saltatoria,_Harlech_forest,_North_Wales,_March_2017_(23819884468).jpg) 191(2); Jorge Núñez (CC-BY-SA-4.0) (https://commons.wikimedia.org/wiki/File:Cydnus_aterrimus_J.N_A_Pobra_do_Brollón.jpg) 242; Jun Souma (CC-BY-4.0) (https://commons.wikimedia.org/wiki/File:Acalypta_marginata_dorsal_habitus.jpg) 190(6); Kelisi (CC-BY-SA-4.0) (https://commons.wikimedia.org/wiki/File:BoxelderBugsBoltonONSep2018.jpg) 103(1); Kleine Halle (CC-0) (https://commons.wikimedia.org/wiki/File:Chlorochroa_juniperina_nymph.jpg) 276(1); Kleine Halle (CC-0) (https://commons.

wikimedia.org/wiki/File:Chlorochroa_juniperina.jpg) 276(2); Line Sabroe (CC-BY-2.0) (https://commons.wikimedia.org/wiki/File:Parent_Bug_female_guarding_eggs_-_Elasmucha_grisea_(21570289224).jpg) 98; Michael J. Raupach, Lars Hendrich, Stefan M. Kuchler, Fabian Deister, Jérome Moriniére, Martin M. Gossner (CC-BY-4.0) (https://commons.wikimedia.org/wiki/File:Coptosoma_scutellatum_ZSM.jpg) 193(1); Michael J. Raupach, Lars Hendrich, Stefan M. Kuchler, Fabian Deister, Jérome Moriniére, Martin M. Gossner (CC-BY-4.0) (https://commons.wikimedia.org/wiki/File:Coptosoma_scutellatum_ZSM.jpg) 312; Michael J. Raupach, Lars Hendrich, Stefan M. Kuchler, Fabian Deister, Jérome Moriniére, Martin M. Gossner (CC-BY-4.0) (https://commons.wikimedia.org/wiki/File:Piesma_maculatum_ZSM.jpg) 190(4); Michael J. Raupach, Lars Hendrich, Stefan M. Kuchler, Fabian Deister, Jérome Moriniére, Martin M. Gossner (CC-BY-4.0) (https://commons.wikimedia.org/wiki/File:Rhopalus_parumpunctatus_ZSM.jpg) 218(5); Michael J. Raupach, Lars Hendrich, Stefan M. Kuchler, Fabian Deister, Jérome Moriniére, Martin M. Gossner (CC-BY-4.0) (https://commons.wikimedia.org/wiki/File:Rhopalus_parumpunctatus_ZSM.jpg) 356(1); Michael J. Raupach, Lars Hendrich, Stefan M. Kuchler, Fabian Deister, Jérome Moriniére, Martin M. Gossner (CC-BY-4.0) https://commons.wikimedia.org/wiki/File:Tingis_ampliata_ZSM.jpg 190(5); Michael J. Raupach, Lars Hendrich, Stefan M. Kuchler, Fabian Deister, Jérome Moriniére, Martin M. Gossner (CC-BY-SA-4.0) (https://commons.wikimedia.org/wiki/File:Graphosoma_lineatum_ZSM.jpg) 189(3), 193(4), 200(2); Michael J. Raupach, Lars Hendrich, Stefan M. Kuchler, Fabian Deister, Jérome Moriniére, Martin M. Gossner (CC-BY-SA-4.0) (https://commons.wikimedia.org/wiki/File:Hydrometra_gracilenta.jpg) 188(5); Michael J. Raupach, Lars Hendrich, Stefan M. Kuchler, Fabian Deister, Jérome Moriniére, Martin M. Gossner (CC-BY-SA-4.0) (https://commons.wikimedia.org/wiki/File:Legnotus_limbosus.jpg) 244; Michael J. Raupach, Lars Hendrich, Stefan M. Kuchler, Fabian Deister, Jérome Moriniére, Martin M. Gossner (CC-BY-SA-4.0) (https://commons.wikimedia.org/wiki/File:Nabis_rugosus_ZSM.jpg) 190(2); Nabokov (talk) (CC-BY-SA-3.0) (https://commons.wikimedia.org/wiki/File:Nesting_Earwig_Chester_UK_2.jpg) 57; phillipengelking (CC-BY-SA-4.0) (https://commons.wikimedia.org/wiki/File:Phloea_corticata_Angra_dos_Reis_Ilha_Grande.jpg) 15(1); Quartl (CC-BY-SA-3.0) (https://commons.wikimedia.org/wiki/File:Coreus_marginatus_qtl2.jpg) 336(3); Rene Sylvestersen (CC-BY-SA-3.0) (https://commons.wikimedia.org/wiki/File:Fossil_Hemiptera.JPG) 132; Robert Flogaus-Faust (CC-BY-3.0) (https://commons.wikimedia.org/wiki/File:Eurydema_ornata_RF.jpg) 307(3); Robert Webster (CC-BY-SA-4.0) (https://commons.wikimedia.org/wiki/File:Narnia_snowi_P1230203a.jpg) 45; Rui Andrade (CC-BY-4.0) (https://commons.wikimedia.org/wiki/File:Micrelytra_fossularum_(10.3897-BDJ.9.e65314)_Figure_1_d.jpeg) 327(3); S. Rae (CC-BY-2.0) (https://commons.wikimedia.org/wiki/File:Gonocerus_cf._juniperi_-_Flickr_-_S._Rae.jpg) 344(2); Siga (CC-BY-SA-3.0) (https://commons.wikimedia.org/wiki/File:Ilyocoris_cimicoides_natur.jpg) 19(2); Siga (CC-BY-SA-3.0) (https://commons.wikimedia.org/wiki/File:Phasia_hemiptera_bl.JPG) 19(1); Slimguy (CC-BY-4.0) (https://commons.wikimedia.org/wiki/File:2017_06_27_Eurydema_oleraceum_nymph.jpg) 304(2); Slimguy (CC-BY-4.0) (https://commons.wikimedia.org/wiki/File:2021_07_27_Eurygaster_austriaca.jpg) 136; Slimguy (CC-BY-SA-4.0) (https://commons.wikimedia.org/wiki/File:2016_10_18_Leptoglossus_occid.jpg) 345(2); Syrio (CC-BY-SA-4.0) (https://commons.wikimedia.org/wiki/File:Pentatoma_rufipes_nymph.jpg) 293(2); Thijsdegraaf (CC-BY-SA-4.0) (https://commons.wikimedia.org/wiki/File:Byrsinus_flavicornis_7_27-10-2015.JPG) 237(2); Thijsdegraaf (CC-BY-SA-4.0) (https://commons.wikimedia.org/wiki/File:Myrmus_miriformis_1_nimf_20-7-2015.JPG) 366(3); Veljo Runnel (CC-BY-SA-4.0) (https://commons.wikimedia.org/wiki/File:Eurydema_dominulus_163295956.jpg) 305; Volkmar Wagner (CC-BY-SA-3.0) (https://commons.wikimedia.org/wiki/File:Baumwanze,_Pinthaeus_sanguinipes_002.JPG) 259(2); Wouter Hagens (CC BY-SA 4.0) (https://commons.wikimedia.org/wiki/File:Thuja_occidentalis_D.jpg) 89; Wouter Hagens (CC-BY-SA-4.0) (https://commons.wikimedia.org/wiki/File:Thuja_occidentalis_D.jpg) 229(2). **Williams, David:** 374. **Wren, Antony:** 144(1,2).

While every effort has been made to trace and acknowledge copyright holders, the author would like to apologise for any errors or omissions, and invites readers to inform the publishers so that corrections can be made to any future editions.

Index

SPECIES INDEX (INCLUDING ORDERS, FAMILIES AND GENERA)

Individual species listed in Appendix 1 have not been indexed. Page numbers in **bold** refer to information contained in figure captions. Page numbers in *italics* refer to the main descriptions of the species in Chapter 8.

GENERAL INDEX

The glossary (pages 401–408) has not been indexed. Page numbers in **bold** refer to information contained in figure captions. Page numbers in *italics* refer to the main descriptions of the species in Chapter 8.

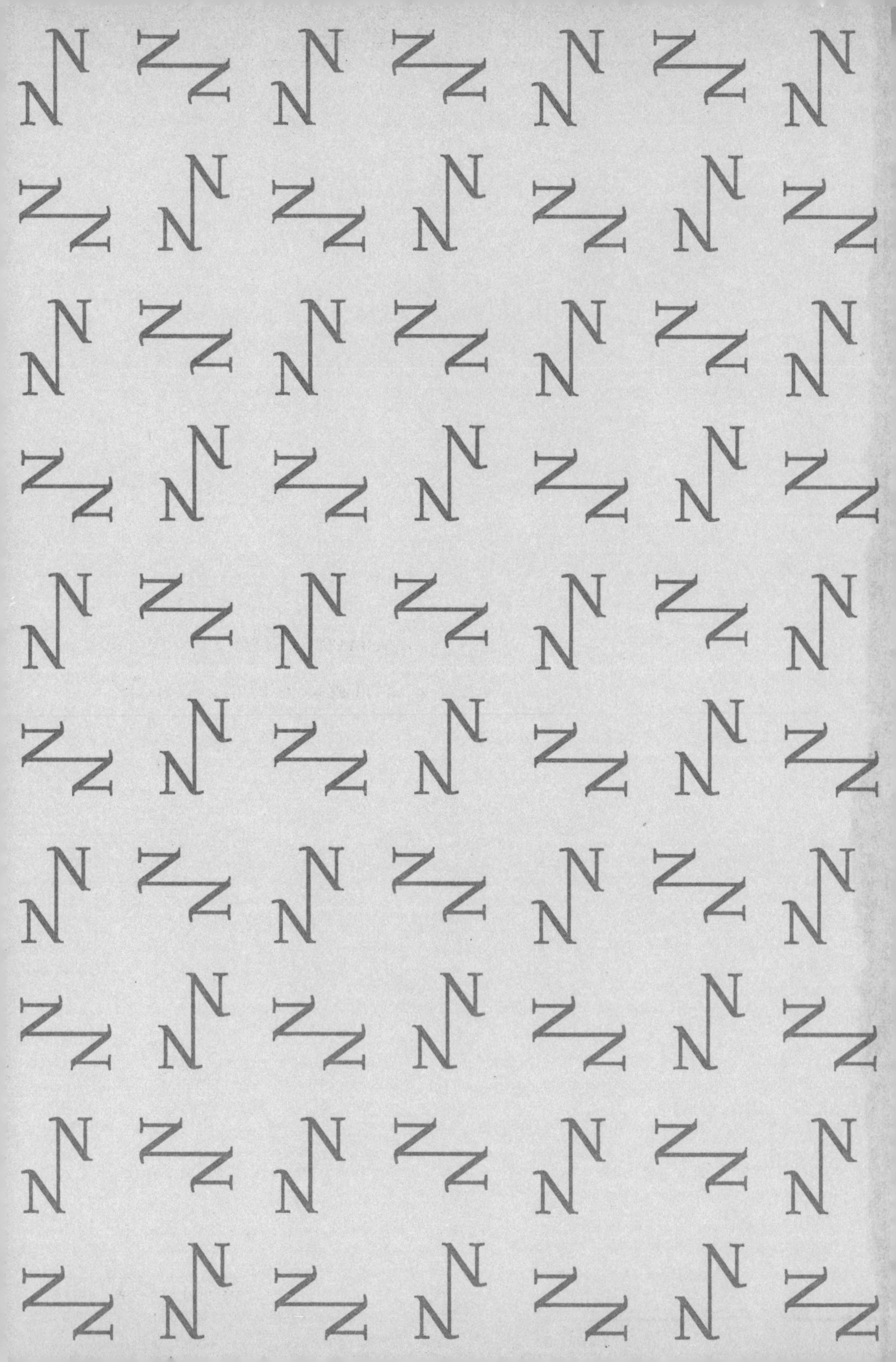